Modern techniques of surface science

Cambridge Solid State Science Series

EDITORS:

Professor R. W. Cahn
University of Cambridge

Professor E. A. Davies
Department of Physics, University of Leicester

Professor I. M. Ward
Department of Physics, University of Leeds

D. P. WOODRUFF AND
T. A. DELCHAR

Department of Physics, University of Warwick

Modern techniques of surface science

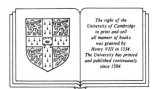

The right of the
University of Cambridge
to print and sell
all manner of books
was granted by
Henry VIII in 1534.
The University has printed
and published continuously
since 1584.

CAMBRIDGE UNIVERSITY PRESS

Cambridge
London   New York   New Rochelle
Melbourne   Sydney

Published by the Press Syndicate of the University of Cambridge
The Pitt Building, Trumpington Street, Cambridge CB2 1RP
32 East 57th Street, New York, NY 10022, USA
10 Stamford Road, Oakleigh, Melbourne 3166, Australia

First published 1986

Printed in Great Britain at University Press, Cambridge

British Library cataloguing in publication data
Woodruff, D. P.
Modern technique of surface science.
1. Solids—Surfaces
I. Title. II. Delchar, T. A.
530.4'1   QC176.8.S8

Library of Congress cataloguing in publication data
Woodruff, D. P.
Modern techniques of surface science.
Includes bibliographies and index.
1. Surfaces (Physics)—Technique. 2. Surface
chemistry—Technique. 3. Spectrum analysis.
I. Delchar, T. A. II. Title.
QC173.4.S94W66 1985   541.3'453   85-10960

ISBN 0 521 30602 7

Contents

Preface

Since the early 1960s or so there has been a virtual explosion in the level of research on solid surfaces. The importance of understanding surface processes in heterogeneous catalysis had been recognised since the early part of the twentieth century, but it was not until the 1960s, with the introduction and development of ultra-high-vacuum techniques, that real advances could be made, even using the 'old' techniques such as low energy electron diffraction (1927) and field emission (1936). The subsequent development of materials science and the growth of the semiconductor industry has added further surface problems for investigation while, at the same time, many new techniques have been introduced and exploited to study surfaces at the atomic level. For someone coming fresh to the field of surface physics or surface chemistry there seems to be a bewildering excess of different techniques, each commonly referred to by its acronym or unpronounceable string of initial latters. Much of the scientific literature in this field is occupied with technique-orientated studies of specific problems in which the strengths and limitations (particularly the latter!) of the technique or techniques used are rarely explained. Quite early in the development of surface science it became evident that surface problems should be tackled using a range of complementary techniques if a proper and complete understanding were to be obtained. Therefore, before one can appreciate the general progress being made in an area of surface science, or select a technique to investigate a particular problem, or understand the results of investigations of one's problem by other methods, it is necessary to understand the basic physical principles, strengths and limitations of the available techniques.

In this book we set out to provide the reader with just this information. The level of presentation and discussion is that appropriate to final-year undergraduates or postgraduate students although the wide scope of the book may well make it useful to many research workers, particularly those working on the periphery of surface science. This wide scope means that we have not attempted to be exhaustive in discussing every application of every technique or the results of all the wealth of published research in the field. What we have attempted to do is to cover all the techniques and to try to illustrate the way in which they can be used. We have also attempted to give some assessment of the value of each technique; these assessments must be, at least in part, subjective, but we hope that they will be seen to be balanced judgements. The presentations of the techniques also include a description of the experimental methods as these often influence or dominate the technique, and in the case of the more longstanding techniques we include some limited historical background. In these instances, where the technique is straightforward, the examples chosen may be from the pioneer workers themselves.

An expert in the field will undoubtedly detect omissions, particularly of some of the most recently developed techniques such as scanning tunnelling electron microscopy, inverse photoemission and Raman techniques. These methods are still quite new and their impact on the field has yet to be assessed. A specialist practitioner of one technique may also feel we have omitted some key application or elegant example; we apologise, but the field is vast and some things have had to be omitted to make this book manageable. We hope, however, that the reader will be able to obtain a clear 'flavour' of the techniques and their applications, and be well armed to delve into specialist review articles on specific techniques to find out more.

Many researchers have kindly given their permission to reproduce their results and in several cases have provided us with more detailed illustrations. Each is acknowledged in the relevant figure caption but we thank them all again for this help and their encouragement.

We should also add a final note on the units used in this book. The normal practice in research papers in surface science, like other research fields, falls short of adopting SI units fully. We have, in order to be consistent with this wider literature, retained at least two non-SI units in this book. These are the Å unit of length (1 Å = 0.1 nm) and the torr unit of pressure (1 torr = 133.3 Pa).

D.P.W.

December, 1984   T.A.D.

Abbreviations

AEAPS	Auger Electron Appearance Potential Spectroscopy
AES	Auger Electron Spectroscopy
APS	Appearance Potential Spectroscopy
ARUPS	Angle-Resolved Ultraviolet Photoelectron Spectroscopy
CHA	Concentric Hemispherical Analyser
CMA	Cylindrical Mirror Analyser
CPD	Contact Potential Difference
DAPS	Disappearance Potential Spectroscopy
EAPFS	Extended Appearance Potential Fine Structure
ESD	Electron Stimulated Desorption
ESDIAD	Electron Stimulated Desorption Ion Angular Distributions
EXAFS	Extended X-ray Absorption Fine Structure
EXELFS	Extended Energy Loss Fine Structure
FEM	Field Emission Microscopy
FIM	Field Ion Microscopy
HEIS	High Energy Ion Scattering
HREELS	High Resolution Electron Energy Loss Spectroscopy
ILS	Ionisation Loss Spectroscopy
IMBS	Inelastic Molecular Beam Scattering
INS	Ion Neutralisation Spectroscopy
IRAS	Infrared Reflection–Absorption Spectroscopy
LEED	Low Energy Electron Diffraction
LEIS	Low Energy Ion Scattering
MBE	Molecular Beam Epitaxy
PSD	Photon Stimulated Desorption
RFA	Retarding Field Analyser
RHEED	Reflection High Energy Electron Diffraction
SEXAFS	Surface EXAFS (Extended X-ray Absorption Fine Structure)
SIMS	Secondary Ion Mass Spectroscopy
SXAPS	Soft X-ray APS (Appearance Potential Spectroscopy)
TPD	Temperature Programmed Desorption
UHV	Ultra-High-Vacuum
UPS	Ultraviolet Photoelectron Spectroscopy
XPS	X-ray Photoelectron Spectroscopy

1   Introduction

1.1   Why surfaces?

The growth in the study of solid surfaces and in the number of techniques available for their study has been enormous since the early sixties. At least one reason for this is the growing awareness of the importance of understanding surface properties and indeed the fact that work on surfaces has had an impact on this understanding and on specific applications in the 'real world'. At a fundamental level surfaces are of great interest because they represent a rather special kind of defect in the solid state. Much of our understanding of solids is based on the fact that they are, in essence, perfectly periodic in three dimensions; the electronic and vibrational properties can be described in great detail using methods which rely on this periodicity. The introduction of a surface breaks this periodicity in one direction and can lead to structural changes as well as the introduction of localised electronic and vibrational states. Gaining a proper understanding of these effects is not only of academic interest, as there is growing interest in the properties of low-dimensional structures in semiconductor devices, and a free surface can represent the simplest case of such a structure.

Perhaps the most widely quoted motivation for modern surface studies is the goal of understanding heterogeneous catalysis. The greatly increased rates of certain chemical interactions which occur in the presence of solid (usually powder) catalysis must result from the modification of at least one of the constituent chemicals when adsorbed on the solid surface and its enhanced ability to interact with the other constituent(s) in this state. One would therefore like to understand what these modifications are, whether there are new intermediate species formed, what are the rate limiting steps and activation energies, what kind of sites on the catalyst surface are active and how these processes depend on the catalyst material. This might lead to better or cheaper catalysts (many such catalysts being based on precious metals such as platinum). The problems of understanding these processes in a microscopic or atomistic way are formidable. Industrial processes frequently operate at high temperatures and pressures (i.e. many atmospheres) and the catalysts are in the form of highly dispersed powders (possibly with individual particles comprising only hundreds of atoms), they frequently involve transition metals on oxide 'supports' which may or may not be passive, and they may include small additions of 'promoters' which greatly enhance the efficiency of the catalysts. The approach which makes the fullest use of the techniques in this book is to study highly simplified versions of these problems. This involves initially taking flat, usually low Miller index, faces of single crystals of the material of interest and studying the adsorption or coadsorption of small quantities of atoms and molecules on them in an otherwise Ultra-High-Vacuum (UHV) environment. The stress of these methods is on characterising the surfaces and the adsorption and reaction processes in fine detail so that the conditions are very well defined. Although it is easy to see reasons why this approach may be too far removed from applied catalytic problems to be of real value, the signs in the last few years have been encouraging and simple catalytic processes are now being understood with the help of these model studies (King & Woodruff, 1982).

Another area of study is in the understanding of corrosion of materials and certain kinds of mechanical failure due to grain boundary embrittlement. One important process in these problems is of the segregation of minority ingredients (often impurities) in a solid to the free surface, or to internal surfaces (grain boundaries) when the temperature is high enough to allow diffusion through the bulk at a reasonable rate. A particular species can find it very energetically favourable to be in one of these surface sites rather than in the bulk so that even a bulk concentration of a few parts per million can lead to surfaces covered with a complete atomic layer of the segregant in equilibrium. Segregation of this kind is now well established as being a cause of intergranular fracture of engineering materials. On the other hand, similar segregation to free surfaces can have the effect of improving resistance to corrosion. Studies in this broadly metallurgical area have proceeded not only by the modelling type of investigation described above, but also by applying some of the techniques described in this book to the study of the surfaces of 'real' materials of interest. In particular, by investigating the composition of the top few atomic layers of a fractured or corroded surface considerable information can be gained. To do this, one requires techniques which are highly surface specific in their analytic capabilities. Coupled with a method of removing atomic layers in a reasonably controlled fashion, usually by ion bombardment, a depth profile of the surface and subsurface composition can be obtained.

The final main area of application of surface studies which lies closest to the fundamental problems mentioned at the beginning of this section is in the fabrication of semiconductor devices. Although there are applications for depth profiling on actual devices for 'trouble-shooting' in production problems (due to contamination or interdiffusion at interfaces) there are also problems of quite fundamental importance which lie naturally quite close to the modelling approach described in relation to catalytically motivated research. For example, the formation of metal–semiconductor junctions with desirable properties is strongly influenced by the tendency for chemical interactions to occur between the metal and the semiconductor. Real devices use well-oriented single crystal samples so this aspect of the modelling is no longer idealised. Moreover, in the case of semiconductor surfaces some of the simplest problems remain far from trivial to solve. Most semiconductor surfaces appear to involve some structural rearrangement of the atoms relative to a simple extension of the bulk structure. For example, the stable structure of a Si{111} surface reconstructs to a 'superlattice' seven times larger in periodicity than the bulk (a (7×7) structure in the notation described in chapter 2). This structure remains essentially unsolved. Even in the case of the {110} cleavage faces of III–V compounds such as GaAs in which there is no change in two-dimensional periodicity of the surface, there is a rearrangement in bond angles influencing the relative positions of the Ga and As layers. Finally, we note that there is increasing interest in the growth of semiconductor devices by Molecular Beam Epitaxy (MBE) using methods very close to those used in surface science generally (UHV and 'adsorption' at very low rates). The surface structures formed during MBE can be very complex and highly sensitive to the stoichiometry of the uppermost layer. Growth studies also reveal that many materials will not grow in a layer-by-layer form on certain other layers. These limitations in 'atomic engineering' need to be understood properly if exotic multilayer devices are to be designed and built.

This book is concerned with the analytical techniques which have contributed, and should continue to contribute, to understanding these problems. It is concerned with the basic underlying physical principles of the techniques and the extent to which those principles constrain their usefulness. As such, it is not intended to be an experimental handbook for surface analysis but the background which allows the techniques to be used and assessed properly. Some experimental details are given, but again with the primary object of understanding the strengths and limitations of individual techniques. In the final section of this chapter a very broad and brief review of the interrelationship of the main techniques is given with reference to their applications. First, however, we consider the need for UHV and define a few of the specialist terms used in surface studies.

1.2   Ultra-High-Vacuum (UHV), contamination and cleaning

If we are to study the properties of a surface which are well characterised at an atomic level it is clear that the composition of the surface must remain essentially constant over the duration of an experiment. This evidently means that the rate of arrival of reactive species from the surrounding gas phase should be low. A reasonable criterion would be that no more than a few per cent of an atomic layer of atoms should attach themselves to the surface from the gas phase in, say, an experimental time scale of about one hour. This requirement can be evaluated readily from simple kinetic theory of gases. Thus, the rate of arrival of atoms or molecules from a gas of number density n per unit volume and with an average velocity c_a is

$$r = \tfrac{1}{4} n c_a \tag{1.1}$$

while equating the kinetic energy of the particle's mass m with a root mean square velocity c_{rms} to their thermal energy determined by the absolute temperature T and Boltzmann's constant k_B gives

$$c_{rms}^2 = 3 k_B T / m \tag{1.2}$$

Finally, using the relationship between the two velocities

$$c_a = (8/3\pi)^{\frac{1}{2}} c_{rms} \tag{1.3}$$

and the fact that the pressure P is given by

$$P = n k_B T \tag{1.4}$$

leads to an expression for the rate of arrival

$$r = P(1/2\pi k_B T m)^{\frac{1}{2}} \tag{1.5}$$

A convenient form of this expression, in which P is expressed in torr (i.e. mm of Hg), T in K and m is substituted by the molecular weight M multiplied by the atomic mass unit gives

$$r = 3.51 \times 10^{22} P/(TM)^{\frac{1}{2}} \tag{1.6}$$

r being in molecules cm^{-2} s^{-1}. For example, N_2 molecules ($M = 28$) at room temperature ($T = 293$ K) at 1 torr have an arrival rate of 3.88×10^{20} molecules cm^{-2} s^{-1}. It is convenient to define a monolayer adsorption time in terms of the pressure. In defining this we assume that a monolayer, i.e. a single complete atomic layer, consists of about 10^{15}– 2×10^{15} atoms cm^{-2} and that all molecules arriving at the surface stick and are incorporated into this monolayer. Thus for the example given the monolayer time is about 3×10^{-6} s at 1 torr, 3 s at 10^{-6} torr, or almost 1 hour at 10^{-9} torr. This means that if all the gas atoms and molecules arriving at a surface in a vacuum system do indeed stick to it, then contamination of only a few per cent of a monolayer in an experimental time of 1 hour requires pressures of 10^{-10} torr or better.

While these are broadly worst case assumptions, some surfaces of interest do react readily with H and CO, the main ingredients of a UHV chamber and so match these conditions. The need for UHV is therefore simply to keep a surface in its clean or otherwise well-characterised condition once produced. Indeed, the need for good vacuum can also extend to the kind of depth profiling study of technical surfaces described in the previous sections. In these cases a sample is initially analysed 'as loaded' so that the surface composition is dominated by contamination from handling in air and uninfluenced by the quality of the surrounding vacuum in the analysis chamber. Once surface layers have been removed in the depth profiling, however, the freshly exposed surface is susceptible to new contamination and must be studied in a good-quality vacuum.

A detailed discussion of the methods of UHV is not appropriate to this book and can be found in many volumes concerned specifically with vacuum technology (e.g. Roberts & Vanderslice, 1963; Robinson, 1968). A few points of general interest are worth noting, however. The first is that a major reason for the development of modern surface science research in addition to those given in the previous section is the commercial availability of convenient UHV components since the early 1960s and their subsequent development. Early work was carried out in glass vacuum systems using liquid N_2 trapped Hg diffusion pumps. The surface science instrumentation had to be incorporated into these sealed glass vessels with electrical connections made through glass-to-metal seals in the containment vessel. Modern surface science studies usually involve the use of many different techniques in the same vessel, each of which may be quite sophisticated and this is achieved by mounting each onto a stainless steel flange which is sealed to a stainless steel chamber using Cu gasket seals. This gives great flexibility and demountability and it is hard to see how this level of sophistication could have been achieved realistically with glass systems. In addition to the development of these demountable metal vessels, great use is now made of ion pumps which require only electrical power to function and do not need liquid N_2 and the regular attention that this implies.

The second general point regarding UHV is the constraints on fabrication methods necessary for instrumentation within the vacuum. Although one must use vacuum pumps capable of operating in the 10^{-10}–10^{-11} torr range, an important ingredient in obtaining UHV is the need to 'bake' the whole system. In the absence of leaks and with suitable pumps, vacua are limited by the 'outgassing' of the inner walls and instrument surfaces within the chamber mainly due to the desorption of adsorbed gases from these surfaces. By heating all of these surfaces, the rate of desorption is increased, the surface coverage decreased, and thus the rate of desorption on subsequently cooling to room temperature is reduced. This reduces the gas load on the pumps and thus allows lower pressures to be achieved. Typically, a stainless steel chamber with all its enclosed instrumentation is baked to 200 °C for 12 hours or so. Obviously this means that all components in the vacuum chamber must be stable and have low vapour pressures at 200 °C. An additional common requirement for the experiments described in this book is that all components must be non-magnetic as many surface techniques involve low energy electrons which are easily deflected by weak electrostatic and magnetic fields. Fabrication methods compatible with these requirements are now well established involving mainly the use of stainless steel and refractory metals with ceramics for electrical or thermal insulation. Many materials acceptable in 'high vacuum' ($\sim 10^{-6}$ torr) such as many adhesives and plastics are not acceptable in UHV.

While UHV guarantees that a surface should not be influenced by the arrival of ambient atoms and molecules on a time scale of the order of one hour or more, a further requirement to studies of the properties of ideal surfaces is to be able to clean them, in the vacuum system, to a level compatible with the contamination constraints we have set on the vacuum, i.e. to be able to produce a surface which contains no more than a few per cent (and preferably less) of an atomic layer of species other than those which comprise the underlying bulk solid. Generally we also require that the surface is well ordered on an atomic scale. The main methods used to achieve this *in situ* cleaning are

   (i)   cleavage,
  (ii)   heating,
 (iii)   ion bombardment (typically Ar ions),
 (iv)   chemical processing.

The first of these is largely self-explanatory; for those materials which do cleave readily (e.g. oxides, alkali halides, semiconductors, layer compounds), and for studies of the surface orientation which comprises the cleavage face, surfaces can be prepared in vacuum which are intrinsically clean. Apart from these limitations the main problem with the method is that it is usually only possible to cleave a single sample (e.g. a long bar) a few times, so that the surface cannot be reprepared many times, and that the cleavage may result in a heavily stepped surface. Large variations in the properties of a surface (particularly the adsorption kinetics) can be obtained from cleave to cleave on many materials. There are examples, moreover, of cases in which the cleavage surface presents a different structure from that obtained by heating to allow the surface to equilibrate; the Si{111} surface is an example of this.

Heating a surface, like heating the walls of a vacuum vessel, can lead to desorption of adsorbed species. However, in most cases some impurities on the surface are too strongly bound to be removed by heating to temperatures below the melting point of the sample. This method of cleaning has been most used for W and similar high melting point materials for which the surface oxides are flashed off below the melting points. Even for these materials, however, it is unlikely that the method can be totally satisfactory due to impurities such as C which form exceeding strongly bound compounds with the substrate material. On the other hand, once these kinds of impurities have been removed, heating alone may be sufficient to regenerate a clean surface following an adsorption experiment using more weakly bound adsorbate species. This surface regeneration by heating may be applicable to many materials for which heating alone is totally ineffective in the initial cleaning process.

The use of Ar ion bombardment of a surface to remove layers of the surface by sputtering is by far the most widely used primary method, particularly for metal surfaces. The actual physics of this process and the yields obtained are discussed in chapter 4. The technique is effective in the removal of many atomic layers of a surface and even if an impurity species is far less effectively sputtered than the substrate it can be removed eventually. One disadvantage of ion bombardment, typically at energies of 0.5–5.0 keV, is that the surface is left in a heavily damaged state, usually with embedded Ar atoms, so that the surface must be annealed to restore the order. This in itself can create problems; as was noted in the previous section, many dilute impurity species in the bulk of a solid segregate preferentially to the free surface and if a sample with a clean surface is heated, the diffusion rates are increased and further segregation can occur. Typical segregants in transition metals of very high average purity are C and S. This then requires further ion bombardment cleaning, further annealing and so on. In practice a number of cycles (possibly tens of cycles) of bombardment and annealing leads to depletion of segregating impurities in the subsurface region and to a clean surface. Far fewer cycles are then required for recleaning the sample after adsorption studies.

The final approach of chemical cleaning *in situ* involves the introduction of gases into the vacuum system at low pressures ($\sim 10^{-6}$ torr or less) which react with impurities on a surface to produce weakly bound species which can be thermally desorbed. It is most widely used for the removal of C from refractory metals such as W which can be cleaned of most other impurities by heating alone. Exposure of such a surface to O_2

at elevated temperatures leads to removal of C as desorbed CO leaving an oxidised surface which can then be cleaned by heating alone.

1.3   **Adsorption at surfaces**

Although many of the techniques described in this book are applicable to a wide variety of surface problems, we concentrate the development and illustration of the techniques in terms of studies on well-characterised low index single crystal surfaces and adsorption of atoms and molecules on them, although not all the techniques are restricted to such studies. It is therefore helpful to define some of the terms and units used in these studies which we will have cause to use in later chapters. The first of these is the definition of a *monolayer* of adsorbate. One way of defining the coverage of a surface at monolayer level – i.e. of a single complete atomic or molecular layer – is in terms of the coverage of a two-dimensional close packed layer taking account of the atomic or molecular size. Such a definition is frequently used in studies of polycrystalline surfaces. However, on surfaces of well-defined crystallography it is generally more convenient to use a definition based on the atomic density of packing of the surface itself. We shall therefore use the definition that a monolayer of adsorbed atoms or molecules involves a number density equal to that of the atoms in a single atomic layer of the substrate material parallel to the surface. In the absence of reconstruction, this is, of course, the same as the number density of atoms in the top atomic layer of the substrate. Frequently, incidentally, saturation of a particular adsorbate species occurs at a coverage of less than one monolayer so the definition implies nothing about the maximum possible coverage which depends on the adsorption system under study.

A second definition concerning adsorption studies is for a unit of exposure. The unit which is firmly established in the literature is the *Langmuir* (abbreviated as L), with $1 L = 10^{-6}$ torr s exposure. A major disadvantage of this unit is that, as may be readily appreciated from equation (1.6), the actual number of atoms or molecules arriving at a surface in 1 L of exposure actually depends on the molecular weight of the gaseous species and its temperature. Table 1.1 illustrates the effect of this variation, showing the number of molecules striking 1 cm^2 of surface in 1 L with a gas temperature of 300 K. Also shown here is the coverage, in monolayers, which would result if all the molecules arriving were to stick on a Ni$\{100\}$ surface, with dissociation assumed for H_2, O_2 and I_2. Despite this disadvantage and the fact that the unit is certainly not an SI one, it has the great advantage of experimental convenience as most researchers performing an exposure are equipped with an ion gauge,

Table 1.1. *Effect of 1 Langmuir exposure of different adsorbates at 300 K*

Incident and adsorbing species	No. of molecules arriving (cm^{-2})	Coverage on Ni{100} with unity sticking factor (monolayers)
H_2 adsorbing as H	1.43×10^{15}	1.80
O_2 adsorbing as O	3.58×10^{14}	0.44
CO adsorbing as CO	3.83×10^{14}	0.24
I_2 adsorbing as I	1.27×10^{14}	0.16

calibrated in torr, and a stopwatch! It provides a convenient unit for characterising the exposures needed to produce certain adsorption states on a surface and allows some transferability between experimenters working on the same adsorption system. It also is a unit of convenient magnitude in that, as table 1.1 shows, 1 L corresponds to of the order of 1 monolayer coverage if all molecules stick to the surface. Although a proposal for a unit based on the actual number of impinging molecules was made a few years ago by Menzel & Fuggle (1978) it has not gained any serious support in subsequent literature. One further point which is worth mentioning in the context of table 1.1 is the question of sticking factors. The final column of table 1.1 is constructed assuming that all impinging molecules stick to the surface (i.e. that the 'sticking factor' is unity) independent of coverage. In fact this would represent a relatively unusual state of affairs. As the coverage increases, some molecules arriving at the surface will strike other adsorbed species rather than the clean surface. Assuming that these also stick (and then may diffuse over the surface to fill clean surface sites) actually involves assuming that second layer adsorption (albeit possibly more weakly bound) is possible. An alternative possibility, that the molecules arriving at occupied sites are not adsorbed, leads to an average sticking factor which falls exponentially with time. This *Langmuir adsorption* is one of several possible forms of adsorption kinetics discussed in many books on adsorption (e.g. Hayward & Trapnell, 1964) and will not be discussed further here. We should note only that in serious studies of adsorption kinetics, exposures given in Langmuirs and based on ion gauge readings of total chamber pressures are unlikely to be very reliable due to difficulties in establishing the pressure at the sample and the need for ion gauge calibration. For the same reason, exposures determined in the same way in different chambers and in different laboratories as those needed to obtain particular adsorption states may well show variations of a factor of 2 or more.

Finally, in this book we shall frequently choose examples of adsorption systems which may be referred to as involving *chemisorbed* or *physisorbed* atoms or molecules. The distinction between these two types of adsorption lies in the form of the electronic bond between the adsorbate and substrate. If an adsorbed molecule suffers significant electronic modifications relative to its state in the gas phase to form a chemical bond with the surface (covalent or ionic) it is said to be chemisorbed. If, on the other hand, it is held to the surface only by van der Waals' forces, relying on the polarisability of the otherwise undisturbed molecule, then it is said to be physisorbed. Clearly physisorption produces weak bonds while chemisorption often produces strong bonds. It is usual to regard the upper limit of the bond strength in physisorption as around 0.6 eV per atom or molecule, or 60 kJ mole^{-1} (1 eV molecule^{-1} = 96.5 kJ mole^{-1} = 23.1 kcal mole^{-1}). Thermal energy considerations such as those discussed in chapter 5 lead to the conclusion that such weakly bonded species would be desorbed from a surface at a temperature much in excess of 200 K. Adsorbates stable on a surface above this sort of temperature are therefore almost certainly chemisorbed. However, the distinction is strictly in terms of the form of the bond, and not its energy, and there are cases in the literature in which electronic modifications characteristic of chemisorption are seen in far more weakly bound species. A low desorption temperature does not, therefore, necessarily indicate physisorption.

1.4   Surface analytical techniques

The rest of this book sets out the main techniques used in the investigation of the detailed properties of surfaces so that there are a limited number of generalisations which can be made regarding the selection of specific techniques and their relative strengths and weaknesses. Broadly, in investigating surfaces we are usually interested in the *structure* of a surface, its *chemical composition* and some information on the *electronic structure* which may be in the form of the chemical state of particular adsorbed atoms or molecules, or may involve a determination of the surface electronic band structure or energy density of electronic states. The listing in table 1.2 of the main techniques discussed in this book, and their abbreviated forms, gives some indication of the kind of information each gives, the ticks indicating that this kind of information is actually extracted from some experiments, while the bracketed ticks indicate a potential or a minor use of the technique in a specialised form. Clearly individual techniques often provide information in more than

Table 1.2. *Summary of techniques and their acronyms and abbreviations*

	Structure	Composition	Electronic structure or chemical state	Chapter
Low Energy Electron Diffraction (LEED)	√			2
Reflection High Energy Electron Diffraction (RHEED)	√			2
X-ray Photoelectron Spectroscopy (XPS)	(√)	√	√	3
Surface Extended X-ray Absorption Fine Structure (SEXAFS)	√			3
Auger Electron Spectroscopy (AES)		(√)	√	3
Appearance Potential Spectroscopy (APS)		√	√	3
Ionisation Loss Spectroscopy (ILS)		√	√	3
Ultraviolet Photoelectron Spectroscopy (UPS)	(√)	(√)	√	3
Ion Neutralisation Spectroscopy (INS)		√	√	4
Low Energy Ion Scattering (LEIS)	√	√		4
High Energy Ion Scattering (HEIS)	√	√		4
Secondary Ion Mass Spectroscopy (SIMS)		√		4
Temperature Programmed Desorption (TPD)		√	(√)	5
Electron and Photon Stimulated Desorption (ESD and PSD)	√	(√)	(√)	5
Field Emission Microscopy (FEM)	√		√	6
Field Ion Microscopy (FIM)	√			6
Work function determinations			√	7
Molecular beam scattering	√	√		8
Infrared Reflection–Adsorption Spectroscopy (IRAS)	√	√	(√)	9
High Resolution Electron Energy Loss Spectroscopy (HREELS)	√	√	(√)	9

one of the three areas and the extent of this information varies considerably.

One generalisation which can be made in the study of surfaces by these techniques is that it is rarely satisfactory to use only one of these techniques. At the very least, some limited information in each of the three areas of structure, composition and electronic structure are needed simultaneously. For this reason most investigations involve several experimental probes on the same chamber and it is this multitechnique approach which has gained much from the development of stainless steel UHV chambers with demountable flanges. For studies on well-characterised surfaces it is probably fair to say that the two most common support techniques to an investigation by other methods are Low Energy Electron Diffraction (LEED) and Auger Electron Spectroscopy (AES). LEED provides a simple and convenient characterisation of the surface long range order while AES provides some indication of chemical composition and, in particular, characterises the cleanness of a surface. Moreover, both can be installed using the same piece of instrumentation (the LEED optics and Retarding Field Analyser (RFA)), although this arrangement falls short of ideal. The wide use of standard characterisation probes such as LEED and AES has greatly improved our ability to compare studies of any particular adsorption system by different techniques performed in different laboratories. Of course the use of these two (or any other) support probes does not totally characterise the surface. A glance at a LEED diffraction pattern shows the basic periodicity of the ordered component of the surface structure but far more study is required to *determine* the surface structure in detail; i.e. to establish the quality of the order or the relative atomic positions on the surface. There is a good chance, however, that two experimenters both working on adsorption of A on B to provide a particular periodicity of the surface and with a particular maximum level of contamination as seen in AES are working on essentially the same surface. In comparing different techniques and investigations in more detail the special aspects of the techniques must be considered. One technique may be dominated by the signal from minority species on a surface while another probes far more 'averaged' properties. We attempt to identify these special aspects, and make some specific comparisons, in later chapters.

One final remark concerns the application of these techniques to 'technical surfaces'; i.e. to the study of surfaces of polycrystalline materials which have typically undergone some high pressure or wet chemical treatment and are loaded into the vacuum chamber and studied without precleaning. Usually it is of interest to 'depth profile' such surfaces to investigate the subsurface as well as the surface layers.

Most of the techniques in table 1.2 are of little use for these surfaces but a few have proved of great value. In particular both AES and X-ray Photoelectron Spectroscopy (XPS) have been widely used to determine surface and subsurface composition and, particularly with XPS, chemical state. Low Energy Ion Scattering (LEIS) and Secondary Ion Mass Spectroscopy (SIMS) have also been used in this way, although SIMS has its greatest application in the analysis of relatively thick films by depth profiling. All of these techniques are primarily concerned with composition and provide no structural information. So far structural studies of technical surfaces have not been pursued with these methods although Surface Extended X-ray Absorption Fine Structure (SEXAFS) may hold some promise for these problems. For these few methods some of their special merits in studies of technical surfaces are also included in the presentations which follow.

2 Surface crystallography and electron diffraction

2.1 Surface symmetry

The classification and description of symmetry properties and structures of bulk (three-dimensional) crystalline materials requires a reasonable understanding of crystallography; notably of the restricted number of types of translational symmetry which crystals can possess (characterised by their associated unit cell which must be one of the 14 Bravais lattices) and the finite number of point and space groups which can define the additional symmetry properties of all possible crystals. Many properties of solids are intimately related to the special symmetry properties of these materials. While a solid surface is intrinsically an imperfection of a crystalline solid, destroying the three-dimensional periodicity of the structure, this region of the solid retains two-dimensional periodicity (parallel to the surface) and this periodicity is an important factor in determining some of the properties of the surface. In particular, it plays a dominant role in allowing electron diffraction techniques to provide information on the structure of the surface, as well as strongly influencing the electronic properties of the surface. For these reasons a proper understanding of surface crystallography is important for a general understanding of many surface effects and is critical for an understanding of the electron diffraction techniques, Low Energy Electron Diffraction (LEED) and Reflection High Energy Electron Diffraction (RHEED).

In discussing the structure of solid surfaces it is helpful to develop a notation which minimises confusion. A surface, in the mathematical sense, cannot have structure, so by 'surface structure' we mean the structure of the solid in the vicinity of the surface. For this reason it is useful to define a region of the solid in the vicinity of the mathematical surface as the *selvedge* (by analogy with the edge of a piece of cloth). A 'surface' can then be thought of as a *substrate* which has the proper three-dimensional periodicity of the bulk, plus the few atomic layers of the selvedge which may adopt atom sites different from those of the bulk; for example, it is probable that the layer spacing normal to the surface will differ slightly from that of the bulk (substrate). It is also possible that reconstruction may take place parallel to the surface in the selvedge (this is known to happen, for example, for some surfaces of Si, Ge, Au and Pt in their 'clean' condition). The selvedge, however, is crystalline in the sense that it retains periodicity parallel to the surface; i.e. it is two-dimensionally periodic. Evidently in reconstructed surfaces this periodicity differs from that of the substrate though it is usually coherent with the substrate periodicity – i.e. both selvedge and substrate have a common periodicity which is greater than that of the substrate (and may be greater than either part in isolation). This notation really only covers clean surfaces; very often the surface structure of interest involves an *adsorbate*. We will use the term adsorbate structure to describe those layers of the surface, usually above the selvedge, which contain a localised excess of some foreign species (whether it arrived from the gas or solid phase). The introduction of an adsorbate may well change the structure of the selvedge and in the presence of a foreign species the outermost layers of the surface, the adsorbate structure, may contain both the new foreign species and the species of the clean substrate.

It is the two-dimensional periodicity of the surface which allows us to classify the possible symmetry elements and the symmetrically different surface structures. It is helpful, however, to clarify the formal reasons for, and extent of, the loss of periodicity in the third dimension. Any surface technique, by definition, penetrates only a small way into the solid; the signal which emerges is likely to contain a large contribution from the top atom layer, a weaker contribution from the next layer, and so on. Evidently the effective depth of penetration varies from technique to technique and is a function of the particular conditions of the experiment. Most techniques, however, penetrate deep enough into the material to contain significant contributions from adsorbate (if present), selvedge and substrate. Thus, from the point of view of classifying the two-dimensional symmetry properties (i.e. those symmetries involving only operations within planes parallel to the surface), it is generally necessary to consider symmetries which are properties of the whole adsorbate, selvedge and substrate complex and not simply one part or one layer of this. The symmetries operate in two dimensions but the surface structure is entirely three dimensional. In some special cases it is possible that the penetration is shallow enough for higher symmetries to be observed because only a small part of this complex is 'seen' by the technique. Of course, even if the whole surface region of the crystal has the same atomic structure as that of the substrate (i.e. there is effectively no adsorbate or selvedge), the technique, by virtue of its surface specificity, 'sees' only a two-dimensionally periodic system; successive atom layers become inequivalent by virtue of the limited penetration itself.

In addition to the translational symmetry parallel to the surface which characterises the crystallinity of the solid surface, it may possess a small number of point and line symmetry operations (which are a small subset of the three-dimensional symmetry operations) which involve rotation or reflection within planes parallel to the surface. This whole subject is dealt with in many undergraduate solid state physics textbooks and is fully classified in detail in the *International Tables for X-ray Crystallography* (1952). Briefly, however, these symmetry operations are (the trivial) one-fold rotation, two-, three-, four- and six-fold rotation axes (note that a five-fold rotation axis or more than six-fold rotation are not compatible with the two-dimensional translational symmetry), mirror reflection in a plane perpendicular to the surface, and glide reflection (involving reflection in a line combined with translation along the direction of the line by half of the translational periodicity in this direction). Consideration of the symmetry properties of two-dimensional lattices (nets) leads to just five symmetrically different Bravais nets. These are hexagonal, characterised by a six-fold rotation axis, square, characterised by a four-fold rotation axis, primitive or centred rectangular which are the *two* symmetrically non-equivalent lattices characterised by mirror symmetry, and oblique which lacks all these symmetries. Note that the centred rectangular net is the only non-primitive net. Centering of any other net leads only to nets which can be equally well classified by primitive nets of the same symmetry. Combining these five Bravais nets with the ten different possible point groups leads to a possible 17 two-dimensional space groups. Thus there are only 17 symmetrically different types of surface structure possible (although, of course, there are an infinite number of possible surface structures). These nets, point group and space groups are set out in figs. 2.1, 2.2 and 2.3 and table 2.1.

It is worth noting that even in the case of unreconstructed clean surfaces the surface unit net is not necessarily a simple projection of the three-dimensional unit cell onto the surface plane. Consider, for example, the {100} surface (i.e. a plane surface parallel to the {100} set of planes) of an f.c.c. solid. Fig. 2.4 shows schematically a view looking down onto such a surface (i.e. a projection of the solid onto the surface). If the x symbols are taken to represent the top layer atoms and all other odd-numbered layers, the o symbols represent all other alternate layer atoms including the layer next to the top one. This surface has square symmetry as expected because both 'surface' and bulk have four-fold rotation axes perpendicular to this surface. However, the surface Bravais net is described by the primitive square unit mesh shown on the right of fig. 2.4. The end or projection of a three-dimensional face-centred unit cell shown on the left forms a centred square unit mesh with twice the area of the true unit mesh. As we have seen, a centred square mesh is symmetrically identical to a primitive square net and is therefore an inappropriate description. This kind of difference in description using two-dimensional mesh and three-dimensional unit cell arises, of course, because of the use of a non-primitive unit cell in three dimensions (which *is* symmetrically distinct), but can lead to confusions of notation, particularly in reciprocal space.

2.2   Description of overlayer structures

If the surface layers of a solid differ structurally from the substrate, either in the form of a reconstructed selvedge, or of an adsorbate, or both of these, then the structure within these layers may be

Fig. 2.1. The five two-dimensional Bravais nets; the specifications are given in table 2.1.

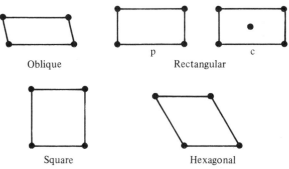

Fig. 2.2. Stereograms of the ten two-dimensional point groups. On the left are shown the equivalent positions, on the right the symmetry operations. The names follow the full and abbreviated 'International' notation.

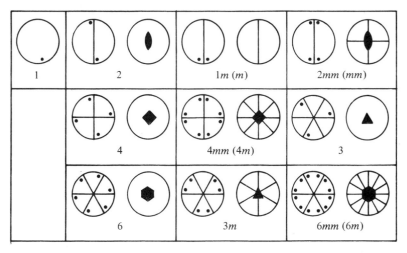

Fig. 2.3. Equivalent positions, symmetry operations and long and short 'International' notation for the 17 two-dimensional space groups.

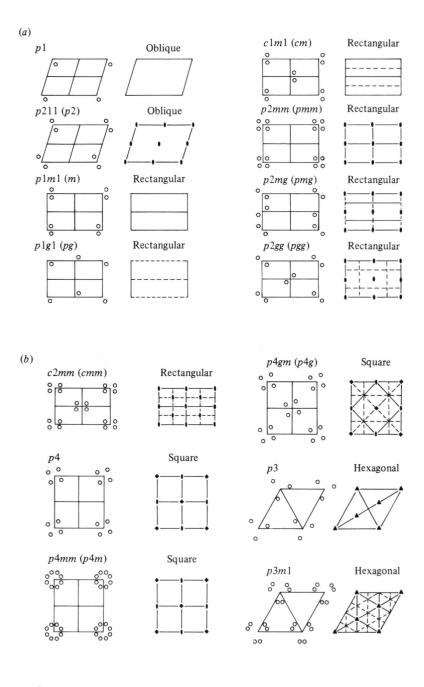

Table 2.1. *The five two-dimensional Bravais nets*

Shape of unit mesh	Mesh symbol	Conventional rule for choice of axes	Nature of axes and angles	Name
General parallelogram	p	None	$a \neq b$ $\gamma \neq 90°$	Oblique
Rectangle	p c	Two shortest, mutually perpendicular vectors	$a \neq b$ $\gamma = 90°$	Rectangular
Square	p	Two shortest, mutually perpendicular vectors	$a = b$ $\gamma = 90°$	Square
60° angle rhombus	p	Two shortest vectors at 120° to each other	$a = b$ $\gamma = 120°$	Hexagonal

either disordered, ordered and coherent with the substrate, or ordered and incoherent with the surface. Evidently the first case is of little interest from the point of view of surface crystallography. Provided order does exist in the adsorbate or selvedge, however, it is convenient to describe this order by relating its Bravais net to that of the underlying substrate. This is usually done in one of two ways, the most general of which,

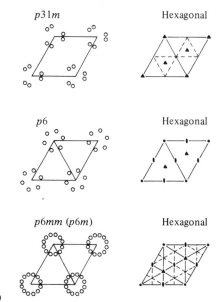

p31m Hexagonal

p6 Hexagonal

p6mm (p6m) Hexagonal

(c)

proposed by Park and Madden (1968) involves a simple vectorial construction. If the primitive translation vectors of the substrate net are **a** and **b** and those of the adsorbate or selvedge are **a′** and **b′** then we can relate these by

$$\mathbf{a}' = G_{11}\mathbf{a} + G_{12}\mathbf{b} \qquad (2.1)$$

$$\mathbf{b}' = G_{21}\mathbf{a} + G_{22}\mathbf{b} \qquad (2.2)$$

where the G_{ij} are four coefficients which form a matrix G

$$G = \begin{pmatrix} G_{11} & G_{12} \\ G_{21} & G_{22} \end{pmatrix} \qquad (2.3)$$

such that the absorbate and substrate meshes are related by

$$\begin{pmatrix} \mathbf{a}' \\ \mathbf{b}' \end{pmatrix} = G \begin{pmatrix} \mathbf{a} \\ \mathbf{b} \end{pmatrix} \qquad (2.4)$$

Another property of this matrix is that, because the area of the substrate unit mesh is given by $|\mathbf{a} \times \mathbf{b}|$ the determinant of G, det G is simply the ratio of the areas of the two meshes and provides a convenient classification system for the type of surface structure involved as follows:

(a) *det G integral and all matrix components integral;* the two meshes are *simply related* with the adsorbate mesh having the same translational symmetry as the whole surface.

(b) *det G a rational fraction (or det G integral and some matrix components rational);* the two meshes are *rationally related.* In this case the structure is still commensurate but the true surface mesh is larger than either the substrate or adsorbate mesh. This surface mesh has a size dictated by the distances over which the two meshes come into

Fig. 2.4. Schematic diagram of a {100} f.c.c. solid surface with the top layer (and odd-numbered layer) atoms shown as crosses, the second (and even-numbered layer) atoms shown as circles. The unit mesh on the left is the projection of the three-dimensional (non-primitive) unit cell. On the right is shown the primitive unit mesh.

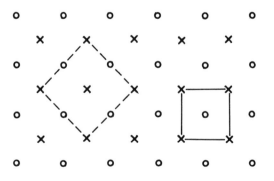

coincidence at regular intervals, and for this reason such structures are frequently referred to as *coincidence lattice* (or more properly *coincidence net*) structures. The true surface mesh now has primitive translation vectors $\mathbf{a}''$ and $\mathbf{b}''$ related to the substrate and adsorbate meshes by matrices P and Q such that

$$\begin{pmatrix} \mathbf{a}'' \\ \mathbf{b}'' \end{pmatrix} = P \begin{pmatrix} \mathbf{a} \\ \mathbf{b} \end{pmatrix} = Q \begin{pmatrix} \mathbf{a}' \\ \mathbf{b}' \end{pmatrix} \tag{2.5}$$

det P and det Q being chosen to have the smallest possible integral values and being related by

$$\det G = \frac{\det P}{\det Q} \tag{2.6}$$

(c) *det G irrational;* the two meshes are now incommensurate and no true surface mesh exists. Such a situation implies that the substrate is simply providing a flat surface on which the adsorbate or selvedge can form its own two-dimensional structure. This might be expected, for example, if adsorbate–adsorbate bonding is very much stronger than the adsorbate–substrate bonding or if the adsorbed species are too large to 'feel the granularity' of the substrate.

A somewhat more convenient, but less versatile, notation for surface mesh structures which is more widely used is that suggested by Wood (1964). In this case the notation defines the ratio of the lengths of the surface and substrate meshes, together with the angle through which one mesh must be rotated to align the two pairs of primitive translation vectors. In this notation if adsorbate A on the $\{hkl\}$ surface of material X causes the formation of a structure having primitive translation vectors of length $|\mathbf{a}'| = p|\mathbf{a}|$ and $|\mathbf{b}'| = q|\mathbf{b}|$ with a unit mesh rotation of ϕ the structure is referred to as

$$X\{hkl\}p \times q\text{-}R\phi°\text{-}A$$

or often

$$X\{hkl\}(p \times q)R\phi°\text{-}A$$

Note that this notation can only be used if the included angles of the surface and substrate unit meshes are the same. Thus while it is suitable for systems where the surface and substrate meshes have the same Bravais net, or where one is rectangular and the other square, in general it is not satisfactory for mixed symmetry meshes. In these cases the matrix notation must be used. As examples of the Wood notation, a clean unreconstructed Ni$\{100\}$ surface is denoted by Ni$\{100\}(1 \times 1)$ while one structure formed by the adsorption of O on this surface is the Ni$\{100\}(2 \times 2)$–O structure; by contrast the Si$\{100\}$ is typically reconstructed to Si$\{100\}(2 \times 1)$ in its clean state, but adsorption of atomic H can 'unreconstruct' the surface to give Si$\{100\}(1 \times 1)$–H. Some examples of surface nets and their matrix and Wood notations are given in fig. 2.5. One particular example of note in this figure is the $(\sqrt{2} \times \sqrt{2})R45°$ structure on the square mesh which is a common one on cubic $\{100\}$ surfaces. It is extremely common to refer to this in terms of a centred unit mesh $\sqrt{2}$ times larger and not rotated relative to the substrate mesh as c(2×2), thus encouraging the use of the notation p(2×2) for a true (2×2) structure. As we have seen, no centred square Bravais net exists as a symmetrically distinct mesh from the primitive square one but this notation is widely used and is thus effectively absorbed into the nomenclature.

Fig. 2.5. Examples of overlayer structures in which the open circles represent the periodicity of the substrate while the crosses show adsorbate or selvedge mesh periodicity. In each case the substrate Bravais net is shown dashed while the full surface Bravais net is shown with full lines. (*a*) shows a $(\sqrt{3} \times \sqrt{3})R30°$ structure on an hexagonal substrate, the matrix notation being $\begin{pmatrix} 2 & 1 \\ -1 & 1 \end{pmatrix}$. (*b*), (*c*) and (*d*) show (2×2) or $\begin{pmatrix} 2 & 0 \\ 0 & 2 \end{pmatrix}$, $(\sqrt{2} \times \sqrt{2})R45°$ or $\begin{pmatrix} 1 & 1 \\ -1 & 1 \end{pmatrix}$ and (2×1) or $\begin{pmatrix} 2 & 0 \\ 0 & 1 \end{pmatrix}$, structures. Note in case (*c*) that the dash–dot net is centred but not rotated relative to the substrate unit mesh so that this structure is often referred to as c(2×2), necessitating the notation p(2×2) for structure (*b*).

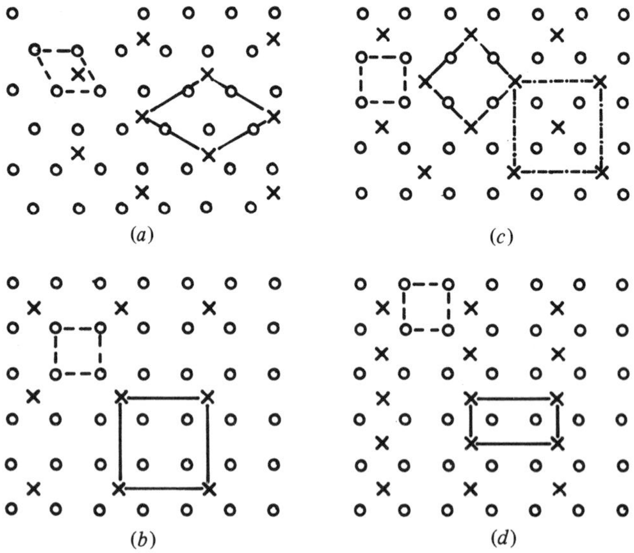

(*a*)   (*c*)

(*b*)   (*d*)

2.3   Reciprocal net and electron diffraction

One of the most important methods of investigating the structure of periodic systems such as bulk solids or solid surfaces is by a diffraction technique (for the bulk solid this can be using X-rays or electrons). Diffraction techniques give information rather readily about the translational symmetry of the system in the form of the 'reciprocal lattice'. For example, diffraction in a bulk solid (three-dimensionally periodic) gives rise to a series of diffracted beams which are readily explained in terms of conservation of energy and conservation of momentum but for the addition of any reciprocal lattice vector. Thus, for a three-dimensional system, if the incident wavevector is $\mathbf{k}$ and the emerging wavevectors are $\mathbf{k}'$ then conservation of energy gives

$$k^2 = k'^2 \tag{2.7}$$

and conservation of momentum gives

$$\mathbf{k}' = \mathbf{k} + \mathbf{g}_{hkl} \tag{2.8}$$

where $\mathbf{g}_{hkl}$ is a reciprocal lattice vector

$$\mathbf{g}_{hkl} = h\mathbf{a}^* + k\mathbf{b}^* + l\mathbf{c}^* \tag{2.9}$$

the primitive translation vectors of the reciprocal lattice $\mathbf{a}^*, \mathbf{b}^*, \mathbf{c}^*$ being related to those of the real lattice $\mathbf{a}, \mathbf{b}, \mathbf{c}$ by

$$\mathbf{a}^* = 2\pi \frac{\mathbf{b} \times \mathbf{c}}{V}, \quad \mathbf{b}^* = 2\pi \frac{\mathbf{c} \times \mathbf{a}}{V}, \quad \mathbf{c}^* = 2\pi \frac{\mathbf{a} \times \mathbf{b}}{V}, \quad V = \mathbf{a} \cdot \mathbf{b} \times \mathbf{c} \tag{2.10}$$

These conditions are essentially an expression of Bragg's law. The diffracted beams are thus characterised by the points of the reciprocal lattice (hkl) and their location in space readily leads to a deduction of the form of the reciprocal lattice and so, by an inverse form of the transformations equation (2.10) to the deduction of the real lattice. A convenient graphical representation of these equations ((2.7)–(2.9)) in reciprocal space used in three-dimensional diffraction studies is the *Ewald sphere* construction. This is shown in fig. 2.6(*a*). The construction, superimposed on the reciprocal lattice, involves drawing a vector $\mathbf{k}$ to terminate at the origin of the reciprocal lattice and then constructing a sphere, radius k, about the beginning of the vector $\mathbf{k}$. For any point at which this sphere passes through a reciprocal lattice point, a line to this point from the centre of the sphere represents a diffracted beam $\mathbf{k}'$. Several such beams are shown in fig. 2.6(*a*) together with one associated reciprocal lattice vector $\mathbf{g}_{hkl}$.

For diffraction processes involving only the surface, a very similar situation exists except that, because the system is only two-dimensionally periodic (parallel to the surface) only the component of the wavevector parallel to the surface is conserved with the addition of a reciprocal net vector. Thus, conservation of energy again gives

$$k^2 = k'^2 \tag{2.7}$$

or, if components parallel to, and perpendicular to, the surface are denoted by suffixes $\parallel$ and $\perp$

$$\mathbf{k}_\parallel{}^2 + \mathbf{k}_\perp{}^2 = \mathbf{k}'_\parallel{}^2 + \mathbf{k}'_\perp{}^2 \tag{2.11}$$

while the conservation of momentum gives

$$\mathbf{k}'_\parallel = \mathbf{k}_\parallel + \mathbf{g}_{hk} \tag{2.12}$$

with

$$\mathbf{g}_{hk} = h\mathbf{a}^* + k\mathbf{b}^* \tag{2.13}$$

and

$$\mathbf{a}^* = 2\pi \frac{\mathbf{b} \times \mathbf{n}}{A}, \quad \mathbf{b}^* = 2\pi \frac{\mathbf{n} \times \mathbf{a}}{A}, \quad A = \mathbf{a} \cdot \mathbf{b} \times \mathbf{n} \tag{2.14}$$

where $\mathbf{n}$ is a unit vector normal to the surface.

Note that $\mathbf{k}_\perp$ is not conserved in this process. Equation (2.12) is applicable to electrons traversing the crystal vacuum interface, whatever their source. It is therefore equally applicable, for example, to Auger or photoemitted electrons created in the surface region. Evidently, as the diffraction conditions are now dictated by a reciprocal net vector, having two components, the diffracted beams are denoted by a two-number index hk (commonly written (hk)). A modified version of the Ewald sphere construction adapted to represent these surface equations ((2.11)–

Fig. 2.6. Ewald sphere construction for bulk (*a*) and surface (*b*) cases. The incident electron wavevector, $\mathbf{k}$, is labelled in each case with possible scattered wavevectors ($\mathbf{k}'$). In the bulk case the reciprocal lattice vector $\mathbf{q}_{303}$ associated with the relevant scattered wave is also shown. The dashed scattered wavevectors in the surface case propagate *into* the solid and are not observable.

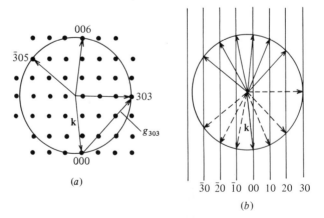

(2.13)) is shown in fig. 2.6(*b*). While the construction remains a three-dimensional one (for the diffracted and incident beams lie in three dimensions), the reciprocal lattice of fig. 2.6(*a*) has been replaced by infinite 'reciprocal lattice rods' perpendicular to the surface and passing through the reciprocal net points. This greatly relaxes the conditions for the formation of diffracted beams; while in the three-dimensional case a small change in the electron energy or direction of **k** will cause the loss of many beams but the appearance of new ones, in the surface case corresponding changes, simply cause slight movement of the diffracted beams. One further consequence of the relaxed conditions in the surface diffraction case is that there are now two diffracted beams associated with each reciprocal net point (reciprocal lattice rod) *hk*. However, half of these, the ones shown dashed in fig. 2.6(*b*), are propagating on into the crystal rather than being backscattered out of it, and so are not observed.

The indexing of the diffracted beams by the associated reciprocal net vector is, by convention, referenced to the substrate real and reciprocal net. This means that if the selvedge or adsorbate structures have larger periodicities, the surface reciprocal net is smaller than that of the substrate alone and the 'extra' reciprocal net points and associated diffracted beams are denoted by fractional rather than integral indices. Some examples of schematic real structures and their associated real and reciprocal nets are shown in fig. 2.7. Note that in the case of fig. 2.7(*b*) the real net is a centred rectangular one. By convention this is therefore described by a non-primitive real net but this leads to a reciprocal net which is too small. In the figure the true reciprocal net is shown (generated by the use of equation (2.14) on the real net *primitive* translation vectors) but by convention the reciprocal net points are indexed relative to the spuriously small net, leading to 'missing' net points and diffracted beams of the type *hk* where $h + k$ is odd. Similar effects occur in three-dimensional crystallography due to the choice of non-primitive unit cells to describe the very common structures having non-primitive Bravais lattices. Fortunately, in the two-dimensional case there is only one non-primitive Bravais net and the problem is much less common (for common, simple surfaces, only the b.c.c. {110} surface possesses this net). Notice, however, that in some early LEED literature other surfaces were described by non-primitive unit meshes leading to similar problems in beam notation. For example, the description of the f.c.c. {100} surface by the centred square mesh of the end of the unit cell (fig. 2.4) leads to spuriously labelled 'missing' beams; fortunately this practice seems to have died out.

By contrast there is one special case in which 'missing' diffracted beams can provide valuable additional information on a surface structure beyond the point of simply establishing surface periodicity. This occurs when the structure possesses glide symmetry lines (other than those associated with centred nets – see fig. 2.3). Only four space groups contain such elements (*pg*, *pmg*, *pgg* and *p4g*), but in these cases it is found that, when both incident and diffracted beam lie in the glide plane perpendicular to the surface, alternate beams (those for which the indexing is odd) along the glide plane are missing. For example, for a *p4g* structure at normal incidence all beams *h*0 or *k*0 for which *h* or *k* are odd are missing. This consequence of glide symmetry can be demonstrated quite generally (Holland & Woodruff, 1973) and is a special case in which the space group symmetry of a surface structure can be readily determined.

Fig. 2.7. Some real and reciprocal meshes, the relevant unit meshes being marked. Reciprocal net points are also labelled using the convention described in the text. Note that examples (*a*) and (*b*) represent unreconstructed surfaces while (*c*) and (*d*) show an adsorbate or selvedge structure (crosses) superimposed on a substrate structure (circles). In (*b*) the *primitive* real and reciprocal unit nets are shown dashed.

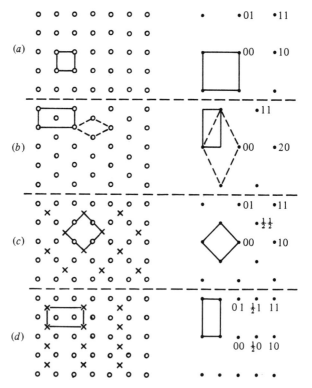

2.4   Electron diffraction – qualitative consideration

In the last section we demonstrated that electron diffraction is able to provide rather direct information on the periodicity or translational symmetry of a surface structure. While this piece of information is valuable, it does not provide any clear picture of the actual location of the surface atoms within the unit mesh. To extract this information we must, as in X-ray crystallography of bulk structures, study the diffracted beam intensities and relate these to a quantitative scattering theory. Unfortunately, under conditions which make the electron diffraction surface sensitive, this scattering theory proves to be complicated and the deduction of the surface structure from the diffracted beam intensities is considerably more complicated than in X-ray diffraction. We will defer discussion of this problem until the next section and first look at the more qualitative aspects of LEED and RHEED.

LEED can reasonably claim to be the oldest of modern surface techniques, as the first LEED experiment by Davisson and Germer in 1927 also provided the first demonstration of the wave nature of the electron. They showed that when a beam of monoenergetic electrons was fired at a single crystal Ni surface the elastically scattered electrons emerged in preferred directions which could be explained by diffraction from the periodic distribution of atoms in their crystalline Ni sample. Despite these early beginnings, however, LEED has, like other surface techniques, only been widely exploited since the development of modern UHV technology in the early 1960s. In the early days of electron diffraction it was soon realised that high electron energies allowed transmission diffraction in thin samples and that this was a far more appropriate method of studying solids; the surface sensitivity of LEED proved to be a considerable disadvantage for bulk studies, not only because the intrinsic surface structure may differ from the bulk, but also because surface contamination is of far greater importance and so introduces stringent vacuum and cleaning constraints. Nevertheless, a small amount of LEED work was performed during this period of relative inactivity, particularly by H. E. Farnsworth and his co-workers, and at least one important instrumental development during the latter part of this period, the introduction of a direct visual display of the diffraction pattern, did much to encourage the rapid growth of interest in LEED in the 1960s.

A schematic diagram of a typical modern LEED 'optics' allowing direct visual observation of the diffraction pattern is shown in fig. 2.8. The electron gun, similar to those used in cathode ray tubes and utilising electrostatic focussing, delivers a beam of typically 1 μA at energies in the 20–300 eV energy range to the target. The energy spread of this beam, determined by the high temperature of the thermionic source, is usually ~0.5 eV. Electrons scattered or emitted from the sample travel in straight lines in the field free region to the spherical sector grids as the first of these grids is set to the same (ground) potential as the sample. The next one or two grids are set to retard all electrons other than those which have been elastically scattered using a potential close to that of the original electron source filament. The elastically scattered, diffracted, electrons passing through are then reaccelerated onto the fluorescent screen by applying about 5 kV to the screen to ensure that they produce a fluorescent image of the diffraction pattern which can be viewed through the grids from behind the sample. Comparison of this figure with that of the Ewald sphere construction under LEED conditions (fig. 2.6(*b*)) shows rather clearly that the observed diffraction pattern is simply a projection of the surface reciprocal net at a 'magnification' determined by the incident electron energy and thus k value. The display nature of this instrument therefore makes it ideal for easy assessment of the translational symmetry of the surface although less well suited to quantitative beam intensity measurements. These can be made in a direct electrical fashion by interposing a movable Faraday cup (with retarding filtering grids fitted) between the sample and grids. Alternatively, the brightness of the fluorescence associated with a diffracted beam can be measured using a spot photometer. A recent variation on this latter approach is to use a TV camera to view the screen and to extract the beam intensities from the TV signal. Interfaced to a small computer this system has the great virtue that diffracted beams can be 'followed' by software rather than by mechanical scanning as the

Fig. 2.8. Schematic diagram of display LEED 'optics'. The potential V_E defines the electron energy as eV_E.

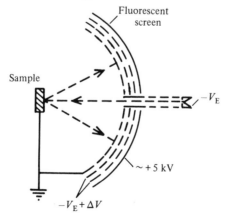

electron energy is changed and thus the magnification of the reciprocal net varies.

The surface sensitivity of LEED, necessary to provide surface rather than bulk information, results from two effects. Firstly in the LEED energy range the mean-free-path for inelastic scattering (by single particle excitations and plasmons) is very short – typically only about 5 Å. The properties of this parameter are discussed more extensively in the following chapter but we note that LEED operates in the energy range in which it is typically at its smallest value, so that electrons penetrating more than two or three atomic layers into the solid have a high probability of losing energy (and coherence) relative to the incident beam and thus being lost from the elastic diffracted flux. A second source of surface sensitivity in LEED is the elastic scattering itself; backscattering is very strong (ion core cross-sections may be as large as 1 $Å^2$) so that successive atom layers receive smaller incident electron fluxes and so contribute less to the scattering. Typically, these two effects contribute about the same amount to the surface specificity.

By contrast the technique of RHEED operates in an energy range (up to ~ 30–100 keV) in which inelastic scattering mean-free-paths are relatively long (~ 100–1000 Å) and the elastic scattering is strongly peaked in the forward direction with very little backscattering. In this case surface sensitivity can only be achieved by using electron beams incident on, and emergent from the surface at grazing angles ($< 5°$). This ensures that the penetration into the surface by the diffracted electrons remains small and that reasonably intense diffracted beams emerge. Rather more details of the RHEED experiment are given later in this chapter. For the moment we simply note that the two diffraction techniques have in common the ability to detect the two-dimensional periodicity of the top few atom layers by displaying a projection of the reciprocal net, and that, in principle at least, the intensities of diffracted beams contain information on atomic locations within the surface unit mesh. We will, however, develop further discussion of the theory and application of electron diffraction in relation to the far more widely used technique of LEED.

2.5 Domains, steps and defects

As we have already noted, there is a rather direct relationship between the real and reciprocal meshes associated with a surface structure and as the diffraction pattern is simply a projection of the reciprocal mesh we might expect the real mesh and associated periodicity of a surface structure to be easily deduced. Moreover, it is easy to show that at normal incidence the diffraction pattern shows the point group symmetry of the surface structure; off normal incidence the direction of the incident beam reduces the symmetry of the diffraction experiment. Finally, we have also noted that the only special symmetry operation involved in two-dimensional space groups which is not contained in the point group or the translational symmetry of the mesh is the glide operation, and that the presence of this symmetry leads to characteristic missing diffracted beams. The full space group of the surface should therefore be obtainable from the diffraction pattern by inspection without any quantitative discussion of the diffraction process.

Unfortunately the foregoing discussion includes a crucial and generally inappropriate assumption that the surface under study consists of a single, perfectly periodic, structural *domain*. Even leaving aside the possibility of defects in the substrate order such as surface vacancies, adatoms and steps, this is unlikely to be true. Consider the formation of an adsorption structure on a perfectly periodic substrate. Adatoms arrive randomly on the surface from the gas phase and generally will adopt preferred local registry sites relative to the substrate, although they may possess considerable mobility to hop between other symmetrically identical sites. At sufficient coverage these adsorbed atoms may display long range as well as short range order and thus give rise to new diffraction features; if there is strong attraction between adsorbate atoms on the surface this may occur by the formation of ordered 'islands' on the surface with rather bare substrate in between. With less interaction the long range order may only become apparent at higher coverages as many of the available sites become filled. In either case, however, we see that if the ordered structure involves a unit mesh larger than that of the substrate there must be more than one identical substrate site per new surface mesh and thus a degree of arbitrariness in the location of occupied sites used to define the origin of the surface mesh. Because the origin of the local surface mesh at different points on the surface is fixed by atoms which arrived independently this arbitrariness of location will lead to defects in the translational symmetry of the surface as a whole. The resulting structure will consist of *domains* of perfectly periodic structure but dislocations in the periodicity occur between the domains at the *antiphase domain boundaries*, so named because these lead to phase differences between the scattered electron amplitude from adjacent, otherwise identical domains. Notice that each domain is symmetrically equivalent with regard to the substrate.

The existence of domain structures leads to a number of features of a real LEED experiment but the simplest of these does not involve coherent interference between the scattering from adjacent domains. In a real experiment it is not only the surface which falls short of our ideal.

The incident electron gun has defects which give rise to diffraction spots of finite size even from a perfect surface. This finite angular size sets limits on the ability to detect faults in the periodicity on a distance scale greater than some *transfer width* on the surface. The effect is similar to that which would occur if the coherence of the source was limited. If this coherence width is significantly smaller than the average domain size then interference across boundaries will be relatively unimportant and the observed diffraction pattern will consist of an incoherent sum of the diffraction patterns of the individual domains; i.e. we may add diffracted intensities rather than amplitudes from the domains. In this case domain boundaries which involve only translational defects in the structure will not affect the diffraction pattern. However, because all domains are symmetrically equivalent as defined by the substrate point group, any new surface structure (adsorption or clean surface reconstruction) which has a lower point group symmetry than the substrate will consist of domains which are not equivalent relative to the new surface point group and so to a mixing of non-equivalent diffraction patterns. This effect is most noticeable when the surface *mesh* has a lower point group symmetry than the substrate (e.g. a rectangular mesh on a 4mm-square substrate). In this case the actual location of diffracted beams differs in some of the domains and the mixed domain diffraction pattern therefore has more beams than the single domain diffraction pattern. This effect is illustrated in the examples of fig. 2.9. In these cases (e.g. the (3×1) rectangular mesh on a square substrate) we see that the resulting diffraction pattern appears to have an associated reciprocal mesh which is smaller than the true one but has many 'missing beams'. Thus, even the determination of the true reciprocal and hence real mesh is not trivial in such cases and must, in complex cases, be determined by a trial and error procedure. A more insidious effect of these domain structures results from a surface structure whose point group is lower than that of the substrate but whose associated mesh has the same point group as the substrate. An example of such a situation would be the formation of a half-monolayer adsorption structure with a $c(2 \times 2)$ $((\sqrt{2} \times \sqrt{2})R45°)$ mesh on a square substrate where the adsorbate adopts a two-fold coordinated bridge site (see fig. 2.10). In this case the surface mesh (square) has the 4mm point group of the substrate but the low symmetry adsorption site reduces the surface structure point group to 2mm. The two symmetrically non-equivalent domains of the structure, related by the 90° rotation of the missing four-fold symmetry operator, therefore lead to two diffraction patterns in which the beam locations are identical but the beam intensities differ (e.g. the 10 and 01 intensities will differ for the individual domains but are mixed in the resulting pattern). As a result, the point group symmetry of the observed diffraction pattern will be that of the substrate even when it differs from that of the individual domains – the 'true surface structure'. This is generally true and sets an important limitation on our ability to determine the surface space group from the observed diffraction pattern. Of course in the case of lower symmetry surface meshes, as we have seen, 'missing beams' always allow the true surface mesh to be found (with perseverance) and also pinpoint at least some domain effects. Moreover, in the case of surface structures containing glide lines, the absence of characteristic missing beams can often permit unique space group determination. Nevertheless, domain effects generally exclude this determination and lead to increased ambiguities to be resolved by full quantitative structural determinations.

One important proviso can be made regarding these general domain effect remarks. In certain cases, such as when a single crystal surface is prepared at a slight misorientation (~ 1–2°) to the desired low index face, the resulting surface may, because of its intrinsically lower symmetry, suppress the formation of some domains to be expected on the low index

Fig. 2.9. Two examples of the effects of multiple domains on a resultant LEED pattern. In each case the real space surface structure, the single domain diffraction pattern and the sum of all equivalent domain diffraction patterns is shown. The upper example of a rectangular mesh on a square substrate involves two domain types. In the lower example three domains of a rectangular mesh on the hexagonal substrate contribute to the final pattern.

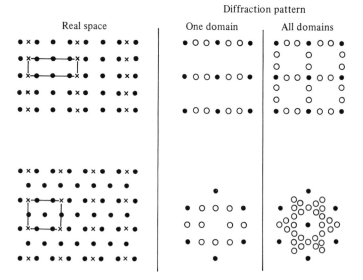

face itself. However, as the surface is likely to be dominated by low index 'terraces', LEED data from the clean unreconstructed surface may appear to have the full point group symmetry of these terraces. Such effects can be invaluable in determining the existence of low symmetry structures if properly appreciated.

We now return to the effect of coherent interference in the electron scattering from adjacent domains. Such effects may be important if the surface structural domains are significantly smaller than the transfer width of the incident electron beam. We should therefore evaluate this parameter first to establish the range of applicability of the arguments we will develop. The instrumental limitations arise from two effects; the energy spread, ΔE, of the incident electron beam of average energy E, and the angular spread $\Delta\theta$ (which may be convergent or divergent). If the electrons left the gun coherent with one another we could ask how this coherence is lost by the deviations in $k_{\parallel}$ at the surface as a result of these two indeterminacies. It would then be appropriate to define a 'coherence width' on the surface. However, the emitted electrons have no specific phase relationships and the diffraction process therefore arises because each electron interferes with itself. In this case the only true meaning of a

Fig. 2.10. Schematic diagram of two domains of a $(\sqrt{2} \times \sqrt{2})R45°$ $(c(2 \times 2))$ structure formed by adsorption at the cross sites onto a square mesh substrate (open circles). The two domains are symmetrically equivalent relative to the substrate and lead to diffracted beams in the same locations, but because the adsorbate–substrate coordination symmetry (two-fold) is lower than that of the substrate (four-fold) the diffracted beam intensities will differ. A sum of the two domains, however, will lead to a diffraction pattern showing four-fold symmetry.

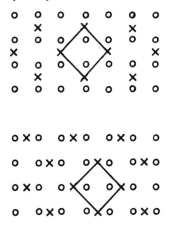

coherence width arises because the electrons originate from a point relatively close to the surface (~ 10 cm) and thus arrive at the surface as spherical rather than plane waves. In such a case a useful definition of coherence width is half the dimension of the first Fresnel zone at the surface which is $r = (D\lambda/2)^{\frac{1}{2}}$ with D the distance to the surface and λ the electron wavelength. Typical values of 10 cm and 1 Å for these parameters give a value of r as 1 μm. This would actually lead to an angular spread of the diffracted beams of about 10^{-4} radians which is negligibly small compared with the widths observed in typical LEED experiments, indicating that this effect is not really significant.

Nonetheless, it is clear that if the defects of the electron gun (notably its angular divergence) lead to some minimum diffracted beam divergence even from a perfect surface, which should give diffracted beams which are δ-functions in angle, then it will be impossible to detect variations in surface periodicity over distances longer than those which would give rise to a smaller diffracted beam divergence. This allows us to define a distance on the surface known as the transfer width (Comsa, 1979; Park, Houston & Schreiner, 1971) which has much the same influence on the experiment as a coherence width, but a different physical origin. The contribution to this transfer width in terms of the experimental parameters is best obtained from a recasting of the diffraction condition. Firstly we note a rather useful relationship from substitution of values in the De Broglie relationship that the electron wavelength in Å, is given by

$$\lambda = (150.4/E)^{\frac{1}{2}} \tag{2.15}$$

with E expressed in eV. If we then define the component of the change in **k** of the electron parallel to the surface in a scattering event as $\mathbf{S}_{\parallel}$ ($= \mathbf{k}'_{\parallel} - \mathbf{k}_{\parallel}$) then for an incidence angle θ_{in} and emergence angle θ_{out}

$$\begin{aligned} S_{\parallel} &= k|\sin \theta_{\text{out}} - \sin \theta_{\text{in}}| \\ &= 2\pi(E/150.4)^{\frac{1}{2}}|\sin \theta_{\text{out}} - \sin \theta_{\text{in}}| \end{aligned} \tag{2.16}$$

Further, we note that the condition for a diffracted beam is that $S_{\parallel}$ is a reciprocal net vector (i.e. 2π divided by a periodicity). Thus if we evaluate the variations in $S_{\parallel}$, $\Delta S_{\parallel}$, due to variations in E and θ_{out}, the transfer width is given by

$$w = 2\pi/\Delta S_{\parallel} \tag{2.17}$$

Taking the partial derivative of $S_{\parallel}$ with respect to θ_{out} gives the appropriate expression for the influence of beam divergence as

$$w_{\theta} = \frac{(150.4/E)^{\frac{1}{2}}}{\cos \theta_{\text{out}} \, \Delta \theta} = \frac{\lambda}{\cos \theta_{\text{out}} \, \Delta \theta} \tag{2.18}$$

while taking the derivative with respect to E given the effect of energy broadening, ΔE as

$$w_E = \frac{2(150.4E)^{\frac{1}{2}}}{\Delta E |\sin \theta_{out} - \sin \theta_{in}|}$$

$$= \frac{2\pi}{S_{\parallel}} \frac{2E}{\Delta E} \qquad (2.19)$$

Typical values of $\lambda = 1$ Å, $\cos \theta_{out} = 1.0$ and $\Delta\theta \approx 10^{-2}$ rad give $w_{\theta} = 100$ Å while for an energy broadening $\Delta E = 0.5$ eV, $E = 100$ eV and $(2\pi/|S_{\parallel}|)$ set at a typical interatomic spacing of 2.5 Å gives $w_E = 1000$ Å. Evidently, it is therefore the angular and not the energy broadening which is important in determining the coherence width. Strictly we should consider individual contributions from different components of the gun which give rise to a finite size of diffracted 'spot' such as aperture sizes and the angular resolution involved in measuring the spot size (Park, Houston & Schreiner, 1971; Wang & Lagally, 1979) but the arguments here identify the main limitation as due to deficiencies in beam size and divergence and not in energy spread, and also indicate that a typical transfer width is of the order of 100 Å. For quantitative measurements of diffracted beam size effects due to structural effects it is clearly necessary to determine these instrumental limitations quantitatively which are convoluted with the 'true response' of the surface to give the observed signal. Procedures for pursuing this in detail are given in the references above. Indeed, a detailed analysis of these procedures suggests that if very careful measurements of the exact beam profile are made the effective transfer width (i.e. the limit of detectable structural effects) may be substantially larger than implied by simple measurements of the overall angular width of diffracted beams (Lu & Lagally, 1980). It is, of course, also possible to refine the instrumentation to some extent if the principal object of an experiment is to study the degree of coherence in surface structures, but even this typical figure of 100 Å is ~ 40 times a typical interatomic spacing on the surface and shows that coherent interference of small domains may be important in LEED.

In order to determine the effects of coherent interference between domains, and of the effect of defects in surface periodic structures, we must develop a simple mathematical theory. So far we have used a model-independent theory which only requires perfect (infinite) two-dimensional periodicity and leads to exact diffraction conditions – i.e. to diffracted beams which are δ-functions in space. Antiphase domain boundaries and other defects break this perfect translational symmetry and must be evaluated with a less general theory. We therefore adopt the single scattering formalism normally used in X-ray diffraction. We assume that the incident electrons form a plane wave of wavevector **k** and that far from the surface the amplitude $A_{\Delta k}$ scattered into an outgoing wavevector **k**′ with

$$\Delta \mathbf{k} = \mathbf{k}' - \mathbf{k} \qquad (2.20)$$

is given by the coherent sum of scatterings from each atom of atomic scattering factor f_j taking into account the phase difference introduced by different path lengths. Thus

$$A_{\Delta k} = \sum_n \sum_m \exp{(\mathrm{i}\,\Delta \mathbf{k} \cdot \mathbf{A})} \sum_j f_j \exp{(\mathrm{i}\,\Delta \mathbf{k} \cdot \mathbf{r}_j)} \qquad (2.21)$$

where the summations are over all atoms in the surface region whose locations are given by the position vector $\mathbf{A} + \mathbf{r}_j$ where $\mathbf{r}_j$ defines the position within a surface unit mesh and $\mathbf{A} = n\mathbf{a} + m\mathbf{b}$ is a real surface net vector. Thus the sum over n and m is over surface meshes and the sum over j is a sum of scatterers within a unit mesh. Some remark should be made concerning the parameter f_j. In X-ray diffraction this is identified with the atomic scattering factor for an isolated atom of the relevant species. It depends on scattering angle and energy (i.e. on $\Delta \mathbf{k}$) and is real (i.e. not complex). In LEED the situation is more complicated. Firstly, even for a free atom the electron scattering cross-section involves a phase shift (also dependent on $\Delta \mathbf{k}$) and so is complex. Secondly, because the mean-free-path for *inelastic* scattering is short (~ 5 Å; a more thorough discussion of this is given in the following chapter) the incident electron wave is attenuated exponentially as it penetrates the solid; one way of incorporating this effect into our simple formulation (equation (2.21)) is to make the f_j smaller for deeper layers, even of the same species. This also ensures that the summation over j, in principle over all atoms falling below a surface unit mesh and therefore effectively infinite, is truncated after a few atom layers because of the rapidly decaying values of the f_j. Finally, however, we should consider the effects of *multiple scattering*. The free atom electron scattering factors can be extremely large (equivalent to a cross-section ~ 1 Å^2 or 10^{10} times larger than X-ray scattering cross-sections) and as a result multiple scattering of electrons between atoms can become important. It is the proper description of this process which presents the main problem in determining surface structures (i.e. actual atomic locations) by LEED and will be discussed further in section 2.6. For our present purposes, however, we note that the effect of multiple scattering is that each atom is presented with an incident electron flux which is not simply the incident plane wave but also has contributions of multiple scattering from other atoms. However, these multiple scattering effects will be identical for all symmetrically equivalent atoms; i.e. for all atoms occupying the same site in the surface unit mesh. This means that these effects can therefore be included in the f_j. The resulting f_j differ for all sites which are symmetrically non-equivalent, even if they are the same species and are in the same atom layer. All of these complications can therefore be separated out into the *geometrical structure factor*

$$F_{\Delta k} = \sum_j f_j \exp{(i\, \Delta \mathbf{k} \cdot \mathbf{r}_j)} \qquad (2.22)$$

so that writing

$$A_{\Delta k} = \sum_n \sum_m \exp{(i\, \Delta \mathbf{k} \cdot \mathbf{A})} F_{\Delta k} \qquad (2.23)$$

the remaining summation in equation (2.23), *the lattice sum*, is independent of all of these complications. Notice that as the lattice sum becomes infinite (i.e. the system becomes perfectly two-dimensionally periodic) the amplitude of $A_{\Delta k}$ vanishes except when

$$\Delta \mathbf{k}_\parallel = \mathbf{g}_{hk} \qquad (2.24)$$

a result presented earlier on the basis of more general arguments. Strictly, this is the only condition under which we can safely separate out all multiple scattering effects in $F_{\Delta k}$; this is because, for a finite summation, the multiple scattering experienced by atoms on the edge of the domain or 'crystal' will differ from that seen by atoms in the centre. This should be a weak effect, however, so that discussions of diffracted beam shapes based on evaluations of the lattice sum should be rather reliable although the actual intensities of these beams can only be determined by proper, multiple scattering, determinations of $F_{\Delta k}$.

Some qualitative features which can emerge from the evaluation of lattice sums for different types of periodic structure and defects can be readily appreciated by one-dimensional models such as those illustrated in fig. 2.11. In this figure relative diffracted beam intensities resulting from the lattice sum alone are shown as a function of $\Delta \mathbf{k}_\parallel$ (note that as the vector $\mathbf{A}$ lies parallel to the surface the only important component of $\Delta \mathbf{k}$ is $\Delta \mathbf{k}_\parallel$). The first three examples show the same well-known effects as in optical interference patterns from single, double and multiple slit sources (excluding the 'single slit' geometrical structure factor – i.e. its Fraunhofer diffraction pattern). The basic periodicity of the diffraction pattern is defined by the spacing of adjacent scatterers but the width of the maxima reduces as the number of coherent scattering events is increased; their half-width relative to their spacing in $\Delta \mathbf{k}_\parallel$ is $1/n$ where n is the number of periodically sited scatterers. The following three examples of fig. 2.11 illustrate the effects of different defects in the periodic structure. Fig. 2.11(*d*) shows the splitting of alternate beams due to periodic antiphase domain boundaries involving separations not of the regular spacing *a*, but an odd half-integral multiple of *a*. If the same antiphase boundaries are retained, but the regularity of their structure is removed, this sharp beam splitting becomes simply a beam broadening. Finally, fig. 2.11(*f*) illustrates the ability of diffraction processes to identify *average* periodicity in structures which are not strictly periodic but are based on a periodic lattice. The loss of true periodicity introduces a background but because this intensity is distributed evenly while the diffracted beams resulting from the coherent interferences concentrate their intensity into distinct beams the dominant effect remains that of the periodic component. While all of the effects illustrated here are

Fig. 2.11. Calculated diffraction intensities, *I*, for one-dimensional scattering models. (*a*) A single atom; (*b*) two atoms with spacing of *a*; (*c*) *N* atoms in a row, spacing *a*; (*d*) several groups of *N* atoms, each with spacing *a* but with the distance between the groups being $(N+\frac{1}{2})a$; (*e*) several groups of atoms of varying size but otherwise as for (*d*); (*f*) *N* atoms randomly distributed over $2N$ sites with regular spacing of *a* (Henzler, 1977).

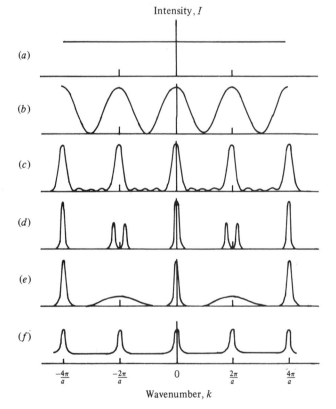

Intensity, *I*

(*a*)

(*b*)

(*c*)

(*d*)

(*e*)

(*f*)

$-\frac{4\pi}{a}$ $-\frac{2\pi}{a}$ 0 $\frac{2\pi}{a}$ $\frac{4\pi}{a}$

Wavenumber, *k*

potentially relevant to two-dimensional surface structure problems, particularly in the study of order–disorder and other surface phase transitions, we will concentrate initially on the effects caused by regular antiphase domain boundaries to illustrate the basic processes involved. For this purpose consider the special case of P domains, each of $N \times M$ regularly spaced scatterers. In this case it is convenient to rewrite equation (2.23) as

$$A_{\Delta k} = \sum_{p=1}^{P} \exp(i\,\Delta \mathbf{k} \cdot \mathbf{r}_p) \sum_{n=1}^{N} \sum_{m=1}^{M} \exp(i\,\Delta \mathbf{k} \cdot \mathbf{A}) F_{\Delta k}$$

(2.25)

where the last two summations are the lattice sum over the individual domains while the summation over p is over the set of domains, the vectors $\mathbf{r}_p$ defining the origins of each domain in the surface plane. Consider now the special case of the centre of each diffracted beam for the individual domains which correspond to the conditions $\Delta \mathbf{k}_{\parallel} = \mathbf{g}_{hk}$. The right-hand pair of summations then simply reduces to the product NM and

$$A_{hk} = \sum_{p=1}^{P} \exp(i\,\mathbf{g}_{hk} \cdot \mathbf{r}_p) NM F_{hk}$$

(2.26)

We can then write $\mathbf{r}_p = l_a \mathbf{a} + l_b \mathbf{b}$ where $\mathbf{a}$ and $\mathbf{b}$ are the primitive translation vectors of the *substrate* mesh, the beam indices h, k also being expressed, as normal, in terms of the reciprocal net vectors associated with the same mesh. Note that we do not require that l_a and l_b are necessarily integral. Then

$$A_{hk} = \sum_{p=1}^{P} \exp[2\pi i(l_a h + l_b k)] NM F_{hk}$$

(2.27)

By setting $l_b = k = 0$ we can reduce this equation to one dimension and analyse the model depicted in fig. 2.11(*d*); l_a is half-integral in this model so for h odd, A_{hk} goes to zero on average because successive domains cancel in amplitude while if h is even these same amplitudes add. Thus alternate diffracted beams split as a result of the zero central amplitude. Of course, if the domains are not all identical in size, these destructive interferences do not exactly cancel and the effects become less dramatic (fig. 2.11(*e*)). Evidently similar beam splitting can occur in two-dimensional cases if $(l_a h + l_b k)$ is half-integral. Consider, for example, a particularly simple, and potentially common situation depicted in fig. 2.12(*a*),(*b*) of adsorption on a square surface mesh in four-fold symmetric 'hollow' sites at one-half of a monolayer coverage to give a c(2 × 2) structure. Because only half such sites are filled, there is, as we have indicated earlier, one degree of arbitrariness in which sites are filled so that antiphase domain boundaries will appear between domains in which either one of these symmetrically identical sites is filled. However, note that the nature of this boundary is such that l_a and l_b are always integral, the identical sites being related, by definition, by translation vectors of the substrate mesh. In such a situation $(l_a h + l_b k)$ can clearly only be other than integral if one or both of h and k are fractions of integers. Thus integral order (non-'super lattice') beams cannot be split by this process, although fractional order beams can be – in this case the $\frac{11}{22}$ beam, for example. Of course, as in the one-dimensional case, exact splitting requires regular domain structures; more generally one might only expect some of the extra beams to be broadened due to domain boundaries (cf. fig. 2.11(e)).

While this simple formalism is valuable in assessing certain types of domain effects on diffraction patterns, it must be used with some caution in that it is only universally applicable to certain kinds of domain interferences. In particular, if adjacent domains are related by a point group operation which belongs to the substrate but not to the full surface structure, important limitations are introduced. A simple example illustrates this. Consider the case of a c(2×2) structure formed on a square substrate mesh not by four-fold coordinated registry, as in fig. 2.12(a) and (b), but by two-fold registry as in fig. 2.12(c)–(f). In this case domains may be related not only by a translation vector of the substrate mesh (but not a translation vector of the surface mesh) as in the four-fold sites, but also some domains involve relative rotation, in this case of 90°. One interesting feature of such domains is that the translation vector relating the adsorbate locations in each domain involves fractional

Fig. 2.12. Different types of antiphase domain boundary caused by adsorption on a square mesh substrate (circles) to form a $(\sqrt{2} \times \sqrt{2})R45°$ (c(2×2)) structure by adsorption (crosses) into four-fold 'hollows' (domains (a) and (b)) or two-fold 'bridge' sites (domains (c)–(f)).

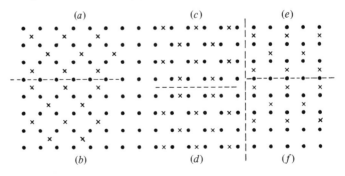

integer values of l_a and l_b. Thus, applying our simple criterion on the value of ($l_a h$ and $l_b k$) we see this may be half-integral, implying split diffraction features not only for fractional order beams, but also, for example, for the 10 beam. This deduction is incorrect because it assumes, implicitly, that the 10 beam of one domain will destructively interfere with the 01 beam of the adjacent beam to produce zero intensity. However, because the surface is only two-fold symmetric, these two beams do not have identical amplitudes (as they did for the substrate alone) and so no longer cancel. In effect our error in applying the simple condition to this situation is our failure to take account of the fact that F_{hk} may be domain dependent. For domains identical apart from a translation this neglect is appropriate but for other types of domain it is generally inappropriate.

A somewhat different, but possibly more commonly important type of domain structure which can be analysed in a similar way is the case of stepped vicinal surfaces. While most LEED studies, and surface investigations generally, are conducted on low index surfaces which are ideally atomically smooth, some important investigations are made on vicinal surfaces. These are surfaces, usually within a few degrees of a low index surface, which can be visualised (see fig. 2.13) as being composed of *terraces* of the nearest low index face, separated by regular arrays of atomic *steps* or *ledges*. For a suitable low index azimuthal tilt from the low index face these steps are smooth; at other orientations the steps themselves contain regular arrays of atomic *kinks*. Thus a vicinal surface is depicted by a regular *terrace–ledge–kink* model. While the exact regularity of such a structure in reality may be questioned, simple theoretical studies have shown that significant variations in terrace size, for example, can be tolerated without destroying the diffraction pattern characteristic of the regular structure. Evidently these regular terraced structures are essentially the same as regular domain structures although one important distinction exists; while for domain structures on a low index face the domains are related by a vector parallel to the surface, adjacent terraces are related by a vector which includes a component normal to the individual domain (terrace) surface. This means that it is not only $\Delta k_{\parallel}$, but Δk itself which influences the interference condition between adjacent terraces.

The rather complex diffracted beam splittings which can occur from vicinal surfaces can be best appreciated using a simple graphical construction. This method relies on the fact that a diffraction pattern, as equation (2.23) shows, is essentially a Fourier transform of the scattering system. Moreover, an important general theorem of Fourier transforms is that the Fourier transform of a convolution product of several

42   *Surface crystallography and electron diffraction*

functions is simply equal to the product of the Fourier transforms of each function. In our case this means that we can split a diffracting system of several periodicities into each of these and the final diffraction pattern will be a product of the diffraction patterns of each component. An excellent example of this idea is the scattering of a {410} f.c.c. surface such as Cu taken from the work of Perdereau & Rhead (1971). Fig. 2.14 shows a ball model of such a surface. We see it is composed of {100} terraces but because there is an even number of atom rows in the terrace the true unit mesh of the surface is oblique (or more strictly centred rectangular covering two terraces). Notice that the {310} surface also shown has an odd number of rows per terrace and can be represented by a primitive rectangular mesh.

Consider now the diffraction pattern to be expected from a single terrace. The location of the diffracted beams will be exactly as from a complete {100} surface but in summing over unit meshes on the surface we see that while one size (say the *n* summation of equation (2.23)) is essentially infinite, the other is small and in this direction the diffracted beam intensity distribution will be as in fig. 2.11(*c*). The principal maxima of this distribution are shown schematically in fig. 2.15 centred on the crosses which are the positions of a {100} surface diffraction pattern. The drawn out shapes are intended to represent the intensity distribution in the direction perpendicular to the step edges. We must now convolve with these terraces the fact that they actually form a regularly repeated unit of (effectively) infinite extent. As this is only a one-dimensional periodicity it is represented by a row of atoms in identical

Fig. 2.13. Schematic diagram of the terrace–ledge–kink (TLK) model of a vicinal surface. Individual atoms are represented by cubes in this diagram so that the ledges and kinks are typically visualised as involving single atom spacing displacements.

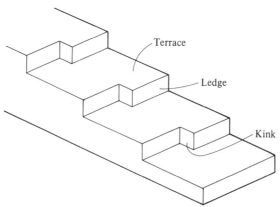

sites in each terrace – say a set of step edge atoms. Notice that this row (like the primitive unit mesh) is oblique to the steps and because it has only one-dimensional periodicity but infinite extent its diffraction pattern is a set of infinitely narrow, infinitely long lines shown dashed on fig. 2.15. The resulting diffraction pattern is therefore the product of these two patterns, shown on fig. 2.15 as the circles. One important feature of

Fig. 2.14. Ball models of two stepped surfaces of an f.c.c. crystal, {410} (*a*) and {310} (*b*). The primitive unit meshes of the two surfaces are shown together with the centred primitive rectangular mesh on the {410} surface.

the pattern, however, which makes this figure only exact at a single energy (which happens to be 118 eV for Cu{410}) is that the two component diffraction patterns have different origins. For example, if *O* is the 00 beam for a {100} terrace and the origin of this component, the terrace row component is centred at the point *P* a distance sin 2α away from *O* where α is the angle between the average surface orientation (and of the line of identically sited atoms) and the {100} terraces. Thus the angle between *OP* is fixed while the dimensions of both diffraction patterns expand and contract with energy about their respective origins. The case illustrated corresponds to alternate diffracted beams associated with {100} terraces being split equally. As the energy is varied so the splittings move through the (moving) {100} 'spots'. Thus spots split and become single in a systematic fashion. Note, incidentally, that for a {310} surface (fig. 2.13) the dashed lines of fig. 2.15 become horizontal and all spots split at the same energy.

Two important and related pieces of information can be extracted from studies of such patterns. The first relates to the size of the splitting when a spot appears as a doublet; as we see from the construction used to form fig. 2.15, this splitting relates directly to the average terrace width which can therefore be measured from this effect. Notice that the effect is quite reliable even when this average terrace width is not an integral number of atom rows; as in other situations, LEED picks out the *average* periodicity (cf. fig. 2.11(*f*)). In addition, knowing the angle between *O* and *P*, it is possible to predict the energies at which the spot splitting

Fig. 2.15. Construction showing the Fourier transform components leading to the diffraction pattern from a Cu{410} surface at 118 eV as described in the text (after Perdereau & Rhead, 1971).

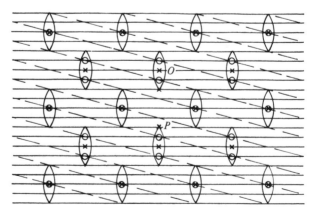

should occur if the step height is known. It is easy to see from fig. 2.15 that the spot splitting criterion occurs when

$$OP = \sin 2\alpha = (n + \tfrac{1}{2})\lambda/D \qquad (2.28)$$

where λ is the electron wavelength and D the terrace width. Fig. 2.16 shows an example of this evaluation in which the energy of splitting (proportional to λ^{-2}) is plotted against S^2 where $S = (n + \tfrac{1}{2})$. The gradient gives, for known α, the value of D, or, alternatively, the value of the step height (the two being related by α). In this case the surface is one of Ni cut on a vicinal orientation $10°$ from a $\{111\}$ face. Interestingly, for this case, results with a sample temperature of $500\ °C$ indicate monoatomic step heights while at $25\ °C$ the steps appear to have coalesced to double atom layer height.

Of course, in dealing with *regular* domain and step arrays some caution is in order regarding the possible effects of multiple scattering. Strictly these regular structures should be treated as large surface mesh structures and the appropriate reciprocal mesh determined. For example, in the case of the $\{410\}$ surface, the primitive net is shown in fig. 2.14 and the reciprocal mesh and diffraction pattern can readily be deduced. In fact, the result is the mesh generated by the intersection of the dashed lines in fig. 2.15 with vertical lines drawn through the streaked terrace spots. The diffraction pattern therefore has spots at the circled points, as described, but with additional features in between. In fact these additional diffracted beams coincide with the weak intermediate diffraction features seen between main maxima in fig. 2.11(*c*); strictly, we

Fig. 2.16. Plot of observed energies of splitting of the 00 beam from a Ni surface cut $10°$ off $\{111\}$ in a $\langle 110 \rangle$ azimuth at $25\ °C$ and $500\ °C$ (after Thapliyal, 1978).

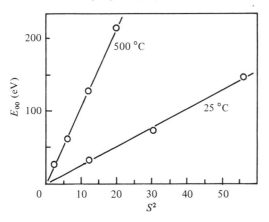

should have already included them, but on the basis of our earlier arguments we assumed they were too weak. This highlights a major feature of these simple single scattering theory methods; they lead to predictions as to which diffracted beams (which should be present on general arguments) are likely always to be weak. We have already argued the case for neglecting multiple scattering in assessing lattice sum effects in diffraction patterns. We should remember, however, that domain or step edge atoms will experience a different multiple scattering environment from those at the centre of a terrace; multiple scattering considerations may therefore become important if these atoms become very numerous (e.g. if the steps or domains become very small). Nevertheless, studies of stepped vicinal surfaces certainly indicate that the simple arguments we have outlined are valuable for terrace widths of only three or four atom rows.

2.6   Surface structure determination using LEED

2.6.1   *General considerations and the failure of single scattering and Fourier transform methods*

The ultimate goal of most diffraction studies of ordered crystalline systems is the determination of the structure of the system; i.e. the exact relative locations of atoms in the material. In principle the diffraction pattern (i.e. the location of diffracted beams) only provides information on the periodicity of the scattering system. We can often infer more information from these basic data. For example, if the diffraction pattern resulting from adsorbing some species on a clean unreconstructed square mesh surface is c(2 × 2) (or indeed if, on any clean unreconstructed surface, the new mesh has twice the area of the original one) we can guess that the structure involves one-half a monolayer of adsorbate species, each atom or molecule occupying a symmetrically identical site. In principle, of course, this periodicity could result from one monolayer of adsorbate in two different sites, or $n/2$ monolayer in n sites. Some independent determination of the coverage by another technique, which in this case can involve errors of up to 1/4 monolayer, can resolve this issue completely. For larger adsorbate net structures this additional coverage information may be more crucial. For example a (2 × 2) pattern, or other structure having a mesh area of four times that of the substrate, could easily be associated with either 1/4 or 3/4 monolayer of adsorbate, still with all adsorbates in identical sites. In the case of the 3/4 monolayer structure an equally possible state is that the adsorbate species occupy two or three different sites. Evidently these high coverage (i.e. close to a monolayer) structures with large surface meshes open up a wider range of structural parameters due to the possibility of the occupation of multiple sites. In such cases, additional information from a technique which can detect multiple site occupation, such as the vibrational state spectroscopies (see chapter 9) may be crucial to reducing the number of possible trial structures to a realistic number.

As we have already discussed, the diffraction pattern may, under favourable circumstances, also provide information on the space group, or sometimes the point group symmetry of the surface structure. This can also be valuable in reducing the range of possible surface structures compatible with the diffraction pattern. However, to proceed beyond this point, and to determine adsorbate–substrate atom registry, we must attempt to interpret not only the diffraction pattern (i.e. the lattice sum of equation (2.23)) but also the diffracted beam intensities which derive from the geometrical structure factor, equation (2.22).

Before considering this problem, however, we should briefly remark on one type of structure for which recourse to intensity analysis is irrelevant. These are incommensurate surface structures. Because the substrate and surface meshes are not commensurate in these structures, no unique adsorbate–substrate registry can exist and so it is clearly futile to attempt to investigate this question. On the other hand, considerable structural information is available from the diffraction pattern alone. For example, if the incommensurate mesh overlayer were to adopt *any* relative location and orientation on the substrate, no sharp extra diffraction features would be seen and, at best, one would observe diffraction *rings* centred about the 00 beam. In fact, exceedingly sharp diffraction features are often seen because the overlayer mesh is *commensurate in one dimension*, but not in the other. Because the one direction in which this commensurate behaviour occurs is usually one of several symmetrically equivalent directions of the substrate, several domains of the overlayer mesh occur and the diffraction pattern can appear quite complex. Fig. 2.17 shows such an example for the case of I adsorption on Ni{100} (Jones & Woodruff, 1981). These kinds of structures are common for the adsorption of the higher halogens. In this case the adsorbate appears to adopt a general centred rectangular mesh and while one side of the rectangle appears to have a constant size and align along a $\langle 100 \rangle$ direction of the substrate, the other dimension varies in a continuous fashion as the average coverage changes. As a result there are certain coverages at which this other dimension is also commensurate with the substrate forming a true commensurate structure (e.g. in the limit a $c(2 \times 2)$ structure seems just possible). These commensurate structures appear to be marginally more stable, as

Fig. 2.17. (*a*) Real space and reciprocal space diagrams of the diffraction patterns seen for I adsorption on Ni{100}. The highest coverage ($\theta = \frac{1}{2}$) corresponds to a c(2 × 2) or ($\sqrt{2} \times \sqrt{2})R$–45° structure. In both diagrams the underlying mesh shown is that of the substrate. At lower coverages the adatom spacing increases in one ⟨110⟩ direction to give a variable size rectangular mesh which is sometimes in coincidence (e.g. (2 × 3)). In the diffraction pattern, obtained by overlapping two domains rotated by 90° relative to one another, groups of four diffraction spots appear to diverge or converge. (*b*) Actual LEED patterns at various coverages; (i) c(2 × 2), $\theta = \frac{1}{2}$; (ii) incommensurate, $\theta = 0.313$; (iii) c(2 × 3), $\theta = \frac{1}{3}$; (iv) incommensurate, $\theta = 0.359$; (v) c(2 × 8), $\theta = 0.375$. LEED beam energies are 70 eV except for (iii) which is at 89 eV (after Jones & Woodruff, 1981).

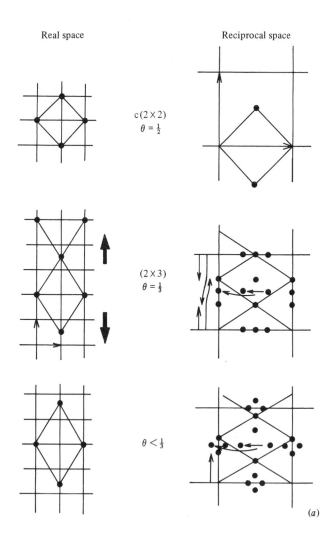

Real space   Reciprocal space

c(2 × 2)
$\theta = \frac{1}{2}$

(2 × 3)
$\theta = \frac{1}{3}$

$\theta < \frac{1}{3}$

(*a*)

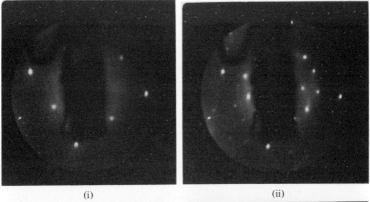

(i)   (ii)

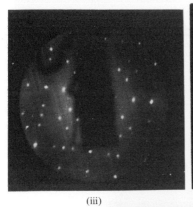

(iii)   (iv)

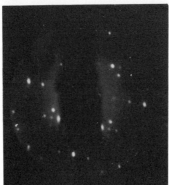

(v)

(*b*)

evidenced by thermal desorption properties, than the intermediate incommensurate structures. Nevertheless, the diffraction patterns clearly show sharp incommensurate overlayer structures with multiple domain beams converging into single beams as the coverage passes through the commensurate structures. In such cases, provided the approximate coverage is known from some other technique, the diffraction pattern provides almost all of the structural information which can be obtained. We could, of course, ask what the average adsorbate–substrate layer spacing is, or whether the commensurate direction imposes special adsorbate–substrate row registry on the structures but, as we shall see, the complexity of the structure lies outside the scope of a full intensity analysis. One final point which might be mentioned with regard to these 'variable dimension' structures arises from the 'split beams' appearance of the diffraction pattern. As we have just remarked, variation in the lattice parameter causes several beams to converge or diverge on a single beam. One alternative explanation of such a structure is therefore that this beam splitting arises from antiphase domain boundaries of the type discussed in the last section. Thus, instead of incommensurate intermediate structures, it is possible to account for the diffraction pattern by regular domains of the relevant commensurate structure in which, as the coverage changes, the domain sizes change. Such a picture requires considerable mobility of atoms around the domain wall boundaries. On the other hand, the variable mesh size model also implicitly assumes ease of adatom motion. Moreover, LEED does tend to sample average periodicities; we have already mentioned that vicinal step surfaces give sharp split beam patterns with splitting appropriate to the average orientation even when the average step size is not an integral number of atom row spacings. It is therefore not strictly possible to distinguish the two models on the basis of the diffraction pattern alone.

Returning to the case of commensurate structures, the question of detailed local structural registry can be answered by intensity analysis, at least in simple cases. The essential problem to be solved in obtaining the contents of a unit mesh is the same as that of finding the contents of a bulk crystal's unit cell using X-ray diffraction but with the important distinction that multiple scattering is an important effect in LEED but not in X-ray diffraction. Because structure determination of simple systems is a very routine and straightforward problem in X-ray diffraction we will outline these methods first in order to highlight the differences between this established technique and that of LEED. If we adopt a single scattering or *kinematical* approximation to the scattering from bulk crystal then the geometrical structure factor for the (hkl)

diffracted beam (i.e. the one defined by $\Delta\mathbf{k} = \mathbf{g}_{hkl}$) is

$$F_{hkl} = \sum_{j=1}^{N} f_j \exp\left[2\pi i(hx_j + ky_j + lz_j)\right] \tag{2.29}$$

$$= \int_{x=0}^{1} \int_{y=0}^{1} \int_{z=0}^{1} V\rho(x, y, z) \exp\left[2\pi i(hx + ky + lz)\right] dx\, dy\, dz \tag{2.30}$$

In the second formulation, equation (2.30), the sum over individual atoms and their atomic scattering factors f_j has been replaced by an integral over the unit cell, volume V, of the electron density $\rho(x, y, z)$ which controls X-ray scattering. In this form it is particularly clear that the scattering amplitude, directly proportional to F_{hkl}, is simply a Fourier transform of the scattering potential so that a direct Fourier transform of the measured amplitudes should provide a map of the structure of the unit cell. Even in X-ray diffraction, however, the problem is not quite so simple, for we must measure not diffracted beam *amplitudes*, but intensities, so that the phase information is lost. However, it can be shown that if F_{hkl} is the complex conjugate of $F_{\bar{h}\bar{k}\bar{l}}$, which is readily satisfied if f_j is real as in X-ray diffraction, then we can define a Patterson function

$$P(u, v, w) = \frac{1}{V} \sum_h \sum_k \sum_l F_{hkl}{}^2 \exp\left[-2\pi i(hu + kv + lw)\right] \tag{2.31}$$

and show that this function is simply a self-convolution of the electron density

$$P(u, v, w) = \int_{x=0}^{1} \int_{y=0}^{1} \int_{z=0}^{1} \rho(x, y, z)\rho(x + u, y + v, z + w)V\, dx\, dy\, dz \tag{2.32}$$

Provided the structure is simple it is rather straightforward to obtain the structure from its self-convolution. Thus in X-ray diffraction we have an established *direct* method of structure determination in which the measured diffracted beam intensities, proportional to $|F_{hkl}|^2$, can be fed into a formula which provides a map of the self-convolution of the structure. One might therefore suppose that the same method could be applied to LEED. There are, of course, some technical difficulties. Because the system under investigation is only two-dimensionally periodic the summation over l in equation (2.31) must be replaced by an integral over $\Delta\mathbf{k}$. One problem that this introduces is that the integral, which strictly should be from zero to infinity, is truncated at both low and high values; the upper (energy) truncation is common also in other techniques and leads to only small errors. The lower energy truncation results from the fact that the lowest possible energy (and hence $\Delta\mathbf{k}$) we can use in LEED corresponds to zero energy *in the vacuum*. However, the energy zero inside the crystal lies at the bottom of the valence band; thus when an electron enters the surface it gains an energy equal to the *inner potential*, roughly equal to the sum of the work function and the Fermi energy and typically 10–15 eV. This lower energy truncation of the integral leads to violent spurious oscillations in the Patterson function which must be removed by complex deconvolution procedures, taking account of the known truncation. Nevertheless, these are only technical difficulties and can be overcome with care. More important is the problem of the nature of the atomic scattering factors f_j in LEED.

We have already mentioned that these factors are complex in electron scattering. It is usual to express the scattering of an electron by an isolated atom in a partial wave representation as a function of scattering angle, θ,

$$f(\theta) = (2\mathrm{i}\,k)^{-1} \sum_{l=0} (2l+1)[\exp(2\mathrm{i}\,\delta_l) - 1]P_l(\cos\theta) \tag{2.33}$$

l being the angular momentum quantum number while P_l are Legendre polynomials and the δ_l are the scattering phase shifts which describe the scattering of a particular species and are, themselves, dependent on k. This complexity of f_j means that a further, more significant, technical difficulty arises in that the condition $F_{hkl} = F^*_{\bar{h}\bar{k}\bar{l}}$ is no longer satisfied. But perhaps more important is the species as well as k and θ dependence of $f(\theta)$. If, for simplicity, we reduce the scattering cross-section to its constituent real and imaginary parts

$$f_j = A_j \exp(\mathrm{i}\,\Delta_j) \tag{2.34}$$

with the A_j and Δ_j being dependent on species, k and θ, and substitute this into the general LEED structure factor, equation (2.22), we obtain

$$F_{\Delta k} = \sum_j A_j \exp[\mathrm{i}(\Delta_j + \Delta\mathbf{k}\cdot\mathbf{r}_j)] \tag{2.35}$$

where, for the hk beam, we have

$$\Delta\mathbf{k}\cdot\mathbf{r}_j = 2\pi(hx_j + ky_j) + \Delta k_\perp z_j \tag{2.36}$$

We note, now, that the fact that a Fourier transform of F_{hkl} (equation (2.30)) leads to the structure, and similarly that the Patterson function leads to a self-convolution of the structure, relies on the fact that the *phase* of F_{hkl} derives entirely from structural registry. Thus the nature of the scattering interferences is controlled only by phase differences due to scattering path length differences. If we now introduce scattering factors such as equation (2.35), however, we find that the phase of F_{hkl} is determined both by these structural components and by the scattering phase shifts Δ_j. Of course, if all Δ_j are the same this component factorises out and becomes irrelevant. If they differ for different atoms within the unit mesh, on the other hand, these scattering differences will be spuriously interpreted as structural effects. In practice *all* atoms within the unit mesh will have different effective Δ_j. This arises from two causes. The most obvious is the possibility that more than one species of atom is involved. Indeed, this is necessary in adsorption systems which form the largest set of relevant structural problems. The fact that the species, energy and angular dependence of both A_j and Δ_j are real and important effects is illustrated by the data of fig. 2.18. Thus uncritical use of a Fourier transform analysis of LEED data for an adsorbate structure would lead to spurious assignments of the adsorbate location. However, even in the case of an elemental solid, where all scatterers are of the same species, multiple scattering ensures inequivalence of both A_j and Δ_j for all non-symmetrically equivalent atoms; i.e. for *all* atoms within the primitive unit mesh. When proper account is taken of multiple scattering, as we have already mentioned, it is not possible to replace the f_j by the relevant isolated atom scattering factors as given in equation (2.33) but account must be taken of the different multiple scattering environments of each atom.

These two problems combine to ensure that *direct* Fourier transform methods cannot be adapted to LEED structure analysis. While the attraction of these simple and direct methods is considerable, substantial efforts to circumlocute the intrinsic difficulties described above have failed to provide any convincing demonstration that the problems can be overcome.

This absence of *direct* methods of handling LEED intensity data means that structural determinations can only be performed by comparing experimental data with the results of theoretical calculations or simulations based on trial structures. When good agreement is obtained between theory and experiment the relevant trial structure is deemed to be the true structure. This trial and error procedure with guessed structures is a major limitation of practical structural investigations. It is one source of limitation on the complexity of structure which can be investigated because the more complex the structure is, the more structural variables there are to test and refine, and thus the more calculations there are which must be performed. The second limitation derives from the complexity and expense of the calculations themselves, the major source of which is the importance of multiple scattering. To illustrate the importance of multiple scattering we can compare the

Fig. 2.18. Calculated modulus A and phase Δ of elastic electron scattering factors for Cu (full line), Ag (dashed line) and O (dotted line) for three scattering angles equivalent to 00 beam scattering with incident angles of 8°, 18° and 33°. The abscissa is in units of Δk of 1.74 Å^{-1} corresponding to the kinematic 'Bragg peak' orders for scattering from Cu{100} (after Woodruff, 1976).

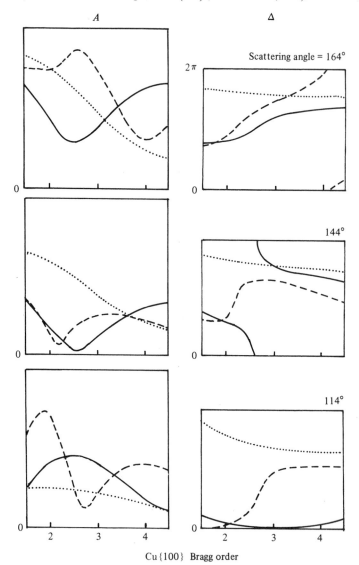

Cu{100} Bragg order

predictions of a single scattering theory with experimental data for the particularly simple case of an unreconstructed low index elemental surface. The most convenient form of intensity data for a LEED experiment is to measure the intensity of a diffracted beam as a function of energy. This is equivalent to fixing hk and scanning through $\Delta \mathbf{k}_\perp$ (the continuous variable related to l) of equation (2.36). For this simple case, in which all atom layers are equivalent, but for the effects of attenuation, we expect peaks in the intensity where $\Delta \mathbf{k}_\perp$ corresponds to the three-dimensional Laue conditions (i.e. the integral values of l). In the absence of attenuation these peaks would be δ-functions; because of attenuation they have a finite width which is inversely related to the number of scattering layers contributing (i.e. to the attenuation length). These conditions are often referred to as Bragg conditions because they correspond to the three-dimensional Bragg's law conditions, and the associated peaks as Bragg peaks. Fig. 2.19 shows some data for the specular 00 beam from a clean unreconstructed Cu{100} surface for a range of angles of incidence. The kinematical Bragg peak locations for normal incidence ($\theta = 0°$) are shown as arrows on the energy axis with no correction for inner potential; a correction for this effect would displace the arrows to lower energies (in the vacuum) by about 14 eV. Two features are clear from this figure. The first is that, particularly near normal incidence, there are many more peaks in the experimental spectra than predicted by kinematical theory, and indeed the largest peaks do not always coincide with these conditions. Secondly, as the Bragg condition depends only on $\Delta \mathbf{k}_\perp$, which for the 00 beam is proportional to $k \cos \theta$ and thus to $E(\cos \theta)^2$, the Bragg peaks should move up in energy as $(\cos \theta)^{-2}$ while in fig. 2.18 the only feature close to a Bragg condition for small θ which appears to display this behaviour is the most energetic one shown. Thus, not only are there many more peaks than predicted Bragg peaks, but also firm assignments of true Bragg peaks within the data are dubious.

In summary, therefore, we see that kinematical theory fails rather badly in its ability to simulate LEED experimental data and that any method of structural analysis which relies on accurate comparison of spectra such as those of fig. 2.19 with theoretically computed spectra requires a theoretical description capable of reproducing most, if not all, of the structure and relative intensities of experimental data. Modern multiple scattering computational schemes do have this ability.

2.6.2 *Basic elements of multiple scattering theories*

The recognition of the importance of multiple scattering in LEED, and the beginnings of the development of a theory of the effect began soon after the original experimental study of Davisson and Germer (e.g. Bethe, 1928). This theoretical interest was revived, along with experiments, in the 1960s and developed along several parallel routes to the production of sophisticated computer programs which, while based on somewhat different formulations of the problem, generally include the same basic physical processes and produce very similar data. Because of this we will not attempt to present an historical survey of this development, nor a detailed presentation of any one formalism, but will briefly discuss the principal ingredients common to all currently used theories. Far more detailed presentation of both the

Fig. 2.19. LEED experimental intensity–energy spectra for the 00 beam from a Cu{100} surface in a ⟨100⟩ azimuth for different incidence angles. The arrows indicate the predicted positions of kinematical 'Bragg peaks' at normal incidence for energies in the crystal.

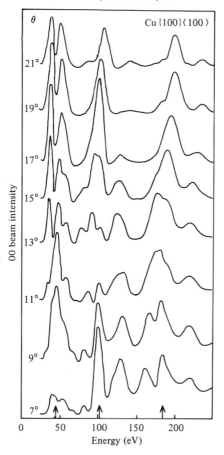

formalism and computer program used in two slightly different
approaches are given in the books by Pendry (1974) and van Hove &
Tong (1979).

The essential ingredients of all theories are attempts to provide
adequate description of

    (a)   the ion core scattering,
    (b)   multiple scattering,
    (c)   inelastic scattering
    (d)   temperature effects.

We have already remarked that the individual atomic scattering
cross-sections (before multiple scattering) are generally written in a
partial wave description (equation (2.33)). We describe the problem as
one of ion core scattering because in electron scattering the valence
electrons of the solid contribute rather little to the process, a fact which is
important to the validation of the methods generally used. In particular
it is usual to describe the scattering potentials by a muffin-tin model (see
fig. 2.20) in which spherical, spherically-symmetric scattering potentials
are truncated at the radii at which they just touch and the intervening
space is assumed to be at a constant potential. Generally the potentials
within the muffin-tin spheres are obtained from spherically averaged
determinations of the free atom charge density distributions, with the
excess free atom charge which lies outside the muffin-tin radius being
distributed evenly throughout the solid surface. Fortunately, it is found
that this cavalier handling of the valence electrons has little effect on the

Fig. 2.20. Plan and sectional views of a 'muffin-tin' potential, V, of a solid near
the surface.

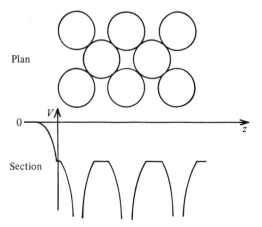

resulting computed LEED spectra. This is fortunate, particularly in the case of adsorbates, because small changes in adsorbate location on a surface imply significant changes in the distribution of valence electrons (i.e. in the chemical state of the adsorbate). If these changes were important it would be necessary to recompute the scattering properties for each geometry, but as they are relatively insignificant this constant property of the relevant ion core need not be recalculated. Once the potential is defined in this way, the phase shifts, δ_l, which characterise the scattering can be obtained from the logarithmic derivative of the radial component of the electron wavefunction at the muffin-tin boundary for an electron at the relevant energy in this potential.

The remaining question which determines the level of complexity of the final computer program required to simulate the LEED experiment adequately is the number of phase shifts necessary for a good description of the scattering. In principle the partial wave sum, equation (2.33), is infinite in l. In practice the higher partial wave phase shifts, δ_l, contribute successively smaller amounts to the scattering cross-section so that at low energies ($\lesssim 150$ eV) typically the first six or eight phase shifts are adequate to describe the scattering, although this number increases as the energy increases. This basic effect is illustrated by some computed phase shifts for Cu shown in fig. 2.21. Because the $P_l(\cos \theta)$ functions become increasingly fine in angular structure as l increases it is evident that high l components are unimportant at low energies and long electron wavelengths. Pendry (1974) has offered a simple argument to indicate that the maximum l value of importance, $l_{\max}$, is given by

$$kR_{\mathrm{m}} \approx l_{\max} \tag{2.37}$$

where k is the usual electron wavevector amplitude and R_{m} is the muffin-tin radius.

While the essential approach to the proper description of ion core scattering is universal, the methods of handling the multiple scattering are more varied. Two general approaches, which are ultimately equivalent, derive from two historically different views of the problem. In one approach a literal, real space, summation of the multiple scattering is made, successive scatterings being described in terms of the partial waves or spherical harmonics emitted from the scattering centres. This angular momentum, or *L-representation* approach, requires summation of events over a finite slab of material over which the higher order scatterings, through inelastic and spherical wave attenuation, fall to inconsequential amplitude. In the alternative approach, sometimes called the 'band structure' method, the two-dimensional periodicity is recognised explicitly by describing the scattering within layers by two-dimensional Bloch waves. The propagation between layers is then dealt with in terms of a finite number of plane waves or beams of well-defined k; this approach is therefore referred to as the *K-representation*. Both approaches have somewhat similar limitations and sources of simplification. For example, the computational complexity, and thus time and expense, as well as potential precision, are related to the number of 'beams' used in the K-representation, or the size of the slab in the L-representation. The finite size of both of these numbers is aided by the damping effect of inelastic scattering which reduces the possible importance of high order multiple scattering. Moreover, both schemes become rapidly more complicated as the size of the unit mesh increases. This arises because the basic unit of scattering summation which lies between the individual atom and the total surface is a *subplane* of atoms consisting of a plane of atoms with one atom per surface unit mesh. If the

Fig. 2.21. Computed electron elastic scattering phase shifts δ_l as a function of electron energy for l values from 0–5.

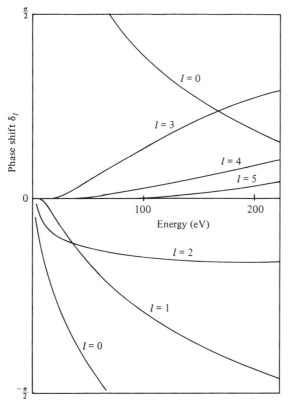

unit mesh is large compared with the substrate unit mesh, each actual atomic layer may have to be described by several subplanes so that, as the number of atom layers required is determined by the depth to which electrons can interact, the number of subplanes typically rises as the area of the surface unit mesh. Essentially the same limitation arises in the K-representation. In this case the number of actual propagating, observed diffracted beams increases also as the area of the surface mesh. The number of plane waves used to sum scattering between layers (which includes damped or *evanescent*, non-propagating beams) therefore increases in a similar fashion. In addition, the number of beams required in a K-representation summation increases as the reciprocal of the square of the subplane layer spacing. One obvious result of this is that, even in a K-representation scheme, coplanar subplanes must be summed in the L-representation. Briefly, therefore, we can state that while there are differences in computational scheme of different approaches to the inclusion of multiple scattering in LEED, and some schemes are better suited to certain problems, they all suffer from the same intrinsic limitation of unit mesh size. This, coupled with the problem mentioned earlier of the number of structural variables involved in complex, large surface mesh structures, imposes the main limitation on quantitative LEED studies. In practice, very little work has been attempted on surface meshes having an area of more than four times the substrate unit mesh (e.g. (2×2)). Moreover, the major expense of LEED computations derives from the need to incorporate a proper description of the multiple scattering and, while simplified perturbation schemes have been developed, no major breakthrough in this problem has been forthcoming. Most of the variation in the computational time taken by different schemes is associated with the extent to which they take account of symmetry in the scattering; these simplifications are only of real value for normal incidence on highly symmetric structures but can reduce the computational time on large-scale computers from of the order of one minute per energy point (for all diffracted beam intensities) to only a few tenths of a second.

The third extremely important ingredient of multiple scattering theories is some description of inelastic scattering. The short (~ 5 Å) mean-free-path for inelastic scattering of electrons in the LEED energy range has a valuable simplifying effect on multiple scattering because it attenuates the incident and scattered electron flux, thus greatly reducing the importance of multiple scattering and the depth of surface into which the incident electrons can penetrate. The existence of the effect, moreover, has the rather direct consequence of determining the minimum width of peaks in LEED intensity–energy spectra such as those of fig. 2.19. In a fully three-dimensional single scattering theory (which therefore has no inelastic scattering) these peaks would have infinitesimal width. Their actual width is inversely proportional to the number of contributing scattering layers and thus to the mean-free-path. In its simplest case the situation is analogous to multiple scattering interferences in an optical Fabry–Pérot interferometer; as the reflectivity is increased in this case, the number of contributing reflections grows and the interference peaks become narrower. Thus, not only does the neglect of inelastic scattering cause a multiple scattering theory to become complex and time-consuming, but it also yields unphysical simulations with peaks which are too narrow and which set too much importance on high order multiple scattering events.

The actual method of including inelastic scattering in LEED calculations is rather simplistic. The damping of an electron wave can be described by a complex k with components $k_r + ik_i$. Thus, for a spherical wave, the wavefunction, ψ, is

$$\psi \sim \exp\left[i(k_r + i\,k_i)r\right] = \exp\left(-k_i r\right)\exp\left(i\,k_r r\right)$$
$$= \exp\left(-r/\lambda_{ee}\right)\exp\left(i\,k_r r\right) \qquad (2.38)$$

showing that the damping effect can be included by an amplitude mean-free-path λ_{ee}. Note that the mean-free-path for *intensity* attenuation (which can be measured by other means – see chapter 3) is equal to $\lambda_{ee}/2$. An alternative approach is to note that the complex k can be seen as a complex energy. As the electron changes its energy on passing from vacuum to solid by an amount equal to the inner potential, due to the shift in energy zero, one way of introducing this complex energy to the electron in the solid is by adding an imaginary component, V_{0i} to the inner potential, attributing the real energy shift to its real part V_{0r}. For the case of electron energies substantially larger than the magnitude of V_{0i} one can then relate this to λ_{ee} by

$$\lambda_{ee} \approx 3.90 E^{\frac{1}{2}}/V_{0i} \qquad (2.39)$$

where E and V_{0i} are expressed in eV and λ_{ee} in Å. Typically, therefore, λ_{ee} of 10 Å at an energy of 100 eV corresponds to a value of V_{0i} of 3.9 eV. Notice that in this formulation the width of peaks in intensity–energy spectra (except where they are widened by overlapping structures) is just $2V_{0i}$.

In LEED computations it is usually convenient to keep either λ_{ee} or V_{0i} constant. These two assumptions imply a difference in the energy dependence of the damping; for example, as may be seen from equation (2.39), constant V_{0i} implies that λ_{ee} is proportional to $E^{\frac{1}{2}}$. As we shall see in chapter 3, neither a constant V_{0i} or a constant λ_{ee} are likely to be strictly accurate in the LEED energy range. On the other hand, it appears that while the inclusion of the basic effect is very important, its exact description is not. For the same reason, probable inhomogeneities in inelastic scattering seem to be of little importance.

The final process included, which can also apparently be included in a fairly crude fashion, is the role of thermal vibrations. These reduce scattering coherence in otherwise periodic structures and thus reduce diffracted beam intensities. It is found sufficient to reduce the individual atomic scattering factors before computing the multiple scattering effects by

$$\exp\left(-\tfrac{1}{2}\Delta k^2\langle u^2\rangle\right)=\exp\left(-M\right) \qquad (2.40)$$

where $\langle u^2\rangle$ is the mean square vibrational amplitude, measured in the direction of Δk for the particular scattering process involved (which may be an intermediate, multiple scattering event), and M is the usual Debye–Waller factor, which in the high temperature limit ($T \gg \theta_D$) can be related to the Debye temperature of the surface, θ_D, by

$$\langle u^2\rangle=\frac{3\hbar^2 T}{mk_B\theta_D{}^2} \qquad (2.41)$$

m being the mass of the scatterer and k_B Boltzmann's constant. This treatment neglects correlations in adjacent vibrations, and it is also usual to assume that $\langle u^2\rangle$ is isotropic. The inclusion of this factor has some effect on relative peak intensities at different energies, through the energy dependence of Δk, but also contributes to reducing multiple scattering by effectively reducing the atomic scattering cross-sections, particularly for backscattering events (involving large Δk).

2.6.3   *Application of multiple scattering calculations*

An important test for the value of these multiple scattering computational methods is their ability to simulate effectively the experimental data from essentially 'known' unreconstructed clean surfaces such as those from Cu{100} shown in fig. 2.19. An example of the quality of match possible in such cases is given by the results shown in fig. 2.22. One outstanding structural unknown in such systems is the exact layer spacing of the solid in its outermost layer (or, in principle, its outermost layers). Fig. 2.22 compares intensity spectra of the 00 beam from a Ni{100} surface under various incidence conditions with theoretical simulations for an ideally terminated solid, and one in which the outermost layer is expanded by 2.5% (i.e. a displacement of 0.05 Å). We see that agreement between theory and experiment is excellent although the differences in the two theoretical calculations are small, so that this spacing change lies on the brink of significance as far as LEED is

Fig. 2.22. Comparison of experimental and theoretical 00 beam LEED intensity–energy spectra for a Ni{100} surface at different angles of incidence θ and azimuth, ϕ. Theoretical results are shown for an ideally terminated bulk solid structure and one in which the top layer spacing is expanded by 2.5% (after Marcus, Demuth & Jepson, 1975).

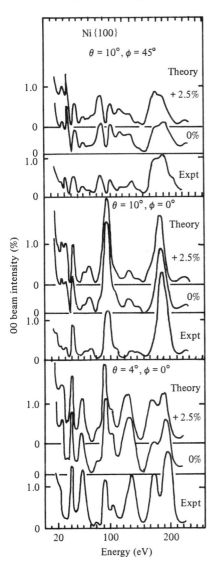

concerned. On surfaces where a larger spacing change is thought to occur (e.g. Ni{110} with a contraction of 5%), more significant conclusions may be drawn.

While agreement between theory and experiment in fig. 2.22 is so good as to leave little doubt as to the essential correctness of both the computational method and the structure, correspondences are not always so good and the plausible range of structures to choose between is often not so clear. In such cases the criterion for 'good agreement' and the 'best structure' are inclined to be highly arbitrary and subjective. Growing concern over this problem has led to the proposal of a number of objective 'reliability factors' to be used to optimise and assess experiment–theory comparisons. While the criteria on which these reliability factors are based remain essentially arbitrary, they do provide a more quantitative and objective basis for structure optimisation. The method proposed by Zanazzi & Jona (1977) is based on considerable experience of attempting to match theory and experiment 'by eye' for different surfaces and attaches most importance to the comparison of peak positions and relative intensities of peaks. Thus they define a reliability factor for a single (one beam) spectrum comparison as

$$r = \frac{A}{E_f - E_i} \int_{E_i}^{E_f} W(E) |cI'_{calc} - I'_{obs}| \, dE \tag{2.42}$$

where E_i and E_f are the initial and final energies of the spectra, I' are the first derivatives of the calculated and observed spectra, c is a normalisation factor chosen to give the same average intensities to experiment and theory

$$c = \left(\int_{E_i}^{E_f} I_{obs} \, dE \right) \Big/ \left(\int_{E_i}^{E_f} I_{calc} \, dE \right) \tag{2.43}$$

and A is a scaling factor to make the function independent of the actual intensities

$$A = (E_f - E_i) \Big/ \left(\int_{E_i}^{E_f} I_{obs} \, dE \right) \tag{2.44}$$

and finally $W(E)$ is a weighting function

$$W(E) = \frac{|cI''_{calc} - I''_{obs}|}{|I'_{obs}| + |I'_{obs}|_{max}} \tag{2.45}$$

which involves second derivatives of the intensity with respect to energy (the doubly primed parameters) and thus enhances the importance of sharp structure. It is found convenient to normalise the resulting r factors to

$$r_r = r/0.027 \tag{2.46}$$

when it is found, subjectively, that r_r values of less than about 0.20 correspond to good agreement while values greater than 0.50 indicate poor agreement. Finally, adding in the results of several spectra and taking account of the beneficial effect of a large data set on reliability, an overall reliability factor R may be defined by

$$R = \left(\frac{3}{2n} + \frac{2}{3}\right)\bar{r}_r \qquad (2.47)$$

$\bar{r}_r$ being the average value of the individual r_r and n the number of spectra incorporated. Using this definition the range of significance of R values is similar to that for r_r; in particular, $R \lesssim 0.2$ suggest a 'good structure', $R \gtrsim 0.5$ a poor one.

An example of the application of this R-factor to a simple LEED structural analysis is provided by fig. 2.23; here a contour map of constant R-factor is shown as a function of top layer spacing and inner potential for a theory–experiment comparison of data from a Cu{311} surface. This shows a rather clear single deep minimum indicating a top

Fig. 2.23. Zanazzi and Jona R-factor contour map of theory–experiment comparison for a study of a Cu{311} surface using inner potential and surface layer spacing change as the variable parameters (after Streater *et al.*, 1978).

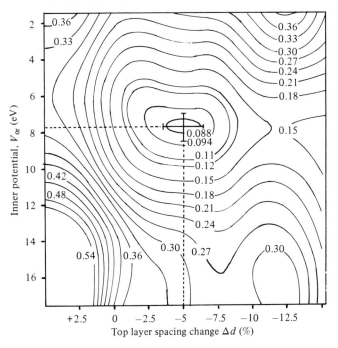

layer spacing contraction for this surface of 5% with an error of about ±1.5%.

An alternative reliability factor, proposed by Pendry (1980) remains essentially arbitrary but avoids the use of double derivatives in the comparison (evidently troublesome in noisy or insufficiently finely-scaled data) and claims to offer a statistical basis for defining the error bars of the structural parameters. It is this parameter which has been used to construct the contour map of fig. 2.24 which relates to an analysis of a c(2 × 2) structure of CO molecules adsorbed on Ni{100}. The 'best' structure results from this map are shown compared with experiment in fig. 2.25, while fig. 2.26 shows schematically the structure itself. Notice that with an absorbed molecule, even as simple as CO, the number of structural parameters in the analysis increases markedly. In particular, we must determine the location of both C and O atoms although we are probably entitled to assume, if other evidence indicates that the molecule does not dissociate, that the C—O bond distance is not significantly different from its value in the free molecule. This does, indeed, appear to be the case and the LEED analysis indicates that the CO bonds linearly to the surface, directly atop Ni atoms in the topmost layer, with the C end 'down' to the surface at a separation from the Ni atoms of 1.71 ± 0.10 Å.

Fig. 2.24. Pendry R-factor contour map for a study of the Ni{100}($\sqrt{2} \times \sqrt{2}$)$R45°$–CO structure using the Ni–C and C–O spacings as parameters but assuming a CO molecule perpendicular to the surface and directly above a top layer Ni atom (see fig. 2.26 – after Andersson & Pendry, 1980).

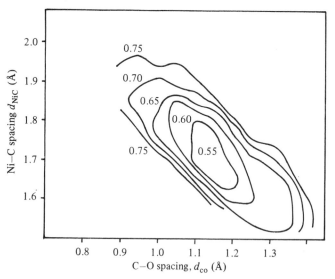

This particular example is interesting for a number of reasons. Firstly, most adsorption structures studies by LEED have been of adsorbed atomic species, generally involving fewer structural parameters, and in almost all cases these adsorbed atoms are found to adopt the highest coordination site possible on the surface, consistent with essentially non-

Fig. 2.25. Comparison of best-fit LEED intensity energy spectra used in the construction of the *R*-factor contour map of fig. 2.24 (after Andersson and Pendry, 1980).

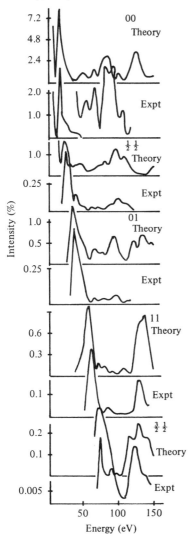

directional bonding. The atop site of the CO molecule thus indicates a more local molecule–atom rather than atom–surface bonding. Moreover, the degree of complexity represented by this example must lie close to the frontier of what is currently possible with LEED structure analysis. Molecules containing more than about two strong scatterers or adopting large surface net structures are likely to present too many structural parameters for the trial and error approach of LEED analyses; moreover they may require computational resources beyond those currently available. Nevertheless, a substantial number of simple adsorption structures have been analysed by LEED and in many cases these studies present the most definitive structural determinations available.

Despite the apparent success of these methods, however, a few words of caution are in order. We have seen that using reliability factors in particular, rather precise structural determinations can be made; in some cases precisions of significantly better than 0.1 Å are claimed. These determinations, however, are based on the comparison of a range of theoretical spectra with a single set of experimental data and so are obviously only as good as this latter data set. In practice, subtle differences in experimental spectra can arise from small angular misalignments, slight defects in magnetic field cancellation and ultimately in the preparation of the surface itself. It is probable that these deficiencies, even in careful experiments, limit the true precision of structure determinations by this method to ∼0.1 Å. In addition, the final structural determination must ultimately be limited by the range of structural models investigated. We have already remarked that using a sufficiently large data set and with reasonably simple structures the problem of uniqueness is not acute. It is clear, however, that if the 'true' structural model is not included in the range investigated, spurious

Fig. 2.26. Schematic plan and sectional view of the structure determination for the Ni$\{100\}(\sqrt{2} \times \sqrt{2})R45°$–CO surface.

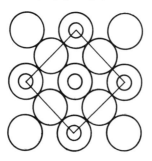

assignments are possible, particularly for surfaces involving several structural parameters.

2.7   Thermal effects

The influence of changing temperature in LEED intensities has already been referred to in the discussion of the essential ingredients of a multiple scattering theory of LEED. In particular we noted that, due to the reduction of scattering coherence brought about by thermal vibrations, a LEED scattering event suffers an amplitude attenuation by a factor of $\exp(-M)$ where M is the Debye–Waller factor (equation (2.40))

$$M = \tfrac{1}{2}\Delta k^2 \langle u^2 \rangle \tag{2.48}$$

where the mean square vibrational amplitude $\langle u^2 \rangle$ is measured in the direction of the scattering vector $\Delta \mathbf{k}$. In principle, therefore, the temperature dependence of LEED intensities allows the temperature dependence of $\langle u^2 \rangle$ to be investigated. Moreover, this parameter is of some interest; it is suggested that vibrational amplitudes of surface atoms, particularly perpendicular to the surface, may be significantly greater than for atoms in the bulk. Indeed, a rather simple theory leads us to this conclusion. If a harmonic oscillator has a force constant E then its vibrational energy $E\langle u^2 \rangle$ can be equated to the thermal excitation energy kT leading to a value of $\langle u^2 \rangle$ which is inversely proportional to E. If E arises from simple pairwise forces between atoms, then as a surface atom has lost roughly one-half of its neighbouring atoms we might expect E to be halved and $\langle u^2 \rangle$ to be doubled, relative to the bulk. Some more sophisticated calculations suggest that even greater enhancements of $\langle u^2 \rangle$ may be expected for atoms in the surface layer.

Somewhat surprisingly, there is some evidence that LEED studies of $\langle u^2 \rangle$ may be interpreted without the use of full multiple scattering calculations. Evidently for a single scattering theory of LEED, the observed *intensities* are simply attenuated by $\exp(-2M)$, the Δk in equation (2.48) now being the vector relating the incident beam and the final diffracted beam. One important result is that some multiple scattering *calculations* indicate that the error involved in interpreting the temperature dependence of the intensities in this way is usually only a few per cent. The reason appears to be that multiple scattering events which are important in LEED usually involve one backscattering event and one or more forward scattering events. Forward scattering implies small Δk and thus little thermal attenuation so that the temperature dependence of the resulting intensity is dominated by the single (large Δk) backscattering event. Indeed, the successive Debye–Waller factors of multiple scattering processes are one factor which greatly reduces the contribution of multiple backscattering events to the observed intensities.

Despite this great simplification in the interpretation of temperature-dependent LEED data, the technique has severe limitations in the investigation of surface vibrational amplitudes. The dominant reason for this is that most theories of the surface enhancement of vibrational amplitudes indicate that it is only in the outermost surface layer that the effect is large, while LEED samples several atomic layers. The value which emerges from the experimental Debye–Waller factors (equation (2.47)) is therefore some $\langle u^2 \rangle_{\text{effective}}$ and not $\langle u^2 \rangle_{\text{surface}}$; $\langle u^2 \rangle_{\text{eff}}$ involves some average over the top three or four atom layers. Because the actual number of layers sampled by LEED is usually smallest at the lowest energies ($\lesssim 50$ eV) as a result of enhanced elastic and inelastic scattering, experiments show the largest values of $\langle u^2 \rangle_{\text{eff}}$ at these lowest energies; at higher energies the value tends towards a bulk value of $\langle u^2 \rangle$. This basic effect is illustrated in the data of fig. 2.27. Typically the maximum value of $\langle u^2 \rangle_{\text{eff}}$ is found to be about twice the bulk value but this evidently implies that the true surface value may be significantly larger. Notice, however, that there is very considerable scatter in the data. This may be due to the fact that multiple scattering can have the effect of varying the sampling depth as a function of electron energy, so that the relative contribution of the surface layer fluctuates with energy.

Fig. 2.27. Temperature dependence of LEED 00 beam spectra at different energies from a Cu{100} surface. The dependence is expressed in terms of an 'effective' Debye temperature which is inversely proportional to the root mean square vibrational amplitude $\langle u^2 \rangle^{\frac{1}{2}}$ (see equation (2.41)). The dashed line shows the bulk value of the Debye temperature (after Reid, 1972).

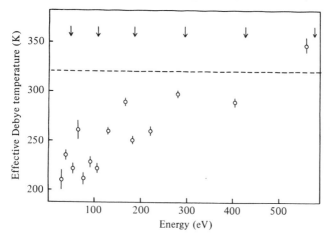

A further complication of temperature-dependent LEED studies in the investigation of surface values of $\langle u^2 \rangle$ is the probable strong anisotropy of this parameter. In particular, while the value of $\langle u^2 \rangle$ perpendicular to the surface is likely to be strongly enhanced, the value parallel to the surface, particularly for highly symmetric surfaces, may approach the bulk value. In principle, this may be investigated by studying different diffracted beams with differing values of $\Delta \mathbf{k}$. In practice the ease of experimental measurements and interpretation has led to almost all studies being performed on the specular 00 beam for which $\Delta \mathbf{k}_\parallel$ is zero so that $\Delta \mathbf{k}$, and thus the appropriate value of $\langle u^2 \rangle_{\text{eff}}$ is perpendicular to the surface.

2.8 Reflection High Energy Electron Diffraction (RHEED)

In LEED we have seen that the choice of a backscattering electron diffraction technique to investigate surfaces necessitates the use of low electron energies both to ensure reasonably strong backscattering cross-sections, and to minimise the inelastic scattering mean-free-path, thus enhancing the surface specificity of the technique. RHEED represents an alternative solution to the same basic problem. Surface specificity and a viable technique can be produced at high energies by using grazing incidence and exit angles in the diffracted beams. In this way, the overall scattering angle is small so that adequate elastic scattering cross-sections are obtained, while the grazing angles imply that the relatively long mean-free-paths for inelastic scattering still keep the elastically scattered electrons in the surface region.

The Ewald sphere construction for this scattering geometry is illustrated in fig. 2.28. This diagram is drawn to scale for a 3° grazing incidence angle on a Cu{100} surface in a $\langle 110 \rangle$ azimuth at an incident energy of 35 keV. Evidently the Ewald sphere is now very large relative to the spacing of the reciprocal net rods (cf. the LEED case in fig. 2.6(b)) and cuts these rods at grazing angles. For this reason RHEED patterns of good surfaces are usually 'streaked' with each reciprocal net point being represented not by a single diffracted beam but by a diffraction streak. Fig. 2.29 shows such a pattern. Notice that the streaks correspond to reciprocal lattice rods lying out of the plane of the diagram of fig. 2.28. The basic experimental arrangement is illustrated in fig. 2.30. Because there is a large energy difference between the elastically scattered electrons and the inelastically scattered background, and because these primary electrons are sufficiently energetic to produce fluorescence on a phosphor, careful energy filtering and post-acceleration are far less necessary than in LEED. Moreover, most

Fig. 2.28. Ewald sphere construction for RHEED. The diagram is drawn to scale for 35 keV electrons incident on a Cu{100} surface in a $\langle 110 \rangle$ azimuth at 3° grazing angle.

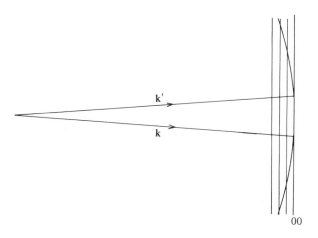

Fig. 2.29. RHEED pattern of a GaAs (100) surface along the [$\bar{1}\bar{1}0$] azimuth at 15 keV taken during MBE in an As_4 flux (courtesy J. H. Neave and B. A. Joyce, Philips Research Laboratories).

applications of RHEED are concerned only with qualitative surface assessment and not with quantitative diffracted beam intensity analysis.

The actual energies and other conditions at which the basic RHEED experiment are conducted vary widely. The most sophisticated instruments derive from conventional electron microscope optics and operate at up to about 100 keV. More recently, however, many applications of the basic technique operate at much lower energies (~ 3–5 keV) using simple electrostatic focus guns such as those designed for oscilloscope and television tube use and frequently used to stimulate the production of secondary electrons in AES (see next chapter). The growth of this use of low energy RHEED or MEED (Medium Energy Electron Diffraction), as it is sometimes called, is associated with the ease of setting up the technique and the space it leaves in front of the sample. Evidently, as fig. 2.30 shows, if a suitable electron gun is already fitted to a chamber for other purposes, the additional resources needed to exploit the technique are minimal. Moreover, as a means of continuously monitoring the growth of epitaxial layers on a surface, it has the virtue of leaving the front of the sample clear for evaporant sources. Current interest in MBE as a means of growing semiconductor device materials has done much to encourage the use of the technique.

Apart from the improved accessibility to a surface provided by the RHEED geometry relative to LEED, the technique has other virtues in relation to the study of epitaxial growth and other multilayer surface processes. In particular, the use of grazing incidence angles clearly makes the technique sensitive to the quality of the microscopic surface. Whereas LEED, typically at normal incidence, picks out the well-ordered parts of the surface with orientations close to that of the average surface, grazing incidence electrons will penetrate into asperities on the surface if it is not microscopically flat. This obviously necessitates more careful sample preparation for RHEED studies but means that the technique can identify changes in surface morphology. For example, if epitaxial growth or corrosion results in the growth of small thick

Fig. 2.30. Schematic diagram of the experimental arrangement for RHEED.

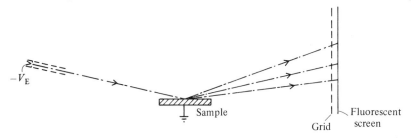

'islands' on a surface the grazing reflection pattern of the flat surface will be obscured, possibly to be replaced by a pattern of sharp diffraction spots due to *transmission* electron diffraction through the tips of these asperities.

By contrast, however, RHEED has certain disadvantages relative to LEED in the study of two-dimensionally symmetric structures in cases where the surface is microscopically flat. For example, the fact that the RHEED pattern is essentially a projection of the reciprocal net (or an approximately planar cut through this net) means that it is necessary to rotate the sample about its surface normal to establish the full two-dimensional periodicity; changes of periodicity lying in the plane of incidence do not lead to changes in the periodicity of the diffraction pattern. More important, however, is the absence of a complete quantitative theory of RHEED such as the multiple scattering theory of LEED, or of comparable experimental intensity data.

One reason for this is quite basic. LEED theory (based on perfect two-dimensional periodicity) predicts diffracted beams which are δ-functions in space; because of experimental limitations the real beams have finite size, but the total intensity within the beam is believed to correspond to the theoretical intensity of these δ-functions. In RHEED, on the other hand, the equivalent of the LEED intensity–energy spectrum is essentially contained in the intensity–position spectrum along a diffraction streak. In this case, however, a strict application of the Ewald sphere construction of fig. 2.28 also leads to diffracted beams as δ-functions (as in LEED) and not streaks. The experimental streaks must arise because of some broadening of the Ewald sphere 'thickness' or the 'width' of the reciprocal net rods so that these almost tangential surfaces and lines meet not at a point, but over an extended volume of reciprocal space. Obvious possible sources of this effect are the limited transfer width or surface order. Estimates of the transfer width based on the arguments outlined earlier (see equations (2.15)–(2.19)), however, indicate this is not likely to be the source of the effect. The influence of energy indeterminacy should be even better than LEED as the primary energy E is much higher while δE is the same. Similarly, incident electron beams from the high energy guns typically have better collimation so that the transfer width should be rather larger than in LEED. Holloway & Beeby (1978) have suggested that the source of the 'broadening' is phonon scattering and have developed the elements of a quantitative theory on the basis of this assumption. Doing so, they suggest that RHEED patterns, and the intensity distribution along the streaks, are rather sensitive to the local structure in an adsorbate overlayer, and could be used to analyse surface structures in as complete a fashion as is currently possible for LEED.

Unfortunately, in the absence of reliable quantitative experimental data, or even of evidence regarding the possible role of phonon scattering in producing RHEED streaks, it is rather hard to assess the future of this development.

Further reading

There are many reviews in the literature in the general area of surface crystallography and electron diffraction from surfaces, mostly emphasising LEED. Two complete books exist on LEED by Pendry (1974) and by van Hove & Tong (1979) and these seem the natural choice for further reading. Pendry's book gives a sound introduction to the basic physics while the later book contains an extensive bibliography of newer original papers.

3   Electron spectroscopies

3.1   General considerations
3.1.1   *Introduction*
A large number of surface techniques involve the detection of electrons in the energy range 5–2000 eV which are emitted or scattered from the surface. A number of features are common to most of these techniques. In particular, all derive their surface sensitivity from the fact that electrons in this energy range have a high probability of inelastic scattering, so that if electrons are detected at an energy which is known to be unchanged by passage through the surface region of the solid, we know that they have passed only through a very thin surface layer; i.e. the techniques are surface specific. Secondly, because this surface specificity derives from a knowledge of the energy of the electrons, some form of electron energy analyser is required by most of these techniques. This piece of instrumentation is therefore common to many techniques.

Of course no classification scheme is perfect. The electron energy analyser can also be used for analysing other charged particles, notably ions, and is also used in ion scattering spectroscopy. Appearance Potential Spectroscopy (APS) is not strictly an electron spectroscopy (in that emitted X-rays are detected) but does depend on electron inelastic scattering for its surface specificity. This classification scheme therefore provides only a rough framework on which to hang some of the basic ideas common to many techniques.

3.1.2   *Electron inelastic scattering and surface specificity*
An electron of energy ~ 5–2000 eV passing through a solid can lose energy in a number of ways. If we neglect phonon scattering (which produces energy changes too small to be detected by most techniques) then the main processes involved are plasmon scattering, single-particle electron excitations involving valence electrons and ionisation of core levels of the atoms which compose the solid. While this last process is the primary process leading to several core level spectroscopies, the associated cross-sections are small compared with the other processes and the mean-free-path for such collisions is generally at least two orders of magnitude longer than the mean-free-path for the other two processes.

Full theoretical calculations of the effects of plasmon and single-particle excitations have only been performed for idealised free-electron materials but are generally performed using electron density parameters for Al. Fig. 3.1 shows the results of one such calculation taken from one of the first papers to consider this subject by Quinn (1962). At low energies (below the plasmon energy) the scattering is obviously dictated by single-particle excitations. Above the plasmon energy, however, the plasmon scattering cross-section rises sharply. At high energies, for this material at least, the scattering becomes dominated by plasmon creation which slowly becomes weaker as the energy increases. The mean-free-path versus energy for the two processes thus shows a steep fall to reach a broad minimum at energies ~ 2–3 times the plasmon energy, followed by a much greater increase with increasing energy (note the logarithmic scale of fig. 3.1).

Quite a large number of experiments have been performed to measure this mean-free-path for a range of materials. These measurements have been reviewed by Powell (1974) and collated in this, and other papers. Fig. 3.2 shows such a collation of experimental values taken from an extensive survey of data by Seah & Dench (1979). Most measurements have been made by an overlayer technique in which a known thickness of

Fig. 3.1. Calculated inelastic scattering electron mean-free-paths in Al showing the one-electron excitation contribution λ_e, the plasmon emission contribution λ_p and the effect of summing these, λ_{tot}. The values are deduced from the calculations of Quinn (1962) taking the Fermi energy, $E_F = 12$ eV, and kinetic energies are shown relative to the Fermi energy.

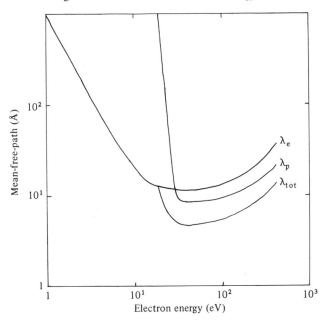

material A is deposited onto a substrate of material B and the attenuation of a discrete energy electron emission from material B (Auger or photoelectron peak) is measured during this process. The degree of attenuation in an overlayer of thickness d is equated to $\exp(-d/\lambda)$ where λ is the mean-free-path for inelastic scattering of electrons of energy dictated by the emission from the substrate (B) passing through the overlayer (A). A number of important experimental aspects of such studies are discussed by Powell; a more fundamental limitation of the technique is the extent to which attenuation in this thin overlayer ($\sim$tens of Å thickness) is representative of the overlayer material in its bulk form. In particular, such thin layers are unlikely to be truly homogeneous, and the scattering may well be influenced by the electronic effects at the substrate–overlayer interface and by its proximity to the overlayer-vacuum interface. Nevertheless, the general trend of results produced by such measurements does follow a curve rather similar to the theoretical data of fig. 3.1. Unfortunately, the overlayer technique is not very easy to apply at very low kinetic energies so most very low energy data are derived much less directly from a very simplified analysis of ultraviolet photoemission data. Evidently there is no reason to expect that the mean-free-path at some energy will be material independent, so the large scatter of data points seen on fig. 3.2 is

Fig. 3.2. Collection of experimental determinations of inelastic scattering electron mean-free-paths as a function of energy above the Fermi level for many different materials (after Seah & Dench, 1979).

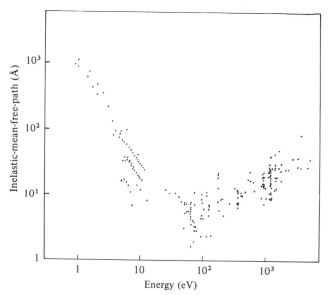

Table 3.1. *Some surface techniques using electrons and photons*

		Detected species	
		Electrons	Photons
Incident species	Electrons	AES LEED/RHEED	SXAPS
	Photons	XPS, UPS	Not surface sensitive

likely to be due, at least in part, to real differences from material to material.

The important result which emerges from both figs. 3.1 and 3.2, however, is that the mean-free-path for inelastic scattering of electrons throughout our specified energy range is of the order of, or less than, a few tens of Å and certainly in the optimum energy range (say, 50–200 eV) is typically less than 10 Å. Any technique involving the analysis of electrons of discrete energy emitted or scattered from a solid in this energy range is therefore highly surface sensitive and samples only the first few atomic layers. This important role of electron inelastic scattering in determining surface sensitivity can be appreciated by considering possible analytic techniques derived from detecting emitted electrons or photons using incident electrons or photons. Table 3.1 lists some techniques in this way and deserves more detailed discussion.

(a) *Electrons in/electrons out.* Two main groups of techniques are listed here: AES for surface compositional analysis and diffraction techniques. AES involves ionisation of core levels by a relatively high energy electron beam ($\gtrsim 1.5$ eV) and detection of emitted Auger electrons of discrete energy resulting from the core hole decay (see section 3.3), usually with energies of $\lesssim 1.5$ keV. The mean-free-path for core ionisation is long ($\gtrsim 1000$ Å) and, as the incident electrons have energy well above the core ionisation threshold, they can still core ionise after losing energy to plasmons and single-particle excitations. Thus the surface sensitivity is dominated by the inelastic scattering of the emerging Auger electrons only, and not the incident electrons. In the diffraction techniques, on the other hand, the emerging electrons have the same energy as the incident electrons because only elastically scattered electrons are studied. Thus inelastic scattering on both ingoing and outgoing electron paths contributes to the surface sensitivity. Indeed, for this reason we might expect the diffraction techniques to be more surface specific than emission techniques at the same energy. LEED is also generally performed in the energy range (~ 30–200 eV) corresponding to the shortest values in inelastic mean-free-paths. RHEED experiments are performed at much higher energies (e.g. 20 keV) where the mean-free-paths are expected to be longer (probably ~ 100 Å) but because RHEED is performed with both incident and emerging electrons at grazing angles to the surface ($\lesssim 5°$) the depth penetration is still small.

(b) *Electrons in/photons out.* Soft X-ray Appearance Potential Spectroscopy (SXAPS or APS – see section 3.4) involves the identification of the threshold energy for incident electrons to ionise a core level of an atom in the solid surface by detecting the onset of X-ray emission associated with electrons falling into the core hole created. These soft X-rays are only weakly absorbed by the solid and so have long mean-free-paths. For this reason, the surface sensitivity of this technique is also dictated by electron inelastic scattering, in this case in the incident beam. Evidently, as in AES, incident electrons with energies substantially greater than the ionisation threshold can cause ionisation of atoms relatively deep in the solid (i.e. over distances much greater than the mean-free-path for inelastic scattering). This technique therefore owes its surface sensitivity to the fact that it involves the detection of the *threshold* of ionisation when electrons which have lost small amounts of energy (e.g. $\sim$ plasmon energy) will not have sufficient energy to ionise the relevant core level.

For this reason the widely used analytical technique of electron microprobe analysis is *not* surface sensitive. This technique utilises (fixed energy) high energy incident electrons (~ 30 keV), and compositional analysis is obtained from energy analysis of emitted X-rays. As the incident electrons can ionise relevant levels after losing substantial amounts of energy, and the escape depth of the emergent X-rays is large, this is essentially a bulk technique and typically samples ~ 1–10 μm of depth into the sample surface.

(c) *Photons in/electrons out.* Photoelectron spectroscopy is generally divided into two techniques by the two types of laboratory photon sources available; gas discharge lamps (fixed photon energy ($h\nu$) in the range ~ 10–40 eV) leading to Ultraviolet Photoelectron Spectroscopy (UPS), and soft X-ray sources ($h\nu$ typically ~ 1200–1400 eV) which are used for XPS. In either case, photoionisation cross-sections are sufficiently small to ensure that the photon penetration depth is large relative to the mean-free-path for inelastic scattering of the emitted electrons which therefore determines the surface sensitivity. In XPS typical escape energies ($h\nu -$ (binding energy)) are ~ 500–1400 eV, so the technique is slightly less surface sensitive than methods using somewhat lower energy emissions. UPS, on the other hand, is most commonly performed with a HeI resonance line ($hv = 21.2$ eV) so emerging kinetic energies are low ($\gtrsim 17$ eV above the vacuum level). The exact degree of surface sensitivity of UPS is therefore difficult to establish, as the emerging electron energies fall in the region of the steep drop of the inelastic scattering mean-free-path with energy of figs. 3.1 and 3.2, and the material dependent variations may be significant. Nevertheless, the surface specificity of UPS is dominated by electron inelastic scattering and experiments do indicate surface sensitivity comparable with other techniques discussed here. It is clear, however, that significant variations in the surface sensitivity may exist in UPS for different materials and in studying levels of different binding energy or by varying the photon energy used.

(d) *Photons in/photons out.* Photons in the vacuum ultraviolet and soft X-ray region of the spectrum (and indeed in the visible region) have long penetration depths in solids relative to the few atom layers of interest to surface physics. All such techniques (e.g. X-ray fluorescence) are therefore essentially bulk techniques.

(e) *Other techniques.* Evidently the combination of probes and detected species above does not provide an exhaustive list of techniques which rely on inelastic scattering for their surface specificity and other references will be found to this parameter elsewhere in this book. For example, in addition to AES using an incident electron probe, Auger electron emission can be studied by incident ion stimulation. This method is, however, rather inefficient and is not used as a primary technique in its own right. It can provide additional information in high energy ($\sim$ MeV) ion scattering experiments when the emerging Auger electron mean-free-path can dictate surface specificity. A further technique, Ion Neutralisation Spectroscopy (INS), appears superficially similar to UPS. Here the stimulating probe is a very low energy ion beam which provides a core hole for Auger de-excitation and emission from surface valence and bonding orbital levels. Typical electron energies in this experiment are similar to those in UPS so the emergent electron escape depth is quite short. However, the actual Auger process is believed to occur with the incident ion outside the surface so that in this technique a very high degree of surface specificity is guaranteed by the exciting probe mechanism.

3.1.3 *Electron energy spectrometers*

In the sections which follow, detailed descriptions of a number of electron spectroscopy techniques will be presented and some attempt will be made to compare their relative strengths and weaknesses in different applications. The historical development of the various electron spectroscopies from rather different roots is at least one reason for the different types of electron spectrometer used in each case. To allow this comparison of techniques to be made, it is therefore necessary to have a basic understanding of the virtues and limitations of the spectrometer used; in this way it may be possible to separate physical limitations of the techniques from the instrumental ones imposed by convention rather than necessity. For this reason a discussion of some of the basic features of different analyser designs is of great value.

The basic purpose of any electron spectrometer is to separate out from electrons entering the spectrometer with a wide range of energies (and angles, due to source divergence) only those electrons in a certain narrow band of energies (independent of their angle). Ideally, this should be achieved with some kind of (energy) band pass filter but can be reached via a simpler high pass filter followed by some kind of signal processing. A form of high pass filter widely used in surface studies is the RFA and this will be discussed first before turning to true band pass, dispersive analysers.

3.1.3.1   *Retarding field analyser (RFA)*

Fig. 3.3 shows a schematic diagram of a set of LEED optics which is the most common form of RFA used in surface studies; indeed, it is the fact that the design of LEED optics (already possessed by many workers) is well suited for use as an RFA which accounts for the popularity of this kind of analyser. The sample sits at the centre of the set of concentric spherical sector grids and, as in the LEED experiment, the first grid (nearest the sample) is set at the same (earth) potential as the sample to ensure that electrons leaving the sample travel in a field-free space to the grids and so maintain their radial geometry. In a LEED experiment, the next grids are then set at a potential slightly less than that of the electron gun filament, so that all electrons having an energy less than those incident on the sample are retarded and do not pass on to the final stage of acceleration to the fluorescent screen. The conventional LEED experiment thus uses the grids as a high pass filter to pass only elastically scattered electrons.

If the retarding grids are set at somewhat lower potential, however, then all electrons having an energy greater than the energy corresponding to this potential reach the fluorescent screen (which is now simply used as a current collector). Thus, if the electron energy distribution is $N(E)$ and the retarding potential is set at V_0, corresponding to a minimum pass energy of $E_0 = eV_0$, then the current arriving at the collector is $\int_{E_0}^{\infty} N(E)\, \mathrm{d}E$, or as the highest energy electrons emitted have the primary beam energy E_p, the current is actually $\int_{E_0}^{E_p} N(E)\, \mathrm{d}E$. Evidently if this current (as a function of E_0) is differentiated in some way, then the resulting signal is the desired energy distribution $N(E)$. A simple way of achieving this is to modulate the retarding voltage V. Consider, for example, a measurement of the current arriving at the collector with retarding voltage V_0, and with retarding voltage $V_0 + \Delta V$. The difference between these two currents is $\int_{E_0}^{E_0 + \Delta E} N(E)\, \mathrm{d}E$ and if $\Delta E = e\,\Delta V$ is small this is evidently equal to $N(E_0)\,\Delta E$ – i.e. it is proportional to the energy distribution required. This is shown schematically in fig. 3.4. Using this kind of modulation leads to a characteristic trade-off between signal and resolution; the resolution, given by ΔE, deteriorates linearly with ΔV, while the signal, $N(E)\,\Delta E$, increases by the same fraction. This conclusion only breaks down at high resolution (small ΔV) when intrinsic limitations due to non-sphericity of the grids and field penetration between them limit the resolution typically to ~ 1 eV although much better resolution is possible if the optics are designed specifically for this purpose. This simple analysis also allows us to appreciate an intrinsic difficulty of a high pass filter used for a band pass application, which is the poor signal-to-background and hence potentially poor noise characteristics of the technique. Suppose, for example, that the noise is limited by electron statistics or 'shot noise' (strictly only true in an electron counting rather than analogue measurement). The noise in the signal $N(E_0)\,\Delta E$ is then given by

Fig. 3.3. Schematic diagram of a set of LEED optics (cf. fig. 2.8) operated as an RFA. Using a modulated retarding voltage as shown, the modulated component of the signal received at the collector is amplified and passed to a phase sensitive detector.

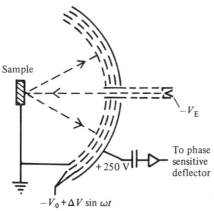

$(2 \int_{E_0}^{E_p} N(E) \, dE)^{\frac{1}{2}}$ while a true band pass detector would display the much smaller noise figure of $(N(E) \, \Delta E)^{\frac{1}{2}}$. Evidently if E_0 is close to E_p these do not differ greatly and the signal-to-noise is good, but with E_0 much less than E_p the situation deteriorates significantly.

In practice the analyser is never used in this digital subtraction mode but the same fundamental limitation exists. Usually the retarding potential E_0 is modulated sinusoidally (i.e. we apply a voltage $V_0 + \Delta V \sin \omega t$). In this case it is easy to show, by a Taylor series expansion, that the current arriving at the collector can be expressed as a sum of

Fig. 3.4. Schematic emitted electron energy distribution $N(E)$ arising from incident electrons of energy E_p. The hatched areas show the total current passed by an RFA operated with retarding voltage (a) of V_0 ($= E_0/e$), (b) of $V_0 + \Delta V$ ($= (E_0 + \Delta E)/e$) and (c) the current obtained by taking the difference between these signals. Evidently if ΔV is small, the difference current in (c) is simply $N(E_0) \, \Delta V$.

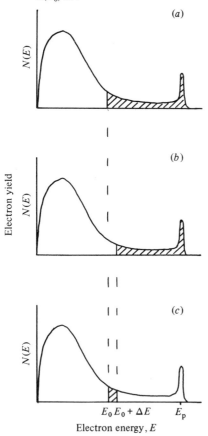

Electron energy, E

harmonics (a d.c. term plus terms in sin ωt, sin $2\omega t$, etc.). The d.c. current is evidently $\int_{E_0}^{E_p} N(E)\, dE$. The first harmonic (i.e. the term in sin ωt) has an amplitude

$$A_1 = \Delta E N(E_0) + \frac{\Delta E^3}{8} N''(E_0) + \frac{\Delta E^5}{192} N''''(E_0) + \cdots \quad (3.1)$$

The second harmonic (sin $2\omega t$) has an amplitude

$$A_2 = \frac{\Delta E^2}{4} N'(E_0) + \frac{\Delta E^4}{48} N'''(E_0) + \frac{\Delta E^6}{1536} N'''''(E_0) + \cdots \quad (3.2)$$

and so on, the primes on the N indicating the order of derivative with respect to E. Evidently, using this sinusoidal modulation, detection of the current at the collector with frequency ω (using a phase sensitive detector) gives a current proportional to $\Delta E\, N(E)$ to a first order as in our digital simulation. Providing that ΔE is kept small ($\lesssim$ a few eV) the higher-order terms can safely be neglected.

In practice, particularly in AES, it is common to measure the amplitude of the *second* harmonic sin $2\omega t$ which can be achieved by using a phase sensitive detector which is referenced by a frequency doubled version of the grid modulation signal. As we see above, the amplitude of this component is, to first order, proportional to the differential of the energy distribution, $N'(E)$. This is because the structure of interest in $N(E)$ is often weak (a small peak on a large background). Differentiating removes the constant background and allows increased amplification. Moreover, a broad peak in particular is more readily seen in the differentiated spectrum because it is changed into a 'double peak', each feature being narrower. This effect is illustrated in fig. 3.5.

There is a further, instrumental reason for detecting the second harmonic signal. The retarding grids and collector form a concentric hemispherical capacitor; there is substantial capacitive coupling between them which leads to a large first harmonic signal being measured at the collector due to the modulation of the capacitively coupled retarding grids. While this effect can be reduced by placing an extra, earthed grid between the retarding grids and collector, and by partially neutralising this capacitively coupled signal with some suitable circuitry (typically a capacitance bridge), the residual effect is still sufficient to make measurements of $N(E)$ directly from the first harmonic signal rather unsatisfactory. To summarise, therefore, the principal virtue of the RFA is its structural simplicity and the fact that LEED optics, already in use by many workers, can be used in this mode. It also has the virtue of a large acceptance angle (typically $\sim \pi$ steradians) and, for truly spherical grids, no aberrations associated with angular divergence of the source (though this is strictly true only for a point source). Its main disadvantage is that, because it is essentially a high pass filter used for a band pass application, the inherent signal-to-noise ratio is poor. Apart from the limitation introduced by the amplitude of modulation, the resolution of the analyser is limited by non-sphericity and the size and separation of grids which influence the degree of field penetration in the high pass filter. Typical conditions using a radius of curvature ~ 50 mm, separation of the grids ~ 2–3 mm and with two adjacent retarding grids to minimise field penetration lead to resolving powers of $E/\Delta E \sim 100$–200. A more detailed discussion of this type of analyser has been given by Taylor (1969).

3.1.3.2 *Electrostatic deflection analysers*

A much more desirable way to measure the number of electrons in a particular energy 'window' is to use an analyser with intrinsic energy band pass characteristics. This can be achieved by passing the electrons through a dispersing field in which the deflection is a function of the electron energy. Such an instrument can be based on either electrostatic or magnetic fields. Here we will only discuss in detail electrostatic deflection analysers. Magnetic analysers are normally used only at very high energies where the strong magnetic fields outside the analyser are not a problem, while electrostatic analysers become difficult to operate

Fig. 3.5. Schematic illustration of the effect of taking the derivative of $N(E)$ with respect to E in the vicinity of a weak peak on a significant background. The derivative spectrum $N'(E)$ is amplified by some large factor M.

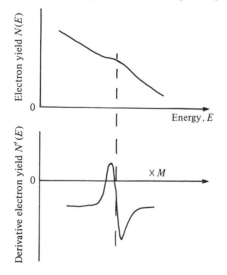

because of the high voltages required and associated insulation problems. At the low energies which are used in most surface electron spectroscopies, electrostatic instruments are simple to operate, compact and UHV compatible.

The simplest possible electrostatic deflection system would be a pair of parallel plates set at different potentials; this produces a field with plane parallel and equally spaced surfaces of constant potential. If electrons are directed into this field they are deflected and the greatest deflection over a given length of travel occurs for those of lowest energy; by putting an aperture into one of the plates, electrons in a specific energy range (the width of the range being a function of aperture size and field strength) will emerge. However, if, for example, the electrons are injected into the field nominally perpendicular to the field but with an angular spread about this mean direction, electrons of the same energy will be deflected by different amounts depending on the angle of injection. This means that the signal passing through the aperture will display an energy spread which is degraded by the angular spread of the incident electrons (see fig. 3.6(*a*)). As a result, both the energy resolution and transmission of the analyser are degraded by the angular spread of the source. Evidently, therefore, a good analyser design should be capable of *focussing* electrons of the same energy but different angles of injection, at the exit

Fig. 3.6. Electron trajectories in a parallel plate capacitor. In (*a*) injection of electrons is along the field lines and perpendicular to the field. Electrons with velocity $v_2 > v_1$ are less readily deflected. However, electrons of initial velocity v_1 may arrive at the same point as the axially injected electrons of velocity v_2 if they have the necessary off-axis injection direction. (*b*) shows a parallel plate device (a 'plane mirror analyser' or PMA) used in a focussing configuration.

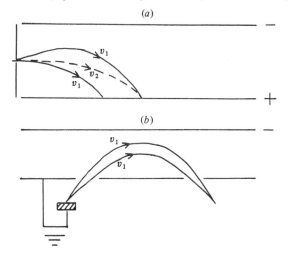

aperture. In the parallel plate analyser this can be achieved by injecting the electrons at a suitable angle. Fig. 3.6 contrasts this focussing domain with the simple conditions just discussed. In fig. 3.6(b) focussing is achieved by choosing a mean injection angle such that electrons which enter at a steeper angle and so need greater deflection to reach the exit aperture, have a larger distance of travel in the region of the deflecting field. Evidently this condition is only well satisfied at certain special conditions and perfect focussing is never obtained. Ideally, we desire a focus position independent of the angle of injection relative to the mean direction. In practice we can write our expression for the image position as a polynomial in powers of the divergence angle α (or in two dimensions, α and β). Geometrical conditions are chosen such that the lowest order terms vanish and we then say that we have nth-order focussing where the lowest remaining power of α or β in the expression has order $(n + 1)$. The remaining higher order terms are referred to as aberrations of the instrument. While the coefficients in this equation are needed to make proper comparison we see that an instrument with second order focussing is likely to be able to operate usefully with a more divergent source than one having first order focussing.

Fig. 3.7 shows some of the more important types of analyser used in surface electron spectroscopies. These are the parallel plate analyser (shown in fig. 3.6(b)) involving a total mean deflection of either 45° (first order focussing) or 30° (second order focussing), the 127° cylindrical analyser (first order focussing), the Concentric Hemispherical Analyser (CHA) with a total mean deflection of 180° and first order focussing, and the Cylindrical Mirror Analyser (CMA); this last instrument consists of concentric cylinders and accepts a conical annulus about a mean angle from the axis of the analyser of approximately 42°: it has second order focussing. We will concentrate our discussion on the last two of these analysers. The parallel plate arrangement has not been extensively used for surface studies, although it has been shown to be convenient for angle-resolved studies because it can be made extremely small and is simple to construct. The 127° analyser has been used particularly for electron energy loss spectroscopy. However, the CHA and CMA are probably the most widely used, particularly in commercial instruments. In some respects they present the extremes of design optimisation and we shall concentrate our discussion on these instruments.

The two main parameters of interest in designing or selecting an analyser for a particular application are the energy resolution and the acceptance angle; both of these also control the sensitivity of the instrument. The energy resolution of all the instruments is controlled by their physical size. If we define a resolving power as $E_0/\Delta E$ then, excluding aberration terms, the resolving power of each instrument is given by the ratio of a physical dimension related to the total electron path length in the analyser, divided by the size of the defining aperture. Thus, for the CHA, this is $2R_0/s$ where R_0 is the radius of the central path through the analyser and s is the size of input (and output) apertures. For the CMA the resolving power is approximately $5.6R_1/s$ where R_1 is the radius of the inner cylinder and s is the size of the defining aperture; for the CMA this is not the size of the aperture in the inner cylinder (which is usually large), but an aperture between the inner cylinder and source and image points. Usually no real entrance aperture exists and so the source size determines the entrance aperture. In fact the detailed design of a CMA is rather complex because the image aperture is not placed on the axis of the analyser; more detailed discussions of CMA design will be found in two general references on analyser design (Sevier, 1972; Roy & Carette, 1977).

Evidently these formulae for resolving power are sufficiently similar to ensure that the resolving power of either instrument is very similar for a similar total size of analyser. The geometry of input acceptance angles, however, is quite different. The CHA is usually fitted with a circular

Fig. 3.7. Schematic diagram of analyser geometry and electron trajectories for a 127° analyser (a) with cylindrical electrodes, a 180° spherical sector or CHA, (b) with spherical sector electrodes, and CMA, (c). The CMA has cylindrical symmetry about the axis.

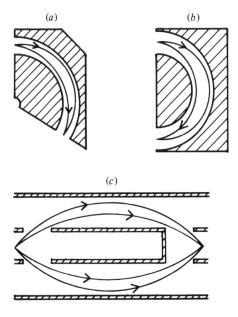

aperture subtending a total angle of less than 5° at the source. The total accepted solid angle is therefore typically less than 10^{-2} steradians, whereas the CMA may accept a range ($\sim \pm 6°$) of incident angles about a mean of 42.3° to the axis, at all azimuthal angles, leading to a total solid angle almost 1 steradian. The total accepted solid angle for the CMA may therefore be 100 times larger than for a CHA. Evidently this larger collection angle should result in larger signal strength and thus improved signal-to-noise characteristics.

The CHA does, however, have certain advantages which cause it to be frequently chosen despite the apparent superiority of the CMA. If electrons are retarded before injection into an analyser, the effective resolving power of the analyser can be improved. This is because, excluding the aberration terms, the resolving power $E_0/\Delta E$ is fixed by the geometry of the system so that if the pass energy, E_0, is reduced, then ΔE is also reduced. If the true energy of the electrons before retardation is γE_0 where γ is some constant ($\gamma > 1$) then the effective resolving power is $\gamma E_0/\Delta E$. Strictly this is a slight oversimplification, because the angular divergence changes during retardation thus increasing the importance of the analyser aberrations. This effect is expressed by the Helmholtz–Langrange equation for any electron optical system which in our case may be written as

$$\alpha_0{}^2(\gamma E_0) = \alpha_1{}^2 E_0 M^2 \tag{3.3}$$

where α_0 is the divergence when at the original energy (γE_0) and α_1 is the divergence after retardation to an energy E_0. The magnification, M, of the optical system (i.e. the ratio of image to object size) is a further variable to be manipulated in the optimum design of pre-retardation stages. Despite aberrations, however, substantial gains in effective resolving power can be obtained by this device. The CHA is particularly well suited to operation in this mode. Because it accepts a circular beam or cone of electrons it is compatible with simple electrostatic aperture and tube lenses, which can be used to produce the necessary pre-retardation, while at the same time imaging the electron source at the entrance aperture of the analyser. A proper choice of electron lenses can provide a considerable working distance between the analyser and specimen. The CMA, on the other hand, has an acceptance geometry which is not compatible with conventional electron lenses; it is, however, possible to operate it in the pre-retardation mode by inserting concentric spherical section grids in front of the analyser and centred on the sample (cf. a set of LEED optics operating as an RFA), and at least one commercial instrument does operate in this way. One disadvantage with the CMA is its rather short working distance and the fact that, because it accepts a large solid angle, most of the space around the front of the sample is obscured by the analyser. The incident exciting beam leading to electron emission must therefore be introduced to the sample at a grazing angle. Fig. 3.8 shows the kind of arrangement possible and also shows a pre-retardation stage as well as second stage of filtering. While it is possible to bevel the ends of the analyser, care must be taken to add special fringing field electrodes to ensure that the field between the cylinders in the region of the electron paths is not influenced by the truncation of the cylinders.

To quantify some of these considerations, a typical CMA with an outer cylinder diameter of about 100–150 mm can operate effectively without pre-retardation with a resolving power ~ 200 and a working distance ~ 5 mm. A CHA of comparable size, fitted with pre-retardation, can be made to operate successfully at resolving powers of 1000–2000 and with a working distance of 25–50 mm.

To summarise, therefore, the CMA is mainly but not exclusively used as a high collection efficiency, low resolution instrument with some associated inconvenience in working distance. The CHA is a low collection efficiency analyser which can be operated at high resolution and, if necessary, with rather long working distances. Of course, not all uses of CMAs and CHAs fit into these simple classifications; for example, at least one commercial CHA is operated in the retarding mode but with a pre-retardation stage of parallel retarding grids rather than proper electron lenses; such an arrangement is simple but discards the potential advantage of large working distance.

Fig. 3.8. CMA fitted with pre-retardation spherical sector grids and showing the angled front cone to permit exciting probes to be directed at the target. In this diagram a second stage of CMA dispersion and analysis is included. This 'double-pass' commercial design is intended to reduce spurious background signal due to secondary electrons generated inside the analyser by having two stages of filtering, the first operated at low resolution (large defining apertures).

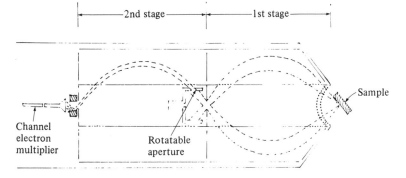

One further important distinction of the analysers relates directly to the acceptance angle which may be needed for an experiment. Evidently, if any angle-resolved experiment is planned, a CHA is much more suitable and experiments of this kind performed with a CMA and aperture can be attributed more to the ready commercial availability of this analyser than to its intrinsic merits for such a purpose. If we briefly reconsider the two other analysers shown in figs. 3.6(*b*) and 3.7, both of these have an acceptance geometry in the form of a slit. They have a collection angle intermediate between a CHA and CMA but are not compatible with conventional electron lenses and must, therefore, be operated in the pre-retardation mode with parallel grids. Their application is therefore similar to the CHA although their electron optics is typically less elegant; a major virtue of these designs, however, is their simplicity of construction and compactness.

Finally, we should remark on the modes of operation of all of these analysers which are important when comparing results. Without pre-retardation (the normal mode of use of a CMA), an energy spectrum is swept by varying the potential on the outer deflection cylinder and, hence, varying E_0. As $\Delta E/E_0$ is constant, this means that ΔE is proportional to E_0 so that if an energy spectrum $N(E)$ is input to the analyser, the output is proportional to $EN(E)$. On the other hand, with pre-retardation, an analyser can either be operated at fixed retardation ratio γ, when the pass energy is again varied and $EN(E)$ measured, or operated at fixed pass energy E_0 and variable retardation γ. In this case E_0 is constant so ΔE is constant (excluding variations in aberrations) so that $N(E)$ is measured at fixed resolution. There are evidently virtues in both modes of operation but the differences are important in comparing data from different analysers. We might also note here that, when using an RFA as described in section 3.1.3.1 in the modulated differential mode, the pass width ΔE is controlled by the (constant) modulation voltage under most operating conditions so that the output is $N(E)$, not $EN(E)$. This distinction is important in comparing Auger electron spectra taken on RFAs and CMAs, the two analysers most used in this technique.

3.1.4   *Electron energy distributions in electron spectroscopies*

If an electron beam is incident on a crystal surface and the energy distribution of emitted electrons is measured, the general form of the distribution is as shown schematically in fig. 3.4. The electrons are generally classified into three groups, elastically scattered, inelastically scattered and secondaries. The peak seen at the incident primary energy consists of the elastically scattered electrons and these form the signal detected in diffraction experiments. This peak is usually taken to include electrons which may also have been phonon scattered as in most instruments such small energy transfers cannot be detected. The exception to this generality is in the study of vibrational energy losses discussed in more detail in chapter 9. The inelastically scattered electrons are generally supposed to be those which have lost energy in at least one inelastic scattering event. In the case of electrons which have suffered several such scatterings or lost energy in continuum excitations they contribute a generally featureless spectrum extending from very low energies to the elastic peak. On the other hand, those electrons which have lost energy in a single process to some discrete quantum excitation produce small peaks on the energy distribution at energies which differ from the elastic energy by this discrete loss energy. The principal types of loss commonly involved are plasmon losses (due to the creation of bulk or surface plasmons) and ionisation losses associated with the ionisation of a core level of an atomic species in the surface region. In addition, continuum electron–hole excitations (interband transitions) may give rise to peaks when there is sharp structure in the initial, filled and final, empty densities of states. These interband transitions usually involve loss energies of a few eV while plasmon losses are typically 10–30 eV and ionisation losses may be from a few tens of eV to 1000 eV or more. Vibrational energy losses (see chapter 9) involve energy vibrational transfers of less than 1 eV.

Most of the 'true' secondaries are situated in the very intense peak lying at very low energies (typically less than 50 eV even for incident energies $\gtrsim 1$ keV) and are supposed to have arisen from a 'cascade' process in the energy loss of high energy primaries. Of course, there is, in general, no way of distinguishing between 'true' secondaries and inelastically scattered electrons when considering the smooth continuum. In addition to this continuum, small peaks can arise due to the emission of electrons associated with the decay of some kind of excited state in the surface region created by the incident primary beam. While such features may occur at very low energies ($\lesssim 20$ eV) associated with interband transitions into the continuum above the vacuum level, or by the decay of plasmons into single particle excitations, the dominant process of this type is Auger electron emission. This process and its study and exploitation is described in detail in section 3.3.

In any electron spectroscopy which involves the release of electrons in the surface region following stimulation by a means other than an incident electron beam, all of the same basic ingredients are to be found in the energy spectrum. The most usual alternative method of excitation is an incident photon beam and all of the same inelastic and secondary

94   *Electron spectroscopies*

processes can occur. However, because of the nature of the excitation, the signal-to-background characteristics can change radically and the general appearance of the energy distributions are, therefore, rather different. In the case of incident electrons, all electrons which are finally detected must have undergone an elastic reflection in the backwards direction (typically less than 1% efficient) or an elastic plus inelastic collision, or result from a secondary process. As a result, below the elastic peak at least, the signal-to-background ratio for the discrete losses and secondary emission processes is generally poor. In a photoemission experiment, however, the photoionisation is followed by electron emission in which typically 50% of this emission is directed out of the crystal without any need for elastic scattering to reverse its direction. Thus, the signal-to-background for these discrete emissions (photo-emission and Auger emission following filling of the photoionised core hole) is usually much better than in incident electron techniques. Indeed, if the incident photon energy is rather low (e.g. 21.2 eV in UPS) then even the low energy secondary electron peak does not totally dominate the spectrum as in incident electron techniques. In XPS, however, with higher photon energies (e.g. 1253 eV) a large low energy 'true' secondary electron peak does cause a severe loss of signal-to-background ratio at low kinetic energies. A typical XPS spectrum (fig. 3.9) also shows another characteristic which is generally present in electron stimulated

Fig. 3.9. 'Typical' XPS spectrum taken using 1253 eV photons (Mg K_α). Note the 'stepped' structure of the background due to inelastic processes. The very high count rate part of the low energy secondary electron emission is omitted.

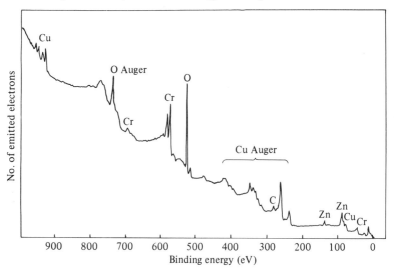

Auger electron spectra, but is not very obvious because of the poorer signal-to-background ratio in this technique. This is the clear inelastic electron tail behind each discrete energy emission; this gives the XPS spectrum a kind of staircase structure as at high kinetic energies there are few primary electrons of sufficient energy to create inelastic electrons while each new discrete emission opens up a new channel of primary electrons to be inelastically scattered. These individual inelastic tails can probably be attributed primarily to photoionisation events occurring at depths below the surface much greater than the mean-free-path for inelastic scattering. Thus, only inelastically scattered electrons from these photoionisation events can be detected outside the surface. This effect is particularly pronounced in photoemission because of the deep penetration of the incident photon beam relative to the escaping electrons.

3.1.5 *Electron spectroscopies: core level spectroscopies*

In all electron spectroscopies, energy analysis of electrons emitted from the solid surface is used to gain information on the electronic energy levels in the surface region. These energy levels can be loosely divided into two groups; core levels and valence levels. In this division the core levels are those associated with electron states localised on a single atom and thus largely characteristic of the atomic species itself; such levels are assumed to be relatively unaffected by the fact that the atom is not free but in the solid surface. They will be expected to have a binding energy of more than, say, 10–20 eV. As such, spectroscopies based on the detection of these levels ('core level spectroscopies') might be expected to be primarily concerned with determining the atomic species present on the surface; in fact, information on the local chemical environment is often also available. Of the core level spectroscopies two stand out because of their exceptional popularity and widespread use; these two, XPS and AES, will be dealt with in some detail in sections 3.2 and 3.3. The remaining techniques of APS, Disappearance Potential Spectroscopy (DAPS) and Ionisation Loss Spectroscopy (ILS) will be grouped together in section 3.4 as they have much in common in terms of the underlying physical principles.

Electronic energy levels of the surface which have lower binding energies will comprise the valence band of the solid and bonding orbitals associated with adsorbed molecules. These states are less well localised and are clearly potentially very sensitive to the local chemical environment in the solid surface. The most widely used technique investigating these levels directly is UPS which is discussed in section 3.5. Another, somewhat less direct, approach is the use of INS. We reserve discussion of this method until the following chapter on incident ion techniques, although it may, in some regards, be seen as an electron spectroscopy comparable with UPS.

3.2   X-ray Photoelectron Spectroscopy (XPS)

3.2.1   *Introduction*

Photoelectron spectroscopy is in principle a particularly simple process. A photon of energy hv penetrates the surface, and is absorbed by an electron with a binding energy E_b below the vacuum level, which then emerges from the solid with a kinetic energy $(hv - E_b)$. In the simplest case, therefore, the energy distribution of photoemitted electrons should simply be the energy distribution of electron states in the solid surface shifted up in energy by an amount hv. Of course, this simple picture is complicated in practice, for example, by the fact that the probability of the photon being absorbed by all the electron states is not the same. However, the relative simplicity of this one electron process makes it a natural choice for discussion first. Any photon whose energy exceeds the work function of the solid $(hv > \phi)$ can be used for photoelectron spectroscopy, which simply excludes the near ultraviolet, visible and higher wavelength radiation. In practice, however, until recently at least, nearly all photoelectron spectroscopy has been performed in two relatively narrow energy ranges defined by convenient intense laboratory sources. The first range is provided by light from gas discharge sources and particularly the intense line emission from He and other inert gases; for He, the two main lines have photon energies of 21.2 and 40.8 eV. For the other inert gases the main emissions have somewhat lower energies. These sources are not capable of accessing a significant range of core levels and provide the means of performing UPS on valance levels. The second readily available photon energy range is usually restricted to two lines (or groups of lines); the Al and Mg K_α X-ray emissions at 1486.6 eV and 1253.6 eV respectively. A small number of experiments have also been performed on the K_α emissions of adjacent atoms in the periodic table. These soft X-ray lines form the basis of XPS. The large gap in energy between these two groups of sources forms a clear division in the two techniques based on essentially the same physical process. Recently, the growing use of the intense continuum source offered by synchrotron radiation from an electron synchrotron or storage ring is leading to a weakening of this division. Synchrotron radiation sources typically provide a continuum for photoemission from the softest ultraviolet to hard $(hv > 10 \text{ keV})$ X-rays, and with suitable monochromators photoelectron spectroscopy can be performed at any energy in this range. In practice, so far most work has been concentrated at energies close to the gas discharge lamps and will be discussed with UPS. A smaller number of experiments have used higher energy radiation (typically $hv \sim 100$–200 eV so far) to study core level effects and as such we will classify these studies as 'XPS'.

3.2.2 *Photon sources*

Before looking at the basic physics and areas of applications of XPS it is useful to consider briefly the experimental constraints which define some of the parameters of the technique. Conventional X-ray sources are created by bombarding a solid target with high energy electrons; the emission from this target consists of characteristic line emissions associated with the filling of core holes created by the incident electron beam, superimposed on a continuum background up to the incident electron energy due to bremsstrahlung. In most cases the electron energy is chosen to be appreciably higher than the K-shell binding energy of the target and lines associated with the filling of K-shell holes dominate the spectrum.

If we are to gain useful information on the occupied electronic energy levels on the surface by energy analysing the photoemitted electrons, our photon source should be as nearly monochromatic as possible. If this is to be obtained directly from an X-ray source, as described, it is important to choose target materials having a low bremsstrahlung background and narrow characteristic line emission, preferably dominated by a single line. A further experimental constraint favours a metallic target; considerable power is fed into the target by the incident electron beam and it is usually necessary to cool the target, a process greatly simplified if the target is a good conductor of heat. This requirement for a cool target is particularly important if the source is to be 'nude' in the UHV analysis chamber vacuum system due to the associated out-gassing of the target which could cause unacceptable pressure rises. These choices favour the use of the two materials previously mentioned, Mg and Al, although some work has also been performed with Na (1041.0 eV) and Si (1739.5 eV). In all of these cases, the emission spectrum is dominated by an unresolved doublet, $K_{\alpha_{1,2}}$ associated with decays from $2p_{\frac{1}{2}} \rightarrow 1s$ and $2p_{\frac{3}{2}} \rightarrow 1s$. In fact, other lines are also present associated with doubly and multiply ionised atoms undergoing $2p \rightarrow 1s$ transitions, all labelled K_{α}, while K_{β} emission lines associated with valence $\rightarrow 1s$ transitions are also present. Fig. 3.10 shows an emission spectrum from Mg with intensities on a logarithmic scale. Evidently using such an unmonochromatised source for XPS will lead to electron energy spectra dominated by emission with the $K_{\alpha_{1,2}}$ photons, although the doubly ionised ($K_{\alpha_{3,4}}$) emission will give rise to photoelectron satellites of about

8% of the main intensities at kinetic energies 10 eV higher. Other lines are generally $\lesssim 1\%$ of the $K_{\alpha_{1,2}}$ emission. The remaining feature of importance in this X-ray spectrum is the width of the dominant $K_{\alpha_{1,2}}$ doublet; full width at half maximum is ~ 0.7–0.8 eV for both Mg and Al.

Most XPS studies are performed with sources of this kind which clearly implies certain limitations for the experiments. Firstly, the existence of satellite X-ray lines will complicate the photoemission spectra and could lead to some overlap of satellite line emission from dominant species and main line emission from minor species in unfavourable cases. Secondly, the intrinsic line width of the dominant X-ray line places a minimum line width on photoelectron spectral features; this will limit our ability to detect changes of width and position of such photoelectron peaks to a few tenths of an eV.

Fig. 3.10. Mg K-shell X-ray emission spectrum. The full line shows the characteristic 'line' emission after subtraction of a constant background as shown by the dashed line. Note the logarithmic intensity scale. Conventional XPS studies rely on the dominance of the $K_{\alpha_{1,2}}$ doublet at 1253.6 eV (from Krause & Ferreira, 1975).

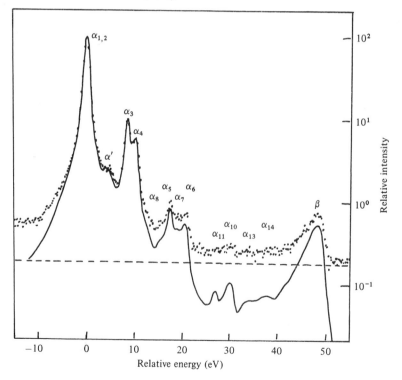

Of course, it is possible to monochromate the X-rays by single or multiple Bragg reflections from suitable crystals to pick out just a part of the dominant $K_{\alpha_{1,2}}$ line; this typically leads to a photon energy spread of ~ 0.2 eV although the loss in intensity is considerable. This improved resolution is vital if useful data are to be accumulated on photoemission core line widths, or on valence band emission density of states. For most purposes, however, the improved photon flux is usually achieved by putting the X-ray target as close as possible to the XPS sample (typically ~ 1 cm). This closeness is partly limited by the need to interpose a thin (~ 10–$30 \, \mu$m) Al or Be window between the X-ray target and XPS sample to prevent detection of secondary electrons from the X-ray source and to isolate the pumping of the outgassing X-ray target from the UHV of the sample chamber.

The other main experimental component, the electron energy analyser design, has been discussed extensively in the previous section. We note, however, that if we are to take advantage of the relatively narrow photon line width in defining shapes and shifts in the photoemission spectrum, the analyser must be of quite high resolving power. Photoelectron kinetic energies may be 1 keV or more, and the analyser resolution should preferably be better than the source line width. This sets a resolving power requirement of around 2000 and has tended to favour the retarded CHA geometries described earlier.

3.2.3 Shapes and shifts

A typical XPS spectrum of an oxidised and partially contaminated Al surface is shown in fig. 3.11. This shows that the spectrum is dominated by a number of sharp emission peaks which have been labelled with the core states of the surface atoms from which they originate. This part of the spectrum follows the anticipated form of a projection of the surface electronic states to energies above the vacuum level. A number of complications do appear, however. The most obvious general feature is the existence of the inelastic loss 'tails' following original photoexcitation of the relevant core level. This feature reflects the fact that while the surface sensitivity shown by the directly emitted ('no-loss') photoemission peaks is defined by the electron inelastic scattering mean-free-path for the escaping electron, the depth of photoionisation is related to the much weaker absorption of the incident X-ray photons.

On a finer scale fig. 3.11(b) shows further structure; notably behind the substrate (Al) emissions are peaks associated with multiple plasmon losses (particularly well defined in Al) and ancillary 'chemically shifted' versions of the substrate emissions themselves associated with Al in an oxide rather than metallic environment. In order to determine the extent to which these ancillary effects may confuse or enhance the usefulness of XPS in surface analysis we shall now consider the processes giving rise to shifts in the photoemission peaks and to the appearance of related fine structure.

In our original very superficial comments about photoemission we suggested that the photoelectron energy spectrum should simply show the density of occupied electronic states in the surface transposed up in energy by an amount hv. Ignoring, for the time being, the question of the relationship between photoelectron peak amplitudes or areas and the occupation of a particular state, we will first look in detail at the location in energy of these peaks. The assumption implicit in this description is that the binding energy 'seen' by the photoelectron, E_b, of the state it leaves is the same as it was before the interaction and hence that all other

Fig. 3.11. Typical XPS spectra obtained from an oxidised and partly contaminated Al sample taken using monochromatic Al K_α radiation. (*a*) shows the overall features with the main core level emissions labelled. (*b*) shows the low binding energy region on an expanded scale; plasmon loss structure and 'chemically shifted' Al emission lines are labelled (after Fadley, 1978).

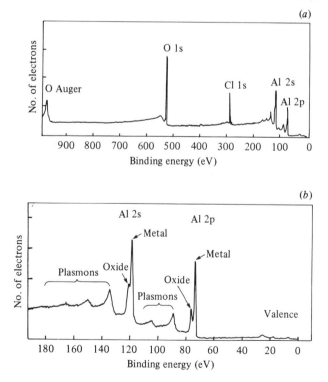

electrons in the system are in the same state as before the photoionisation event. Such a situation obeys Koopman's theorem and this energy, E_b, is therefore referred to as the Koopman's energy. In this case, if E_b is referred to the vacuum level, the emergent kinetic energy is indeed given by

$$KE = hv - E_b$$
(3.4)

In reality, this Koopman's energy is never observed. The main reason for this is the so-called relaxation shift. When the core hole is created by photoionisation, other electrons relax in energy to lower energy states to screen this hole partially and so make more energy available to the outgoing photoelectrons. If we first consider the case of a free atom then this difference in energy from the Koopman's energy may be represented by the intra-atomic relaxation shift E_a and we have

$$KE = hv - E_b + E_a$$
(3.5)

This description would be adequate if photoionisation and photoemission were slow processes such that the system was always allowed to reach a stable equilibrium. In reality the process is rapid and a more usual and apparently valid assumption is the 'sudden' approximation in which the perturbation is switched on very rapidly. The result is that the final state may be one in which an electron is in an excited bound state of the atom, or in which another electron is ejected into the continuum of unbound states above the vacuum level. Such processes leave less energy for the emitted photoelectron and this gives rise to lower kinetic energy satellites; these are usually referred to as shake-up features (when excitation is to a bound state) and shake-off (when excitation is to the continuum) – see fig. 3.12.

Of course, this complex energy spectrum is one which simply manifests the difference in energy between the initial (neutral unexcited atom) state and the final state ion. Such states must be represented by many-electron wavefunctions which take account of the interaction of the different electrons with each other and with the nucleus. Relaxation, shake-off and shake-up are intrinsically many-electron or 'many-body' effects. However, one way of trying to compute the form of the emission spectrum is to represent the final state as a linear superposition of the $(N-1)$ electron states of the final ion originating from the N electron initial neutral atom. One can then try to determine the coefficients in this linear superposition leading to the minimum total energy. Such a procedure is similar to the standard procedures for computing molecular orbital hybridisation character. and leads to terms indicating the relative magnitude of the different shake-up and shake-off satellites.

This procedure has led to the effect being known as Configuration Interaction (CI) (or strictly Final-State Configuration Interaction (FSCI)) and this name is sometimes used instead of shake-up and shake-off in describing the origin of the lower kinetic energy satellite fine structure.

Now let us consider how this situation is changed when the free atom is placed in or on a solid surface. If we revert once again briefly to the adiabatic approximation in which the photon interaction is slow and the ion is, therefore, left in its ground state, the main difference lies in the energy states of the more weakly bound valence electrons. In the case of a metal, in particular, these are very mobile and so will screen the core hole efficiently; this leads to an additional 'interatomic' relaxation shift and to an emitted kinetic energy which is higher than from the atom in its free state; thus,

$$KE = h\nu - E_b + E_a + E_r \tag{3.6}$$

where E_r is the extra relaxation energy associated with the solid environment. As E_b is typically several hundred eV and E_a and E_r are only of the order of a few eV or less, it is easier to measure experimentally, and compute theoretically, the relaxation shifts directly, rather than to expect perfect agreement in the total absolute values of the kinetic

Fig. 3.12. Energy level diagram and schematic photoemission spectra for a core level emission from an atom (*a*) in the sudden approximation, and for the same species in a solid showing the adiabatic (*b*) and sudden (*c*) limits. Note the intra-atomic and interatomic relaxation energy shifts E_a and E_r.

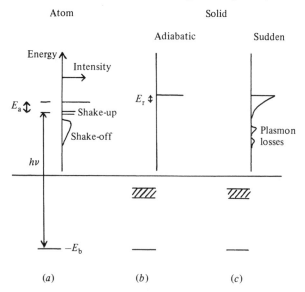

energies (although even this approach has met with success). The only experimental difficulty in this process lies in the proper determination of the energy zero; we will return to this point later.

If we now remember that the 'sudden' approximation is much more nearly valid than the adiabatic one, we expect to see satellite features associated with shake-up and shake-off. In the solid environment we might expect that the free atom picture of the final ion configurations, being in discrete excited states, is not appropriate, as these states normally form a continuum and, although some solids show a remarkable degree of atomic-like behaviour, this is usually true. In this case we expect to be able to excite electron–hole pairs around the Fermi level giving rise to a low kinetic energy 'tail' to the observed photoemission peak; this leads to a characteristic asymmetric line shape for XPS photoemission peaks. Another solid state form of excited state is the possibility of creating bulk or surface plasmons during the photoemission process, giving rise to plasmon satellites at energies lower than the adiabatic peak by amounts which are integral multiples of the plasmon energy. Such effects are clearly observable in materials (such as Al, Mg, Na) in which the plasmons are weakly damped. Features of this kind ('plasmon losses') are seen clearly in the Al spectrum of fig. 3.11. However, these losses and indeed the whole 'inelastic tail' behind each photoemission peak can actually be attributed to two separate processes. One is the 'intrinsic' one just discussed in which the losses are incurred in the photoemission process itself and may be regarded as resulting from excited states of the residual ion and its local environment. The second is the possibility of 'extrinsic' losses which can be incurred by the electrons in the adiabatic peak (and others) during their transport through the solid from the emitter to the surface. Such electrons can suffer inelastic scattering leading to the creation of electron–hole pairs or plasmons; these losses during transport to the surface can therefore also lead to inelastic tails and discrete plasmon satellites of the kind observed.

Quite a number of attempts have been made to try to determine the relative importance of intrinsic and extrinsic excitation processes in determining the loss structure (particularly the plasmon losses) associated with the photoemission peaks. The problem is inherently difficult because the time scale of the transport of the electrons through the surface region is not long when compared with the time scale of the photoemission process, so that the intrinsic and extrinsic processes may interfere coherently; i.e. they are not actually separable in general except in an essentially arbitrary, theoretical way. However, in so far as this work has been conclusive, it appears that, at XPS energies at least, the most important loss process is probably extrinsic. This separation may actually be quite important in determining the usefulness of XPS as a quantitative technique for the analysis of surface composition. This is because an important sum rule tells us that the integrated intensity of the adiabatic peak, plus all intrinsic loss structure, is characteristic of the initial state and final ground state ion. If considerable intensity is diverted from the adiabatic peak to the loss terms, then the intensity of the readily identified adiabatic peak would become dependent on the electronic environment of the atom through its associated shake-up structure.

The foregoing discussion allows us to understand in general the positions and shapes of the structure in the spectrum shown in fig. 3.11(*b*) with the exception of the shifted substrate photoemission peaks labelled as associated with the oxide state. The existence of these 'chemical shifts' associated with different local chemical and electronic environments is of considerable practical value in XPS. These chemical shifts originate from the sum of two effects; the first of these is the final state effect of relaxation already discussed. We have seen that a metallic environment modifies

Fig. 3.13. Comparison of experimental XPS C 1s binding energies with those calculated via Koopman's theorem for C in a range of molecules. Although experimental and theoretical values differ by 15 eV (associated with relaxation effects) the systematic comparison is excellent as indicated by the straight line of unity gradient (after Shirley, 1973).

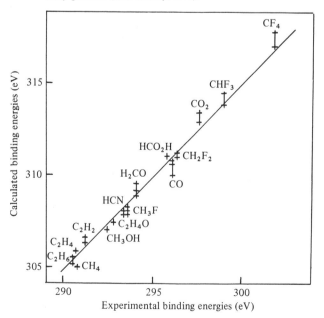

the energy due to an associated relaxation shift and a further change of nearest neighbour atoms, say from metal to O atoms, can modify this intra-atomic relaxation shift. The second effect is an initial state effect which might properly be called a true chemical shift; this is the shift in the original binding energy due to the changed electronic environment of the atom. The relative size of these two effects can vary substantially. For example, fig. 3.13 shows a comparison of the observed binding energy of the C 1s level in a range of carbon-containing molecules with the Koopman's theorem calculated values for the same molecules. While the two energies differ by some 15 eV due to the neglect of relaxation shifts, the points all fall close to the straight line of unity gradient, indicating that the chemical shifts are well described by this model which only includes the initial state effects. On the other hand, calculations of the relaxation and true (initial state) chemical shifts incurred by various atoms adsorbed onto jellium shown in fig. 3.14 show that in these cases the changes in relaxation shift are similar in magnitude to the 'true chemical shift'.

These results show that the exact origins of observed chemical shifts may vary; nevertheless, for practical studies it is their existence which is important. Even without a quantitative description of these effects, valuable work can be performed by using the chemical shifts observed in known systems as a 'fingerprint'. Thus, extensive tabulations of observed

Fig. 3.14. Calculated extra-atomic relaxation shifts and chemical shifts for O, Na, Si and Cl adsorbed on jellium – an ideal smooth free electron surface having an electron density appropriate to Al. Metal–adatom separations were ~ 1.32 Å, 0.85 Å and 0.80 Å for Na, Si and Cl while O was placed inside the metal (after Williams & Lang, 1977).

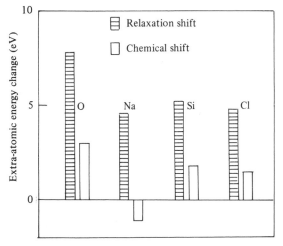

chemical shifts have been compiled, and a common approach to the XPS study of absorption of species A on solid B is to try first to characterise absolute peak positions of XPS peaks in all known bulk compounds of A and B to provide a reference for the various known chemical shifts to be expected. In this way chemical shifts observed during the adsorption may be related to different stages in the chemical reaction.

One problem of this approach and all approaches involving the determination of absolute binding energies in XPS, is that of defining a fixed reference energy. Ideally, the reference should be the vacuum level which is the most relevant parameter for a free atom. Experimentally, however, it is usual to reference to the Fermi level of the solid; this is because the determination of the vacuum level relies on accurate knowledge of work functions and contact potentials in the experimental system. The Fermi level, on the other hand, is usually a reference point *within* the XPS spectrum because, for a metal at least, it represents the most weakly bound occupied energy state and is thus seen in the spectrum as the most energetic emission. It is therefore a straightforward matter to reference the apparent binding energy of a core level emission to this feature in a reliable fashion. The basic problem of proper referencing of observed binding energies is readily seen by considering the effects of adsorbing species A on a surface of (metallic) species B. The 'true' chemical shift between the clean (1) and adsorbed (2) states is given by the difference in the binding energies relative to the vacuum level which is equal to the difference in kinetic energies. Thus,

$$\Delta E_b^V(1,2) = (E_b^V)_1 - (E_b^V)_2$$
$$= (E_{kin})_2 - (E_{kin})_1 \qquad (3.7)$$

If the binding energies are referenced to the Fermi levels then

$$\Delta E_b^F(1,2) = (E_b^F)_1 - (E_b^F)_2$$
$$= (E_{kin})_2 - (E_{kin})_1 + (\phi_{spect})_2 - (\phi_{spect})_1 \qquad (3.8)$$

That is, the new shift is influenced by any possible change in the work function of the spectrometer, deflection and retardation system due to the adsorption of A on these components. Even if $(\phi_{spect})_2 - (\phi_{spect})_1$ is zero, we have

$$\Delta E_b^V(1,2) = \Delta E_b^F(1,2) + (\phi_{surf})_1 - (\phi_{surf})_2 \qquad (3.9)$$

i.e. the difference in the chemical shift relative to the vacuum level such as might be calculated or observed in the gas phase and that relative to the Fermi levels is the work function change of the surface of interest (which might be as much as 1 eV or more). Moreover, we have neglected in this discussion the possible effects of surface charging; insulator surfaces can charge up as a result of the electron emission and while, in several cases, this can lead to drifting of peaks, stable charging shifts can be several eV (i.e. similar to, or greater than, the chemical shifts of interest). While the use of chemical shifts in XPS, even in the purely empirical form, is therefore valuable, it is not without its difficulties. Various approaches have been adopted to try to minimise these problems; in insulators, for example, thin Au films can be deposited on the surface to reduce charging and provide a reference in the form of the Au 4f photoemission (a common XPS 'standard'). It remains a problem, however, in difficult cases or where extreme accuracy is required.

3.2.4   *XPS as a core level spectroscopy*

Strictly, the object of a core level spectroscopy is to provide a compositional analysis of a surface. Because core levels are essentially characteristic in their energy of the atomic species (despite the energy shifts just described), the observation of certain binding energy peaks in an XPS spectrum can be taken as an indication of the presence in the surface region of a particular elemental species. Thus an XPS spectrum should present information from which the composition of the surface region may be determined; the additional information on *exact* peak positions may then be capable of indicating the chemical state of some of these component elements. In principal, therefore, XPS well deserves its alternative name of ESCA – Electron Spectroscopy for Chemical Analysis – first attached to the technique by Siegbahn and his co-workers at Uppsala in Sweden, who can claim primary responsibility for developing the technique originally (see, for example, Siegbahn *et al.*, 1967). The second name has generally fallen out of favour and become less appropriate because of the development latterly of so many other electron spectroscopies with the same object; for that reason this acronym is not generally used in this book.

The usefulness of XPS (or any other analytical technique) for compositional analysis, once the basic ability is established, depends upon two factors. How sensitive is the technique (what is the minimum detectable surface concentration) and how easily can the technique be made quantitative (and how quantitative)?

A primary component of both questions is the photoionisation cross-section for different energy levels of different atomic species. Additionally, we should question the availability of accessible energy levels. Fig. 3.15 shows a plot of the binding energies of filled atomic energy levels across the periodic table; it is clear that with a photon energy in excess of 1 keV, photoemission from some energy levels of all elements is possible and in most cases several levels are accessible. The character of these levels (principal (n) and angular momentum (l) quantum number) varies, however, as does the proximity of the photon energy to the photoemission threshold for the various levels. These are the main parameters affecting the photoionisation cross-sections. The photo-ionisation cross-section is normally computed using the Golden rule that the emission rate is proportional to the square of the matrix element $\langle f|\mathbf{H}'|i\rangle$ where $\mathbf{H}'$ is the interaction between the electron or electromagnetic field and $|f\rangle$ and $|i\rangle$ are the final and initial states. Strictly, these states are many-electron wavefunctions and the interaction must be summed over all electrons. While many-body effects can be important in XPS, it is usual to calculate these cross-sections assuming one-electron wavefunctions only.

The interaction Hamiltonian may be written as

$$\mathbf{H}' = -\frac{e}{2mc}\,(\mathbf{p}\cdot\mathbf{A}+\mathbf{A}\cdot\mathbf{p}) \tag{3.10}$$

Fig. 3.15. Core electron binding energies of filled levels of the elements. Note that the majority of levels lie below about 10^3 eV and are therefore accessible to conventional laboratory source XPS (from Wertheim (1978) based on the tabulated values of Siegbahn *et al.*, 1967).

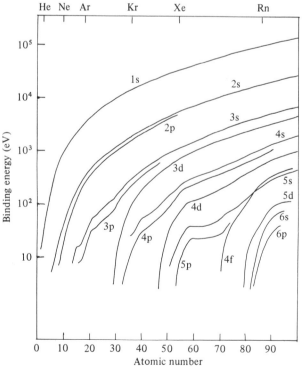

where **p** is the electron momentum operator $(-i\mathbf{V})$ and **A** is the vector potential of the field. Generally, it is possible to ensure that $\mathbf{p} \cdot \mathbf{A} = 0$; a spatial variation of the vector potential may occur in the surface region but such effects will be small if the photon frequency greatly exceeds the plasmon frequency of the solid. Such a situation should certainly hold for XPS.

A further approximation involving the neglect of the spatial variation of the vector potential is almost universally applied. This spatial variation can be written as

$$\exp(i\mathbf{k}_v \cdot \mathbf{r}) = 1 + i\mathbf{k}_v \cdot \mathbf{r} - (\mathbf{k}_v \cdot \mathbf{r})^2 \tag{3.11}$$

with $\mathbf{k}_v$ the photon wavevector; all except the first term (unity) are neglected. Such an approximation should be satisfactory for XPS as the core levels are localised so that the average value of **r** is small and $\mathbf{k}_v$ is small; for these circumstances the error is probably $\sim 2\%$. This is usually known as the 'dipole approximation'. With these approximations the matrix element, apart from constant factors, becomes $\langle f|\mathbf{V}|i\rangle$. This can be written in three equivalent forms

$$\langle f|\mathbf{V}|i\rangle = \frac{i}{\hbar} \langle f|\mathbf{p}|i\rangle$$

$$= \frac{m\omega}{\hbar} \langle f|\mathbf{r}|i\rangle$$

$$= \frac{1}{\hbar\omega} \langle f|\mathbf{V}V|i\rangle \tag{3.12}$$

where ω is the photon angular frequency and V the potential. These three forms are known as the 'dipole velocity' or 'dipole momentum' form, the 'dipole length' form and the 'dipole acceleration' form. All are equivalent but, because the wavefunctions used to describe the initial and final states are not exact, it is common to compute more than one form and use the differences as an indication of the errors involved. In some cases one form may be preferred for certain reasons; for example, the length form which involves multiplying the wavefunctions by r before integrating will tend to exaggerate the effects of errors in the wavefunctions at large r. The acceleration form, on the other hand, will be sensitive to the quality of the mathematical description of the potential near the core where its gradient is large.

If the initial state is described by a one-electron wavefunction with quantum numbers n, l, m of the form $U_{nl}(r)Y_{lm}(\theta, \phi)$ and the final state is written in a similar form (but with no value of n, being in the continuum) as $U_{l'}(r)Y_{lm'}(\theta, \phi)$ then evaluation of the angular integrals in the matrix elements leads to the selection rules

$$l' = l \pm 1$$

$$m' = m, m \pm 1 \qquad (3.13)$$

Thus, for any particular l value of the initial state the final state will contain a coherent sum of two angular momentum states of values $l + 1$ and $l - 1$ (frequently referred to as the 'up' and 'down' 'channels'). The total cross-sections then become proportional, for an initial state characterised by quantum numbers n and l, to $lR_{n,l,l-1}^2 + (l+1)R_{n,l,l+1}^2$ where $R_{n,l,l'}$ are the radial integrals. In the dipole velocity form, for example,

$$R_{n,l,l'} = \int_0^\infty U_{nl}(r)r\left(\frac{\mathrm{d}}{\mathrm{d}r} \pm \frac{2l+1\pm1}{r}\right)R_{l'}(r)r\,\mathrm{d}v$$

$$\text{with } l' = l \pm 1 \quad (3.14)$$

Evidently the cross-sections are dictated particularly by the character of the radial wavefunctions which in turn are influenced by the n and l values. The character of these functions can lead to interesting effects in the photon energy dependence of the cross-section for particular levels but we will discuss this briefly in section 3.2.5. For conventional XPS our main concern is the variation of cross-section at a *fixed* photon energy for different levels. Fig. 3.16 shows the results of a relevant calculation under these conditions by Scofield (1976) for a wide range of accessible levels at a photon energy of 1487 eV corresponding to the Al K_α emission. These calculations were based on a single-particle Hartree–Slater atomic model. The cross-sections are given in barns ($= 10^{-24}$ cm^2); note that, as a typical monolayer of atoms contains $\sim 2 \times 10^{15}$ atoms cm^{-2}, a cross-section of one Mb (10^6 barns – the largest values shown) corresponds to roughly 1 photoelectron emitted out of the surface per 1000 photons incident for each monolayer of the species contributing to the XPS signal. Fig. 3.16 shows that most materials not only have accessible levels but have some level in the top decade of possible cross-sections; on the other hand, of course, a factor of 10 difference in cross-section may influence substantially the sensitivity of XPS to materials at the extremes of the decade. A comparison of figs. 3.15 and 3.16 shows that valence levels, being shallow, generally have low cross-section, another factor making XPS studies of such levels difficult. This seems to be associated primarily with the fact that the photon energy is far above the photoionisation threshold; generally the levels of high cross-section are rather close to threshold.

Although these cross-sections are computed assuming that the initial and final states can be described by one-electron wavefunctions involving only the ionised level, many-body effects in the final state are known to be appreciable in determining the line shapes observed. An important sum rule exists within the sudden approximation in this regard, which states that the one-electron cross-section equals the total cross-section from summing the adiabatic peak and all shake-up and shake-off components. This has important repercussions for the capability of performing quantitative chemical analyses with XPS. Generally, it is only easy to measure the no-loss peak area in a spectrum, as the intrinsic loss contributions are mixed in with the general inelastic 'tail' behind each peak. This area is strictly not the value appropriate for a coverage determination if the calculated cross-sections are used for calibration. Problems also arise if the same species is to be compared in

Fig. 3.16. Calculated cross-sections for photoemission from occupied levels of the elements for 1.5 keV photons (from Wertheim (1978) based on the calculated value of Scofield, 1976).

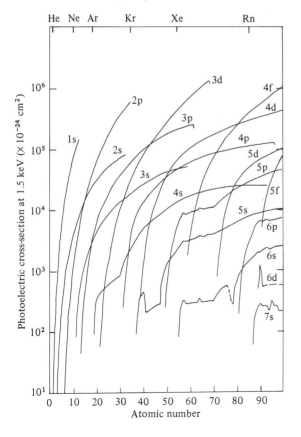

different electronic environments (where the observed many-body effects may differ).

Despite these problems, quantitative studies are possible using XPS as a means of compositional analysis, and by taking great care in the mode of calibration, analyses accurate to even 5% or 10% are claimed. In any such attempt it is usual to regard the photoemission in the 'three-step model' of photoionisation, transport of the photoelectron to the surface and transmission through the surface. The last of these three is rarely a matter of issue for XPS; when the emergent kinetic energies are low (as in UPS) total internal reflection can occur for electrons approaching the surface at grazing angle, but in XPS the energies largely exclude this problem. An alternative important third step, however, is the collection and detection by the electron analyser. If the total intensity in a photoelectron peak N_i is composed of components dN_i originating from positions, x, y, z, then we can write

$$dN_i = (\text{X-ray flux at } x, y, z) \times (\text{no. of atoms of i at } x, y, z)$$
$$\times (\text{differential cross-section of relevant level of species i})$$
$$\times (\text{probability of no loss escape of electrons from } x, y, z)$$
$$\times (\text{acceptance solid angle of electron analyser})$$
$$\times (\text{instrumental detection efficiency}) \qquad (3.15)$$

The last two of these are instrumental in nature and we will not discuss them in detail but note that these will depend on the emergent electron kinetic energy and on the area illuminated by the X-ray source. Normally the penetration depth of the X-rays is very long compared with that of the escaping photoelectrons so that the spatial dependence of the first term can also be neglected. However, the spatial distribution does influence the total signal through the fourth term, primarily associated with inelastic scattering. Thus, a large concentration several atom layers below the surface may give less signal than a much smaller concentration in the top layer. In some cases it is possible to gain information about this depth distribution by studying the angular dependence of the emission. If the only angular effects are associated with the inelastic scattering of the photoelectrons (a question which we will discuss in section 3.2.6.1) then, for an emitter at a depth z below the surface, the signal emerging from the surface at an angle θ from the surface normal will be given by $\exp(-z/\lambda \cos \theta)$ where λ is the mean-free-path for inelastic scattering, assuming isotropic emission. This yields a distribution peaked towards the surface normal and most strongly peaked for a deep emitter. Often, however, this spatial distribution is not considered and some average value for the top few monolayers is derived. The state of knowledge of the photoionisation cross-section has already been reviewed. Obviously a computation of all these parameters is difficult and large errors can creep in; it is therefore usual to try to keep as many of the variables as possible constant and then to *compare* signals from several components. Moreover, it is common to calibrate the signals of species from 'known' surfaces in order to avoid the use of calculated photoionisation cross-sections. Using these procedures typical errors of 10–50% in the inferred concentrations are possible, the size of this error itself depending on the species, coverage and degree of care taken in correcting for variations in the various terms of this equation. Lower limits of detectable surface concentrations in the range 1–10% of a monolayer are possible. Unfortunately some of the more interesting and common light elements such as C fall in the range of low cross-sections and poor detectability (see fig. 3.16).

3.2.5   *Synchrotron radiation studies*

So far we have discussed almost exclusively the use of X-ray sources based on the Al and Mg K_α lines. While there are many suitable target materials providing higher photon energies it is clear from fig. 3.15 that there is little advantage in doing so as this simply increases the kinetic energy of emergent photoelectrons causing a reduction in surface sensitivity and increased difficulty in achieving suitable energy resolution in the detector system. The only advantage is in gaining access to certain further levels but, as fig. 3.15 shows, all materials already have available levels. Lower photon energy sources, on the other hand, do offer some virtues. While they reduce the number of accessible levels, the range of electron kinetic energies is also reduced, leading to shorter inelastic scattering mean-free-paths, thus enhancing surface sensitivity. Some work has been performed on line sources constructed using higher levels in more massive targets (especially Y, which provides a useful line at 132.3 eV) though, so far, little work has been performed using these sources to study solid surfaces and they generally lack intensity. The alternative approach is to use synchrotron radiation. A detailed discussion of the characteristics of synchrotron radiation is not appropriate here and can be found reviewed elsewhere (e.g. Doniach & Winick, 1980; Koch, 1982). The main interest in this form of light source for surface studies lies in the fact that the radiation emitted from charged particle beams accelerated around a synchrotron is in the form of an intense polarised continuum narrowly defined in the plane of the accelerator. The radiation is a necessary part of accelerating charged particles in circular accelerators (being essentially bremsstrahlung) and the main source of energy loss in such a machine. This radiation can therefore be extracted from machines otherwise devoted to high energy physics, in a paratisic mode, although many electron 'storage rings' are now coming into use as dedicated sources of synchrotron radiation.

The output curve from this type of machine is shown in fig. 3.17 and clearly demonstrates the broad continuum; the exact range of this is machine dependent but most machines are capable of high emission throughout the range relevant to photoemission (i.e. from photon energies of a few eV to a few keV). If the output is passed through a monochromator, a tunable photon source results. Because of the need to monochromate the light it is often difficult to make fair comparisons of the relative intensities of line sources used in XPS and UPS with synchrotron radiation; for example, it may not be necessary (especially in UPS) to operate with a monochromator band pass width as narrow as the line width of a laboratory source so that intensity and resolution can be traded. In general, however, under typical operating conditions, the synchrotron radiation intensity from a 'good' accelerator is similar to the best available line sources. The great virtue, of course, then becomes the photon energy tunability. So far, most synchrotron radiation photo-emission has been performed with photon energies less than about 50 eV and will be discussed with UPS; it is also in these studies that the plane polarised nature of the radiation has been exploited most. More recently,

Fig. 3.17. 'Universal' spectral curve for synchrotron radiation. The photon flux is normalised to 1 mA of circulating electron current at 1 GeV in the accelerator and measured in 1 mradian of horizontal divergence. The emitted wavelength λ is normalised to λ_c which is characteristic of the synchrotron or storage ring. Expressing the electron energy E in GeV and the magnet bending radius R in metres, $\lambda_c = 5.6R(E)^{-3}$ in Å. For a typical machine $E \sim 2$ GeV, $\lambda_c \sim 4$ Å so with 100 mA of operating current peak emission is $\sim 10^{14}$ photons s^{-1} mrad^{-1} (0.1% band width)$^{-1}$.

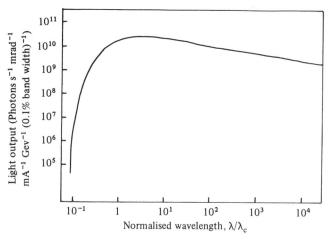

however, photoemission studies in the energy range from 50 eV to several hundred eV have been performed, primarily on core levels and hence falling into our definition of XPS. Mainly, these have attempted to choose the photon energy to optimise the mean-free-path of the outgoing photoelectron and hence enhance surface sensitivity. By varying the outgoing photoelectron energy, it is also possible to measure or utilise the change in mean-free-path and the associated change in sampling depth. So far the full range of possibilities of such work has yet to be explored.

One particular feature of a tunable source, in XPS, however, is that it permits exploitation of the known energy dependence of photoionisation cross-sections. A particularly marked example of this has been used to enhance sensitivity to valence levels but utilises an effect in the photoionisation cross-section of a core level. Generally, above threshold, the photoionisation cross-section of a core level rises to a peak value close to (or at) threshold and then falls monotonically as the emergent kinetic energy increases. However, in the case of initial states characterised by wavefunctions having radial nodes, the behaviour is more complex. For example, states with a single radial node in their wavefunctions have a principal quantum number $n = l + 2$ where l is the angular momentum quantum number (e.g. 2s, 3p, 4d), and show a particularly pronounced 'Cooper minimum' in their cross-sections at a certain energy above threshold. They are therefore characterised by a cross-section which rises, falls to near zero, and then rises again before falling monotonically in the usual way. The effect can be attributed to a sign change in the 'up channel' ($l + 1$ final state) matrix element at a certain energy. Fig. 3.18 shows the radial part of the bound state wavefunctions for a state of this kind (the 3p level in Ar) and for a state not showing a radial node (2p in Ne). Also shown are the radial part of the $l + 1$ continuum wavefunctions at zero kinetic energy. Evidently, in the case of the 3p Ar level, the radial integral in the matrix element is negative under these conditions, whereas for the 2p Ne level it is positive. As the kinetic energy of the outgoing electron increases, the spatial periodicity of the final state wavefunction is reduced and in due course the radial integral in the 3p Ar level also becomes positive and so must pass through zero. This effect is seen in the calculated cross-sections shown in fig. 3.19. The ability to vary the photon energy with a synchrotron radiation source means that conditions can be chosen to maximise or minimise the sensitivity to a particular level. For example, the 4d valence band of Pt retains this core state type of behaviour and by choosing a suitable photon energy it is possible to suppress the emission from this region at its Cooper minimum and hence enhance the sensitivity of the photoelectron spectra to other (overlapping) shallow levels of different n, l character associated with adsorbed molecules.

3.2.6   *Structural effects in XPS*

In our discussion so far we have indicated that XPS from solids is dominated by atomic rather than solid state effects; indeed, this is the essential requirement of all core level spectroscopies. Nevertheless, the local electronic environment in the solid can influence the observed peak positions and line shapes through both initial and final state effects. Effects also exist through the final state in which the local *structural* environment can influence various aspects of the photoemission process. Such effects arise because other atomic centres in the environment of the emitter can elastically scatter the outgoing emitted electron wave, leading to coherent interference effects in the final state. These effects can be manifested in two ways; in the observed angular distribution of emitted photoelectrons outside the solid, and in variations in the photoionisation cross-section (EXAFS or Extended X-ray Absorption Fine Structure) due to electrons coherently backscattered onto the emitter ion core. We will deal with these two effects separately.

Fig. 3.18. Radial part of the 2p (Ne) and 3p (Ar) atomic wavefunctions together with the d continuum (final state) wavelengths at zero kinetic energy (i.e. at photoionisation threshold) (after Fano & Cooper, 1968).

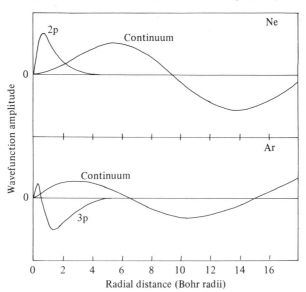

3.2.6.1 *Angle-resolved XPS*

Fig. 3.20 shows schematically an electron emission process from an adsorbate atom on a surface and some of the elastic scattering from neighbouring atoms. As these scattering processes are coherent the amplitude of emission seen by a detector at a fixed angle to the surface involves the sum of the amplitudes of all these scattered terms as well as the directly emitted wave, and the result will be dependent on the emission angle. Moreover, the interference of scattered and directly emitted waves is determined by the position of the emitter relative to the substrate. Measurements of this angular variation or 'photoelectron diffraction' should therefore contain information on the adsorbate–substrate registry.

Fig. 3.19. Theoretical computation of the Ar 3p atomic photoionisation cross-section by Kennedy and Manson (1972) showing the 'Cooper minimum' around 50 eV above threshold. L and v correspond to the 'length' and 'velocity' forms of the matrix element used on the computation (cf. equation (3.12)). These computations include both d and s final states so that the minimum is not identically zero as it is in the d channel alone.

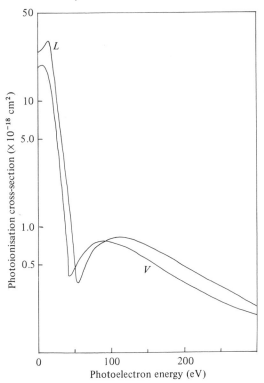

Angular effects associated with this final state scattering are well known for XPS from substrate core levels and the observed structures are very like the so-called Kikuchi patterns seen in high energy electron diffraction studies of bulk materials. Of more interest to the surface scientist is the possibility of studying adsorbate emission effects as a means of determining surface structures. The quality of this structural information is dependent on the strength of the scattering from the substrate, which in turn depends on the energy and geometry of the experiment. If the photoelectron kinetic energy is in the LEED energy range (say, less than about 150 eV) then we know that backscattering is strong. Effects should therefore be observable over the whole range of emission angles. Furthermore, at low kinetic energies the wavelength of the emitted electrons is much larger than the thermal vibrational amplitudes of the atoms so that the incoherency introduced by these vibrations (and the associated Debye–Waller factor) should be small. Of course, we know that at LEED energies multiple scattering is important so the theory could be complex, but at least established computational procedures of LEED theory should be readily adaptable to this problem. The principal experimental complication arises from the fact that low kinetic energies of photoemission are normally only achieved using synchrotron radiation.

One way of retaining substantial scattering interferences using conventional sources and their attendant high photoelectron kinetic energies is to concentrate on grazing angles of emergence from the surface. Under these conditions the scattering angles of some of the

Fig. 3.20. Schematic sectional view through a surface showing directly emitted electrons from a localised state of a surface adsorbed atom and some trajectories involving elastic scattering from surrounding atoms which can interfere coherently with the directly emitted wave.

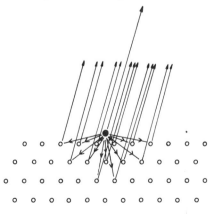

substrate scattering events are small and the scattering strength and relative insignificance of thermal effects can be retained. Thus, while the synchrotron radiation version of the experiment is a kind of 'internal LEED', the grazing emergence XPS technique would be more akin to 'internal RHEED' (although the kinetic energies are somewhat lower than in RHEED and so the range of usable emergent angles may be greater).

The exploitation of both versions of these studies is still in its infancy, although early results from both are promising, as may be seen in figs. 3.21 and 3.22. To appreciate fully both the theoretical and experimental constraints of the technique some comments on the purely atomic aspect of angular effects in XPS are worth noting. As we have already seen, the dipole selection rule in photoemission requires that the final state be comprised of states with $l+1$ and $l-1$ angular momentum. If we consider first the special case of an s-character ($l=0$) initial state it is clear that the final state must be entirely p-character ($l=1$) and the emitted intensity therefore is proportional to $\cos^2 \theta_p$ where θ_p is the angle between the photon polarisation (**A**) vector and the collection direction. This leads to a strong anisotropy of emission (with no emission

Fig. 3.21. Radial plots of the azimuthal dependence of adsorbate core level photoemission at photon energies of 80, 90 and 100 eV from the Te 4d level in a Ni{100} c(2 × 2)–Te structure and from the Na 2p level in a Ni{100} c(2 × 2)–Na structure. The polar emission angle relative to the surface normal was 30°. The outer dots are data points while the inner curves are enhanced data plots obtained by subtracting the minimum value from all points (after Woodruff *et al.*, 1978).

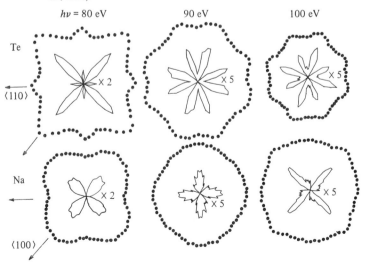

120   *Electron spectroscopies*

perpendicular to the photon **A** vector). For other states the two possible outgoing channels must be included and these interfere coherently to determine the final angular distribution. For photoemission from a filled state of a free atom we can average over all *m* states as there is no preferred axis (other than the photon **A** vector) and it can then be shown that the resulting angular distribution of the emitted intensity can be written as

$$I(\theta_p) = \frac{d\sigma}{d\Omega} = [1 + \beta P_2(\cos\theta_p)] \tag{3.16}$$

Fig. 3.22. Radial plots of the azimuthal dependence of O 1s photoemission for a Cu{100}c(2 × 2)–O structure at a photon energy of 1487 eV (Al K_a) and at the *grazing* emergence angles shown. Inner plots are minimum subtracted data as for fig. 3.21 (after Kono *et al.*, 1978).

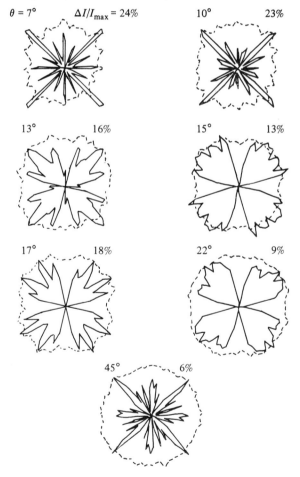

where $d\sigma/d\Omega$ is the derivative photoionisation cross-section and $P_2(\cos\theta_p)$ is a Legendre polynomial $(=(3\cos^2\theta_p - 1)/2)$. β is known as the asymmetry factor and

$$\beta = \frac{l(l-1)R_{l-1}^2 + (l+1)(l+2)R_{l+1}^2 - 6l(l+1)R_{l+1}R_{l-1}\cos(\delta_{l+1} - \delta_{l-1})}{(2l+1)[lR_{l-1}^2 + (l+1)R_{l+1}^2]}$$

(3.17)

where $R_{l\pm1}$ are the radial matrix elements for the up and down channels and $\delta_{l\pm1}$ are the phase shifts associated with the two outgoing waves being scattered by the emitter potential. In general, β can take any value from -1 to 2 (note that if $l=0$ then $R_{l-1}=0$ and $\beta=2$, the most asymmetric result). Although the total photoionisation cross-section is usually dominated by the $l+1$ channel, the β factor is rather more sensitive to the relative values of the two matrix elements. Because of this, measurements of β are commonly made in atomic physics to apply more stringent tests to theoretical descriptions of the photoionisation event. In the solid state, of course, the two outgoing partial waves are then scattered by the surrounding atoms. This leads to a significant increase in computational complexity because the elastic (coherently adding) scattering must be performed for the $l\pm1$ channels (also coherently adding) but separately for each m value. This is because the m waves add incoherently and so can only be summed in intensity *outside* the crystal. In the case of the free atom this summation over m can be performed analytically and leads to the simple result above. The existence of large anisotropies (from $\cos^2\theta_p$ $(\beta=2)$ to $\sin^2\theta_p$ $(\beta=-1)$) due to purely atomic effects also imposes certain constraints on the experimental methods used to study solid state structural effects, as it is desirable to choose a mode of data collection which avoids the possibility of obtaining data dominated by atomic effects. One way around this is to keep the photon incidence and electron analysis directions constant and rotate the sample about its surface normal (a so-called ϕ-plot or azimuthal plot). In such a study the collection angle is unchanged relative to the photon polarisation vector (i.e. constant θ_p) so the atomic physics remains unchanged and all anisotropy observed is due to final state effects in the solid surface. The data shown in figs. 3.21 and 3.22 were obtained in this way. One alternative method of achieving the same aim is to hold all incident and emergent directions constant and vary the photon energy; the scattering variable is thus the photoelectron kinetic energy. Strictly, this is not a measurement of angular dependence, but an angle-resolved measurement. On the other hand, the basic effect investigated is the same in that the variations (other than those due to variation in the total photoionisation cross-section) are due to changes in the interferences between directly emitted and scattered waves. This mode of data collection is referred to as a CIS (Constant Initial State) spectrum and is conceptually very similar to a LEED intensity energy spectrum except that, because the source of electrons is at the site of interest, the sensitivity to adsorbate–substrate registry should be greatly enhanced. Of course, this mode of data collection is only appropriate to a tunable photon source such as synchrotron radiation.

3.2.6.2   *Extended X-ray Absorption Fine Structure (EXAFS)*

In the considerations of elastic scattering of the outgoing photoelectron we have so far concentrated on scattering events which cause electrons to emerge from the crystal surface. One other special kind of scattering event is backscattering to the emitter itself. These elastically scattered contributions arriving back at the emitter will add coherently to the outgoing wave and thus directly modify the final state at the emitter which appears in the photoionisation cross-section matrix element. As a result this backscattering causes the photoionisation cross-section to oscillate as the photoelectron energy increases and its associated wavelength passes through values which are submultiples of the total distance to and from nearest neighbour atoms. This effect forms the basis of the technique of EXAFS. Because of this physical background we discuss EXAFS here in the context of XPS, although it is generally regarded as a quite separate technique and is not monitored by the photoemission itself. In order to separate out the variation of the photoionisation cross-section associated with atomic effects from that resulting from backscattering from neighbouring atoms, it is usual to define a 'fine structure function', χ, as

$$\chi = \frac{\sigma - \sigma_0}{\sigma_0} \tag{3.18}$$

σ_0 being the cross-section of the free atom and σ the cross-section in the solid state. If we now simply consider single scattering events back onto the emitter atom one can show that

$$\chi(k) = -k^{-1} \sum_i A_i(k) \sin\left[2kR_i + \phi_i(k)\right] \tag{3.19}$$

k being the photoelectron wavevector amplitude while the summation is over surrounding 'shells' of neighbouring atoms. R_i is the distance from the emitter to the ith shell and the $2kR_i$ term is the phase factor associated with the outward and return paths of the photoelectron back to the emitter. The essentially simple relationship between $\chi(k)$ and the unknown R_i is complicated by the phase factor ϕ_i; this is composed of a phase shift associated with the emergence of the outgoing wave through the emitter potential and its return through the emitter potential and also with its backscattering off the atoms at R_i. It is clear from equation (3.19) that if we can neglect the ϕ_i a Fourier transform of $\chi(k)$ will lead to a function with peaks at R_i which is essentially the radial distribution function about the photoabsorbing atom species. This *direct* structure determination from the experimental data contrasts strongly with the usual indirect trial-and-error approach used in other surface structural techniques such as LEED. While we cannot, in fact, neglect the ϕ_i we note that in taking a Fourier transform any component of ϕ_i which is independent of k is lost and it is the k dependence which leads to distortion of the structural information. Fortunately, $\phi_i(k)$, like scattering phase shifts in LEED, is dominated by the ion cores and not by the valence electron so that these phase shifts are generally 'transferable'. Thus $\phi_i(k)$ is characteristic of a particular emitter–scatterer species pair, but insensitive to the chemical nature of this pair. This means that rather accurate comparisons of R_i are possible between known and unknown structural combinations; for example, if A and B form a known bulk compound AB, the EXAFS from this may be compared with a study of A on a surface of B to obtain accurate A–B distances for this surface adsorption system.

The amplitude of $\chi(k)$ for the ith shell, A_i, is given by

$$A_i(k) = (N_i/R_i^2)|f_i(\pi, k)| \exp\left(-2\langle u^2 \rangle_i k^2\right) \exp\left(-2R_i/\lambda\right)$$

(3.20)

N_i being the number of atoms in the ith shell having mean square thermal vibration amplitudes of $\langle u^2 \rangle_i$ and backscattering amplitudes $|f_i(\pi, k)|$. Notice that the phase factor of the complex f_i is separated out into the ϕ_i. While this amplitude is proportional to the number of atoms in the ith shell it is attenuated not only by the Debye–Waller term (in $\langle u^2 \rangle_i$) but also by the $\exp\left(-2R_i/\lambda\right)$ and the $(R_i)^{-2}$ term which simply accounts for the fact that the emission is from a point source. It is this additional attenuation which helps to suppress the importance of multiple scattering so that the Fourier transform forms a useful basis for data analysis in a way which it fails to do for LEED (see previous chapter). As a result, provided that the very near edge data ($\lesssim 50$–100 eV) are neglected, multiple scattering effects are weak.

Experimentally, EXAFS in a bulk material is measured, as the name implies, by measuring the X-ray absorption coefficient directly from the attenuation of the X-ray flux passing through a thin film of material. This method has also been applied to studies of highly dispersed metal powder on supported high area catalysts which comprise light element species (e.g. graphite of Al_2O_3) which are weakly absorbing to X-rays. In this case essentially all the metal atoms are 'surface' atoms and reasonable surface sensitivity is possible using a bulk technique. This approach is not viable for studies on well-characterised single crystal surfaces; even if these samples could be prepared sufficiently thin, the very small concentration of surface atoms would preclude measurements of the X-ray absorption associated with them.

One obvious possibility for studying photoionisation effects in a surface specific way is to study the photoemission itself. Unfortunately, as we have already seen, other final state scattering processes lead to variations in measured photoemission yields so that such measurements are likely to comprise a confusing mixture of EXAFS and photoelectron diffraction. It is possible that this is true even if the photoelectrons are measured over a large collection angle, as this is necessarily much less than the 4π steradians of initial emission. As an alternative to measuring the cross-section via the emission itself or by the absorption of light, we can measure it through the decay of the core hole produced. As we will discuss in more detail in the next section, this core hole decay can occur either by X-ray emission or by Auger electron emission in which the energy is given up to another weakly bound electron. This Auger process forms a convenient method of detecting photoionisation events in the surface region. The Auger yield thus displays EXAFS when the photon energy is varied. Alternatively, it is found that measurements of the *total* yield also show this effect. The total yield is dominated by the low energy secondary electrons but these, in turn, arise from a collision cascade initiated by a high energy emission. Thus, as the photon energy passes through an absorption edge (photoionisation threshold) the newly opened channel of photoelectrons and Auger electrons leads to an increase in the total yield. While the signal-to-background of this method is worse than direct Auger electron detection, the crystal surface acts as an electron amplifier and the increased yield can lead to improved signal-to-noise. Of course, the inelastic cascade process leads to reduced surface specificity, but the technique is still quite surface sensitive.

Fig. 3.23 shows an example of the data from a SEXAFS study of I adsorption on a Cu{111} surface to form a $(\sqrt{3} \times \sqrt{3})R30°$ structure. Panel (*a*) shows raw spectra, in this case of the total yield, for photon energies in the vicinity of the I L_3 edge. Spectra are shown for the surface adsorption system (labelled S) and for bulk CuI (labelled B). Panel (*b*) shows the extracted fine structure functions $\chi(k)$, multiplied by k^2 to enhance the higher k values. Panel (*c*) shows the Fourier transforms of this processed information. The main peak in both transforms is associated with the nearest neighbour Cu–I distance although they are displaced in R due to phase shift effects. Filtering out this peak and back-transforming leads to the results of panel (d). The small displacement associated with a different Cu–I distance is clearly seen, and taking the spacing in bulk CuI as 2.617 ± 0.005 Å leads to a determination of the Cu–I distance for the adsorption structure as 2.66 ± 0.02 Å.

Notice that while this rather precise nearest neighbour distance determination compares very favourably with LEED determinations of site position (to ~0.1 Å), these data, as shown, do not determine the adsorption *site*. In principle, this can be obtained either from second nearest neighbour distances or from the EXAFS amplitudes. Until

Fig. 3.23. SEXAFS data for I in a Cu{111}($\sqrt{3} \times \sqrt{3}$)$R30°$–I structure compared with bulk CuI. (a) shows raw data, (b) the extracted fine structure function, (c) Fourier transforms of these and (d) filtered and back-transformed data (after Citrin, Eisenberger & Hewitt, 1980).

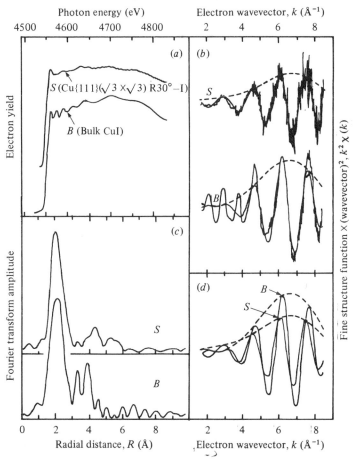

recently, the signal-to-noise quality of SEXAFS data has not permitted the application of the first approach; it remains to be seen how useful this will be in the future. The second approach relies on the dependence of the amplitude A_i on the number of nearest neighbours N_i. Unfortunately, while phase shifts are transferable between different chemical states, amplitudes are not, partly because of the variation in the inelastic scattering mean-free-paths λ. Nevertheless, polarisation dependent amplitudes and amplitude ratios have been applied with some success to site determination. This approach relies on the partial directionality of the emitted photoelectron partial waves which therefore sample certain directions more than others in the nearest neighbour environment. By varying the direction of the polarisation vector, $\mathbf{A}$, some control over this directionality can be imposed.

3.3   Auger electron spectroscopy (AES)

3.3.1   *Introduction – basic processes*

When an atom is ionised by the production of a core hole, either by an incident photon as in XPS, or by an incident electron of sufficient energy, the ion eventually loses some of its potential energy by filling this core hole with an electron from a shallower level together with the emission of energy. This energy may either appear as a photon, or as kinetic energy given to another shallowly bound electron. These competing processes are dominated by the photon emission only when the initial core hole is deeper than about 10 keV. Indeed, this is the physical process used in a conventional laboratory X-ray generator. The alternative radiationless emission of the energy as electron kinetic energy is the Auger effect, named after its discoverer, Pierre Auger. Fig. 3.24 shows a schematic of the two processes. In the case of photon emission we have (ignoring relaxation, etc.)

$$h\nu = E_A - E_B \tag{3.21}$$

and in the case of Auger electron emission

$$KE = E_A - E_B - E_C \tag{3.22}$$

In each case the emitted particle is characteristic of some combination of atomic energy levels of the emitter and so forms the basis of a core level spectroscopy. This is still true if either, or both, of the levels B and C are valence band levels as at least one core level (E_A) is involved which is characteristic of the atomic species alone and is likely to be dominant in defining the emitted particle energy. However, photon emission is of little interest in surface science; these emerging photons have long mean-free-paths in solids and, as the exciting probe is also generally deeply penetrating (except at an ionisation threshold – see the discussion of APS in section 3.4), it gives rise to a bulk analytic tool (microprobe analysis).

By contrast, Auger electron emission is an efficient means of filling core holes of low binding energy, thus giving rise to relatively low kinetic energy Auger electrons of short mean-free-path. Their detection outside the solid therefore provides a surface sensitive probe of chemical composition. While the initial core hole may be created by either incident photons or incident electrons, the relative ease of producing sufficiently energetic ($\sim$ 1.5–5 keV) electron beams of high intensity (1–100 μA) means that AES is invariably performed with incident electron beams. Of course, Auger electrons are produced in XPS and their study in conjunction with direct photoemission can give some special complementary information which we will discuss later.

Although Auger electron emission, being a three level process, is intrinsically more complex than photoemission, its strength lies in the fact that it can be generated by incident electron beams and that the production, focussing and deflection or scanning of electron beams is a well-developed technology. Most AES studies simply use the Auger spectrum as a fingerprint of the chemical composition and are not concerned with a detailed understanding of the basic processes. In order to make a proper comparison with XPS, however, we will discuss some of the fundamentals before going on to look at some of the problems and strengths of AES as a tool for the determination of surface chemical composition.

Fig. 3.24. Energy level diagram showing the filling of a core hole in level A, giving rise to (X-ray) photon emission on the left, or Auger electron emission on the right. The levels are labelled with their one-electron binding energies.

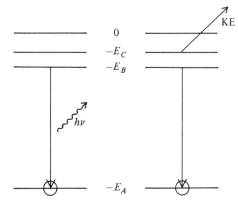

3.3.2   *Energy levels, shifts and shapes*

Equation (3.22) provides an approximate formula for the kinetic energy of an Auger electron based on the one-electron binding energies of the core levels involved. These energies can be obtained empirically, for example, from X-ray absorption studies, and it is usual to label the Auger transition using the X-ray level notations. Thus the transition shown in fig. 3.24 is an *ABC* transition where *A*, *B* and *C* are the X-ray level notations *K*, L_1, L_2, etc. For example, a transition involving a 1s level as *A*, and 2p levels (not spin split) as both *B* and *C* levels (which may have the same energy) would be labelled as a $KL_{2,3}L_{2,3}$ transition. If the shallow levels are in the valence band of a solid the atomic level notation for this state is often replaced by the symbol *V*. In the example above, therefore, the transition might be *KVV* if the 2p levels form the valence band.

Equation (3.22), however, does not give a very exact description of the energy as it takes no account of the fact that the true energy is the difference between a one-hole binding energy state and a two-hole binding energy state. One very approximate method of taking account of this is to replace the binding energy of the shallow level *C* for an atom of atomic number *Z*, with the corresponding binding energy for an atom of atomic number *Z* + 1, thus

$$\text{KE} = E_A{}^Z - E_B{}^Z - E_C^{Z+1} \tag{3.23}$$

or, recognising that the roles of the levels *B* and *C* are indistinguishable (i.e. the *ABC* and *ACB* transitions are the same) one might use

$$\text{KE} = E_A{}^Z - \tfrac{1}{2}(E_B{}^Z + E_B^{Z+1}) - \tfrac{1}{2}(E_C{}^Z + E_C^{Z+1}) \tag{3.24}$$

These approximate, highly pragmatic formulae were widely used in establishing the basic features of Auger spectra in their early use in surface science. Nowadays they are unnecessary as collections of fingerprint spectra for essentially all elements exist for the purposes of species identification (e.g. Davis *et al.*, 1976). For a proper description of the kinetic energy, however, we may write,

$$\text{KE} = E_A - E_B - E_C - U \tag{3.25}$$

where *U* is the hole–hole interaction energy and E_A, etc., are once again the normal one-electron binding energies, normally taken to be those measured in XPS which therefore include one-hole relaxation effects. The *U* therefore lumps together the two-hole energy effects within the material under study; it may be divided into two terms

$$U = H - P \tag{3.26}$$

where H is the hole–hole interaction energy in the free atom and P takes account of the extra-atomic or screening polarisation or relaxation effects of the solid state environment. Written as U, however, we have a readily determined parameter from experiment. Evidently 'chemical shifts' due to different local electronic environments can arise either through the individual E_A, etc., as in XPS, or through U. So far relatively little effort has been devoted to investigating this problem in any systematic fashion for conventional AES although the Auger line shapes and energy positions of a number of specific species are well known to indicate different chemical environments. Fig. 3.25 shows gas phase Auger electron spectra containing the C KLL (or KVV) emission from C in singly, doubly and triply bonded states. The differences are pronounced and demonstrate clearly the existence of chemical effects although, in common with photoemission involving transitions from valence states, we may expect some broadening in the solid state. In fact, the KLL C spectrum from solid surfaces is also known to show 'chemical' effects; fig. 3.26 shows an example of this, the Auger spectrum

Fig. 3.25. Gas phase electron excited C KVV Auger electron spectra from CH_4, C_2H_4 and C_2H_2 characteristic of single, double and triple C—C bonds (from Rye *et al.*, 1979).

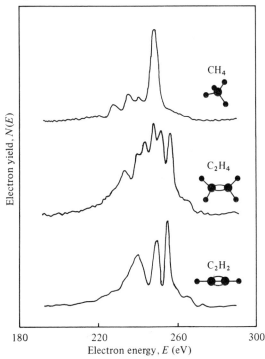

now being shown in the more usual form involving a differentiation with respect to the energy. This mode has the virtue of suppressing the large secondary electron background and has the side effect of turning a simple peak into a positive and negative excursion (see section 3.1.3). The different 'fingerprint' of the C *KLL* spectrum for carbide and graphitic forms shown in fig. 3.26 is one of the best-known examples of chemical effects in Auger electron spectra. Note, however, that the main differences are seen in the positive excursion part of the spectrum which is almost entirely absent from some of the spectra. A comparison with the (undifferentiated) spectra shown in fig. 3.25 shows how this effect may be explained; some of the main 'chemical' effects are in the loss structure (intrinsic and extrinsic) behind the main peak. The peak may therefore be expected to always show a sharp high energy edge leading to a large negative excursion in the differentiated mode. On the other hand, if there is a broad loss 'tail', then in the differentiated spectrum the positive

Fig. 3.26. Derivative C *KVV* Auger electron spectra from two carbides, graphite and diamond (after Haas, Grant & Dooley, 1972).

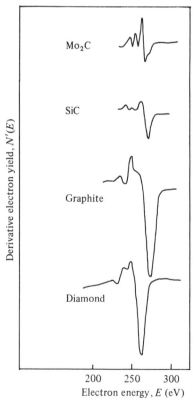

excursion part will be relatively insignificant. For this reason it is common to label peak energies in AES as the energy of largest negative excursion in the differentiated spectrum; this is not the true 'peak' energy but is the most readily compared energy in identifying surface species from the spectra. Finally, fig. 3.27 illustrates another commonly cited example of chemical effects; Auger electron spectra (in the differentiated mode) are shown for a Si sample covered with surface oxide and contaminants, and in a cleaned state. The spectra include various impurity species but show in particular that, while for the clean surface the main low energy Si peak (an $L_{2,3} VV$ transition) is seen at an energy of ~ 92 eV, on the contaminated surface only a very small peak is seen at

Fig. 3.27. Derivative Auger spectra from clean and contaminated Si samples. Auger peaks for the impurity species are labelled. Note the difference in the Si $L_{2,3}VV$ Auger transitions. The contaminated surface shows this peak dominated by the line shape characteristic of SiO_2.

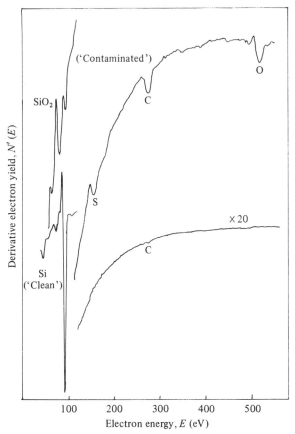

this energy while a strong feature is seen at ~ 70 eV. This lower energy feature is characteristic of the Si $L_{2,3}VV$ transition from SiO_2. Such large shifts, however, are not common.

Note that all of these illustrations involve valence level transitions which might be expected to show chemical effects not only due to intrinsic relaxation, screening and chemical shifts but also due to changes in the valence levels themselves which should be reflected in the energy and shape of peaks. Auger electron emission lines yielding good intensity usually involve at least one valence level and frequently two. Some discussion of the line shape of core–valence–valence (CVV) transitions is therefore appropriate. While the initial core level has a well-defined, discrete energy, the valence levels occupy a band, so if any energy level within this band can be involved in the transition, the probability of some particular energy level being involved is proportional to the valence band density of states at this energy. For a CVV transition, therefore, which involves two such valence states, both of which can be freely selected, the observed kinetic energy line shape should reflect the self-convolution of the density of states in the valence band. In some systems this does seem to be approximately true. In particular the Auger electron line shape is broad (~ twice the width of the valence band due to the self-convolution) and shows a similar shape

Fig. 3.28. Theoretical simulations of the Al $L_{2,3}VV$ Auger electron line shape assuming a simple self-convolution of the density of states but using a localised density of states derived from a 'pinned free-electron model' (Gadzuk, 1974). Note that the main peak position shifts towards that of the experiment and away from the bulk self-convolution as the surface enhanced damping length is reduced, but fails to match experiment well even with this parameter set to 0.4 Å.

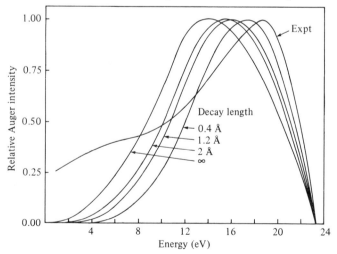

to a calculated self-convolution although the comparison is always hindered by the problems of removing inelastic loss effects (both extrinsic and intrinsic – cf. section 3.2.3) from the experimental data. Moreover, it has been suggested that the density of states sampled may be distorted due to the technique's surface specificity. Fig. 3.28 shows some calculations (with different damping lengths for the decay of surface enhanced bulk states) for the $L_{2,3}VV$ transition in Al compared with experiment and the self-convolution of the bulk density of states. Agreement is never excellent but the trends in the physical effect are clear. In this and subsequent figures we will revert to the undifferentiated mode of spectral display which is more appropriate to comparisons of peak shapes.

In other cases, however, such as the $L_{2,3}VV$ transition in Si, the simple self-convolution of the density of valence states provides a very poor description of the observed line shape (see fig. 3.29(a)). In this case, at least, the primary reason for the failure can be attributed to a failure to include the transition matrix elements for the different components. While a self-convolution of the density of states is satisfactory when the transition rate is independent of which states are involved in the band, a transition matrix element of the form $\langle \psi_f | 1/r_{12} | \psi_i \rangle$ must normally be included (cf. XPS) where the $(1/r_{12})$ represents the Coulomb interaction at a separation r_{12} and ψ_i and ψ_f are the initial (one-hole) and final (two-

Fig. 3.29. Calculated line shapes for the Si $L_{2,3}VV$ Auger electron peak from a Si{111} surface compared with experiment. The local density of states in this surface was calculated and (a) shows a self-convolution which also includes some weighting to take account of transition multiplicities. (b) shows the result of including transition matrix elements (Feibelman, McGuire & Pandey, 1977; Feibelman & McGuire, 1978).

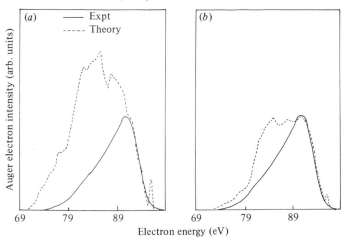

hole plus emergent electron) states. In particular, these matrix elements may vary substantially depending on the angular momentum character of the states. The selection rules in the Auger process are generally much weaker than in photoemission, but certain transitions are nevertheless forbidden. In the case of the Si $L_{2,3}VV$ transition proper account of these matrix elements (fig. 3.29(b)) provides much better agreement with experiment. The principal difference between the self-convolution and the full calculations in this case is the much greater importance of transitions involving p-character electrons near the top of the band relative to the s-character electrons near the bottom. Indeed, even better agreement is found by suppressing the importance of the s-character electrons even further as shown and discussed by Feibelman & McGuire (1978).

While considerations of selection rules and matrix element effects account for the main features of the Si $L_{2,3}VV$ transition in Si, similar calculations for the L_3VV transition in Cu, for example, prove totally unsatisfactory (fig. 3.30). The observed Auger line shape is much narrower than the self-convolution of the density of states or the full calculated line shape. Moreover, the experimental data shows rather narrow satellite structure reminiscent of free atomic transitions. In this case the deficiency in the model is more basic and relates to the effect of

Fig. 3.30. Comparison of experimental and calculated line shapes for the Cu L_3VV Auger electron peak. (a) shows a calculated self-convolution of the density of states while (b) shows the effect of including transition matrix elements (Feibelman & McGuire, 1977). The inset in (a) shows the density of states in its unfolded form.

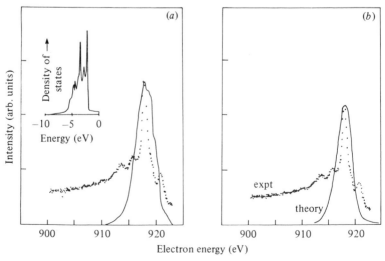

the hole–hole interaction term U discussed earlier. In the simplest cases of CVV transitions we may take the view that the holes in the valence band are delocalised and that U should be essentially zero (particularly with screening). In this case we expect a line shape based on a self-convolution of the two-hole density of states. However, if U for the creation of the two holes at the same atom is large, this simplicity is lost.

In this case the two-hole density of states with the two holes localised becomes displaced in energy from that associated with two separated holes. If the hole–hole repulsion term U is large compared with the band width of the one-electron density of states W, these localised hole states split off the delocalised state band such that the two holes become trapped or bound. In this case it is also found that most of the weight in the Auger transition is taken up by these bound states. This basis effect is illustrated by the results of model calculations of Sawatsky & Lenselink (1980) shown in fig. 3.31. The one-electron density of states of the simple s-band used in the calculation is shown in panel (*a*), while subsequent

Fig. 3.31. One-electron density of states (*a*) and two-hole density of states for model calculations of Auger spectra using different values of the parameter (U/W) as follows: (*b*) 0, (*c*) 0.33, (*d*) 0.67, (*e*) 1.0, (*f*) 1.33. Note that the energy scale is in units of the one-electron bandwidth W and the two-hole densities of states are plotted with increasing hole energy (and thus decreased Auger electron kinetic energy) from left to right (from Sawatsky & Lenselink, 1980).

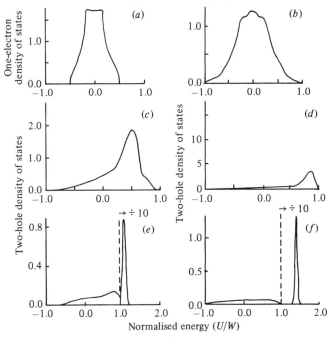

panels show the two-hole density of states for various values of the dimensionless parameter U/W. Notice that the energy is given in units of W and, because these are plotted relative to the hole state energy, the anticipated Auger electron line shape will be the same as these curves but with kinetic energy increasing from right to left. With $U = 0$, the two-hole density of states is a self-convolution of the one-electron density of states, but for increasing U the line shape becomes distorted by the transfer of weight to the displaced bound hole state energy; this becomes most pronounced when $U > W$ when nearly all the weight is taken up by the bound atomic-like state. Notice that while we refer to this state as localised the two holes can move through the solid but are bound together like an exciton. The essential reason for the sharp line shapes in CVV transitions from materials such as Cu and Zn is then readily appreciated by compounding this result with the data presented in fig. 3.32. This shows a comparison of experimental values of U, determined from Auger spectra on comparison with photoemission data, with the experimentally determined value of $2W$ for some materials from the first

Fig. 3.32. Comparison of experimentally determined values of U and W (3d band width) for a series of 3d materials (after Antonides, Janse & Sawatsky, 1977).

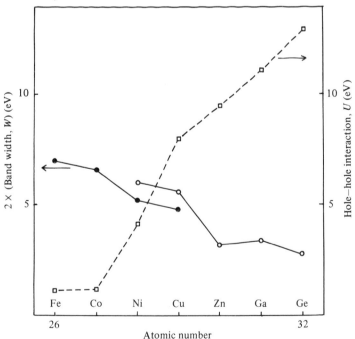

transition row of the periodic table and beyond. The data are derived from $L_{2,3}M_{4,5}M_{4,5}$ Auger spectra. Strictly the critical value of U/W at which we expect the transition from broad solid state to narrow atomic line shapes depends on the nature of the band and the definition of its width; fig. 3.32 is based on an early assertion that the critical state occurs when $U = 2W$. Evidently, however, fig. 3.32 shows that we expect the species Cu and beyond to show atomic-like spectra and the earlier elements to show broader structure even if we had compared U with W rather than $2W$. A proper analysis of Fe, Co and Ni is more difficult as the relevant 3d-band is unfilled while the theory of Sawatsky outlined is strictly valid only for filled bands. Nevertheless, the experimental $L_3M_{4,5}M_{4,5}$ Auger spectra for Co, Ni and Cu shown in fig. 3.33 display the predicted narrowing effect rather clearly.

In the general area of energy shifts and line shapes we see, therefore, that AES is considerably more complicated than the one-electron core level photoemission process involved in XPS. All the complexities of true chemical shifts and interatomic and intra-atomic relaxation effects present in XPS also occur in AES, but with the complicating factors of hole–hole interactions in the final state and a great multiplicity of lines

Fig. 3.33. $L_3M_{4,5}M_{4,5}$ Auger electron spectra of Co, Ni and Cu (after Yin, Tsang & Adler, 1976).

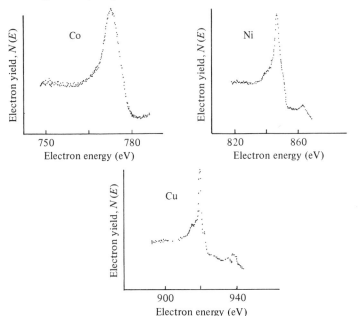

(or complex shape distortion in valence level related emission) due to the weak selection rules. Some understanding of line shapes is emerging but it is not clear that this understanding will lead to these shapes being very useful as a means of determining surface electronic structure. Moreover, in contrast to XPS, rather little effort has been devoted in AES to systematic studies of 'chemical shifts' in emission lines for use as 'fingerprints', despite the fact that such shifts can be larger than in XPS due not only to the same one-electron effects as in XPS, but also due to changes in U with changing electronic environment. Indeed, the main systematic studies which have been performed on Auger electron peak shifts have emerged from XPS experiments. Studies of the difference in energy between a photoelectron peak and its associated Auger electron have the distinct advantage of showing up 'chemical shift' effects without the need to make accurate absolute energy measurements. Some tabulations of changes in this 'Auger parameter' have been made using core–core–core (CCC) Auger electron peaks. Of course, one important reason for the relative lack of systematic study of chemical shifts in AES is that most of the intense peaks in a typical spectrum are not CCC but CCV or CVV transitions. These may have line widths as great as twice the band width so that small shifts (of a few eV) are not always easy to detect, particularly if they involve only part of the signal.

3.3.3   *AES for surface composition analysis*

As we have already remarked in section 3.2, the criteria for a good core level spectroscopy for surface composition analysis are availability of a good range of spectral peaks for all elements, at least some of which can be relatively intense, and an ability to make the technique quantitative. Casting an equation in the same form as that for XPS (equation (3.15)) we can write down the Auger yield for an ABC Auger transition from species i on the surface located at a site x, y, z as

$$\mathrm{d}N_i = (\text{incident electron flux of energy } E_p \text{ at } x, y, z)$$

$\times$ (no. of atoms of i at x, y, z)

$\times$ (ionisation cross-section of level A of
species i at energy E_p)

$\times$ (backscattering factor for energy E_p and
incident direction)

$\times$ (probability of decay of A level of species i
to give ABC Auger transition)

$\times$ (probability of no loss escape of electrons
from x, y, z)

× (acceptance solid angle of analyser)

× (instrumental detection efficiency)

Cast in this form we can see that many of the factors are similar to those in XPS. In particular the first two terms are simply the incident flux and atomic concentration and the last three terms are also essentially as in XPS; note, however, that effectively contained in the last two instrumental terms is an effect (if any) due to intrinsic angular dependence in the Auger electron emission (see later). The third term, the ionisation cross-section, is also comparable with the photoionisation term in XPS although we note that in AES the incident energy beam energy E_p may be varied easily (typically from 1 keV to 5 keV but sometimes to much higher energies) so that an additional parameter is involved. Moreover this term is really convoluted with the following (fourth) term covering backscattering in a rather complex way, as we will show below. Finally, there is an extra term relating to the probability that the core hole in level A created by electron ionisation then decays by the specific Auger transition studied. At least for relatively shallow levels this term is close to unity; X-ray emission is very improbable and the CVV transition is often the only Auger transition energetically possible.

Electron ionisation cross-sections of core levels have been investigated extensively in contexts other than Auger electron spectroscopy. For example, electron impact is often more convenient than using incident photons in the study of energy levels in atoms and molecules. Electron ionisation cross-sections are also needed for quantification in the technique of electron microprobe analysis in which the radiative X-ray emission is used for bulk composition determination. The general trend of all electron ionisation cross-section data as a function of incident electron energy is of a rapidly rising curve above the threshold energy (equal to the core level binding energy) to a peak at about 3–4 times this energy, beyond which the cross-section falls off again slowly. Typical results and some theoretical calculations are shown in fig. 3.34 for some K-shell data of first full row elements. Note that to superimpose the results from several elements two normalisations have been applied to the data; the energies E_p are scaled to the threshold energy E_K for the particular element so the abscissa is (E_p/E_K) and the cross-sections are multiplied by $E_K{}^2$. These scalings are rather effective and lead to a 'universal curve' from which other K-shell cross-sections may be predicted with reasonable precision. For other shells the overall energy dependence of the shape of the cross-section curve remains but the absolute cross-sections scale by slightly different powers of the binding energy. This is indicated in fig. 3.35 which shows a logarithmic plot of peak cross-sections for various core levels of different character and different species. Note that not only are the binding energy dependences for the K- and $L_{2,3}$-shells similar, but the absolute cross-sections are also very comparable. These data therefore show that most elements may be expected to show Auger electron peaks of comparable intensities and that useful rules exist for estimating ionisation cross-sections. Notice, however, that the scales of fig. 3.35 are logarithmic and that small deviations of points from the lines can lead to substantial errors in quantitative yields. Several other important limitations also exist. In particular, core holes leading to particular Auger emission peaks can arise by methods other than direct ionisation of the incident electron beam. One such method is the occurrence of Coster–Kronig transitions. For example, if an incident electron creates a hole in the L_1-shell (2s-state) one possible mode of decay is an Auger process $L_1 L_{2,3} X$ where X could be any level sufficiently shallow to make the transition energetically possible. This Auger-type transition involving the creation of a final-state hole in the same shell as the initial hole is known as a Coster–Kronig transition. Generally such transitions occur at a high rate and so

Fig. 3.34. K-shell ionisation cross-sections using primary electrons of energy E_p for C, N, O and Ne. The incident electron energies are scaled by the K-shell ionisation energy E_K while the cross-sections are relative values multiplied by $E_K{}^2$. The three lines relate to three theoretical descriptions of the cross-section energy dependence, details of which may be found in Glupe & Mehlhorn (1967).

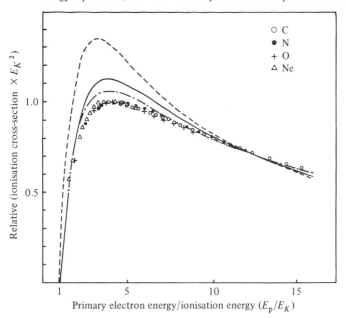

are very probable (and may lead to very broad emitted electron lines due to the short lifetime). If has the effect of making an $L_{2,3}$ hole available for $L_{2,3}XY$ Auger transitions so that, typically, the $L_{2,3}XY$ Auger yield is dictated by the sum of L_1 and $L_{2,3}$ ionisation cross-sections. Fortunately in this case the L_1 cross-section is much smaller than that of the $L_{2,3}$-level so that the correction is not large.

An even more important source of core holes for Auger decay, however, is ionisation by backscattered or secondary electrons in the solid surface generated by the incident beam. This is the origin of the fourth factor in equation (3.27). The overall shape of the ionisation cross-section versus electron energy curve of fig. 3.33 indicates that maximum Auger yields should be obtained using incident primary electron energies of at least three times the binding energy of the deepest core levels of interest. A rather low incident electron energy is therefore far

Fig. 3.35. Experimental electron ionisation cross-sections at an incident electron energy of four times the ionisation energy E_i for various K-shells (+), $L_{2,3}$-shells (o) and $M_{4,5}$-shells ($\triangle$). The K-shell data are those shown in fig. 3.34; the remaining data are of Vrakking & Meyer (1975).

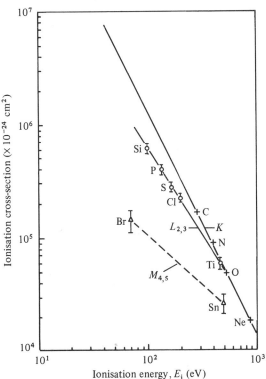

more deleterious to the Auger yield from high binding energy states than a high incident energy is to the yield from shallow states. Because levels up to 1.0–1.5 keV are generally of interest, it is therefore normal to use incident primary energies in the 3–5 keV energy range. This ensures that all levels of interest are excited with a cross-section close to their maximum values. However, it also means that many inelastically scattered electrons also have sufficient energy to ionise core levels and thus give rise to Auger electron emission. For shallow levels in particular, this backscattered flux can be the dominant source of ionisation and must be accounted for in any attempts to quantify the surface composition using equation (3.27). This is a complex problem as the ionisation effect of the backscattered electrons is a convolution of the ionisation cross-section versus energy curve and the inelastic scattering spectrum. Moreover, the backscattering will depend on the substrate or matrix in the surface and subsurface region. For a reliable quantitative determination of surface coverage which takes proper account of backscattering, it is therefore necessary to perform a calibration experiment with the particular system of interest or one very closely similar to it. Before discussing such experiments, however, we should comment briefly on one other physical effect loosely included in backscattering which can lead to difficulties in Auger quantification, particularly in the context of equation (3.27). This effect is that of diffraction of the *incident* electron beam in the crystal surface. Such diffraction can lead to marked variations in the penetration of the incident beam as a function of crystallographic direction and to large variations in the local electron flux 'seen' at different sites in the surface region. The effect is essentially the inverse of 'Kikuchi patterns' seen in high energy electron diffraction, and in the thermal diffuse scattered background of high energy LEED experiments; these patterns consist of dark and bright bands of emission of pseudoelastically scattered reflections along major crystallographic directions and planes originating from diffuse scattering within the solid followed by diffraction or 'channelling' during transport to the surface. The result of similar effects in the incident electron beam used in AES is that marked changes in the absolute Auger electron current from a particular species are found as a function of incident electron direction. An example of this effect is seen in fig. 3.36 which shows variations in Auger yields from a clean Cu substrate by factors of 2; fig. 3.36 also shows that the elastic scattering (and, indeed, the whole secondary emission spectrum) shows similar qualitative variation, demonstrating that the effect is associated with the incident beam and is not specific to the Auger process. This kind of effect is present in any spectroscopy relying on well-defined incident electron

beams as the initial stimulant. Fortunately the effect appears to give rise
to comparable changes in both substrate or matrix Auger peaks and
those due to surface adsorbates or contaminants, so that some
normalisation of the Auger yields can be undertaken to minimise the
errors caused by these diffraction effects.

Finally, in the context of incident beam direction effects we should
note that a further non-crystallographic effect occurs. As an incident
electron beam becomes more grazing to the surface, the probability of
ionising species in the surface region sampled by AES increases, simply
because the incident beam path length in this region increases by a factor
$\sec \theta$ where θ is the angle between the surface normal and the incident
beam direction. This same factor must also occur in XPS, but the broad
X-ray beam used in most instruments prevents this effect from being
utilised. In AES, however, the effect tends to favour the use of grazing
incidence electron beams in the absence of other constraints. One

Fig. 3.36. Elastically scattered and Auger electron emission from Cu using
1.5 keV incident energy as a function of incident electron beam direction. The
data are taken from a cylindrical single crystal surface so that the incidence
plane always lies along the cylinder axis and includes the surface normal
although the incidence direction is at 15° to the surface normal in the plane
(from Armitage, Woodruff & Johnson, 1980).

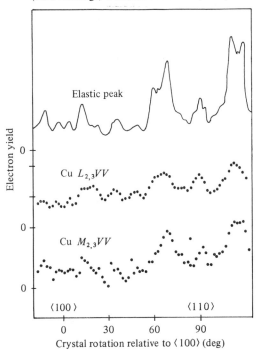

important effect of the sec θ dependence, however, is that on rough surfaces marked variations in Auger yield may occur due to local variations in sec θ.

In summary, therefore, it is clear that there are significant difficulties in the proper absolute quantification of AES on the basis of equation (3.27). In particular, while the difficulties associated with ionisation cross-sections, escape probabilities and instrumental effects are comparable with those in XPS, the effects we have lumped in the general heading of backscattering are potentially very troublesome and can lead to Auger yield variations of at least a factor of 2 depending on circumstances. In practice, reasonable levels of quantification are achieved in AES by less absolute approaches. At the most pragmatic level a handbook of Auger spectra of most elements exists (Davis *et al.*, 1976) using a standard instrument (a CMA) under well-defined incidence electron energies, and comparisons of these spectra automatically include most of the effects of equation (3.27) without the need to separate out individual terms. As much AES is performed with similar instrumentation, these spectra form valuable standards. On the other hand, one effect certainly not included in comparisons of these spectra is the role of backscattering in different environments. If this is to be accounted for properly it is usually necessary to perform some calibration experiment or use an existing, related calibration. Favourable situations do occur in which this may not prove necessary; if we are studying a system where two surface species have Auger transitions of similar character and similar binding energy, then the ratios of these signals, possibly corrected by a scaling factor of the ratio of the two binding energies squared, should provide an accurate assessment of their relative concentrations as both signals should be subject to essentially the same backscattering effects.

For many systems calibration experiments can be performed in a relatively simple way by depositing species A at a fixed (and ideally known) rate onto a clean surface of species B. In the most favourable case the adsorbed species will grow layer by layer on to the substrate. During the formation of the first atomic layer the Auger signal from A will therefore grow linearly with coverage and exposure, while the signal from B will attenuate linearly due to inelastic scattering in the layer of A of fixed thickness but varying coverage. Only when this first layer is complete will the second layer form. The Auger signal from A will then continue to grow linearly but this linear rate will change. This is because the second layer of A not only leads to additional Auger yield but also attenuates the emission from the first layer due to inelastic scattering. Thus a plot of Auger yield versus coverage will show a break in gradient to a new lower gradient. An example of the results of such an experiment is shown in fig. 3.37. If the layer growth persists, further gradient breaks will occur at each successive layer. It is the presence of one or more of these breaks of 'knees' in the gradient which provides an absolute calibration as these are known to correspond to complete monolayers. Unfortunately, this method does depend on their being a linear correspondence between exposure and coverage (or an alternative measurement of coverage), on the growth proceeding beyond one monolayer and on the growth mechanism being such that one layer is completed before growth of the next commence. Of course, if the growth does not proceed in a layer-by-layer fashion, no knee will be found and so no spurious values will be deduced. The failure to obtain multilayer growth or to have a linear exposure–coverage relation (i.e. a constant sticking factor), however, is a common one applying to many interesting gaseous adsorbates for example. In such cases it may only be possible to obtain coverages relative to some saturation value (obtained at very long exposure); the absolute value of this saturation coverage may often be deduced from other considerations (e.g. a LEED pattern). In some special cases, however, a very elegant version of the layer calibration experiment may be performed. In this method of Argile & Rhead (1975) a species C is adsorbed on the surface of B to some (sub-monolayer)

Fig. 3.37. Auger peak height ratio of 102 eV Bi peak to Cu 62 eV peak for Bi grown on an evaporated Cu substrate. Note the break or 'knee' corresponding to monolayer coverage (from Powell & Woodruff, 1976).

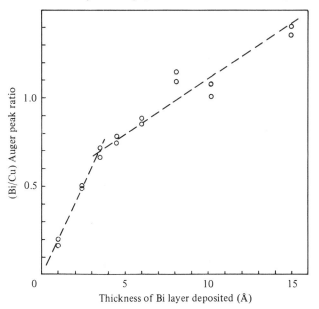

coverage and is found to permit subsequent adsorption of species A (which does satisfy all the growth criteria) only on those parts of the surface not covered by C. Thus, by performing layer growth studies of A and B with various prior exposures of C the knee in the A growth curve is displaced in coverage by an amount related to the coverage of C. Fig. 3.38 shows the result of one such experiment in which Pb (species A) was deposited on Cu (species B) after various exposures of S (species C). The shift of the knee in the Pb and Cu Auger signals to lower coverage of Pb due to the prior S exposure is clear and permits the coverage of S to be deduced, thus allowing the S Auger signal strength to be calibrated.

Fig. 3.38. Auger electron signal versus time plots for deposition of Pb at a constant rate onto a Cu{100} surface previously contaminated with S. The dashed lines show the result expected for an initially clean surface, the Cu plot being adjusted for the attenuation due to the presence of the S (from Argile & Rhead, 1975).

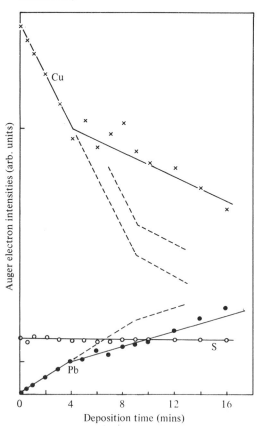

Despite the problems with quantification in AES which make absolute signal levels difficult to interpret, and which can lead to errors of factors of 2 in concentration determinations where careful calibration experiments are not possible, the technique remains, by surface science standards, a simple one to use which can rapidly produce approximate surface composition analyses. Moreover, in common with XPS, but in contrast to some other techniques (such as LEIS, SIMS (see chapter 4) or some of the core level spectroscopies discussed in the next section) the sensitivity variations for different atomic species are not large (typically no more than a factor of 10). In AES it is usually easy to detect 1% of a monolayer of almost any impurity species. One of the greatest strengths of AES, however, particularly in comparison with XPS, is the fact that the stimulating beam is an electron beam of intermediate energy which can readily be focussed and moved or scanned across a surface by electrostatic or magnetic fields. This means that it is easy to produce quite high spatial resolution in the surface analysis and that surface composition imaging using Auger electron emission is possible. Imaging is achieved using the methods of scanning electron microscopy, and the well-advanced technology of this conventional microscopy can be taken over, subject only to the more rigorous vacuum constraints of the surface analysis technique. An example of Auger electron imaging is shown in fig. 3.39; the size of an Auger peak from a particular species is used to define the intensity on a cathode ray tube rastered in phase with the rastering of the electron beam on the sample. The image obtained can be compared (simultaneously if necessary) with a conventional secondary electron emission scanning electron microscopy image. Of course, the yield of Auger electrons from some species at a point is much less than the secondary electron current, so this imaging is necessarily slower and of poorer signal-to-noise than the conventional image. The technique is at least comparable with X-ray microprobe methods, however, and in favourable cases scanning Auger images may be obtained in a few seconds. This chemical imaging in AES, particularly combined with ion sputtering to produce depth profile images into a sample, has proved of great value in the study of 'real' problems in metallurgy and semi-conductor device fabrication. As such, AES is probably the surface analytic technique most widely used outside the domains of well-characterised single crystal surface studies. Of course, an ultimate consideration in the use of high current electron beams of small size (to permit rapid data collection with good spatial resolution) is that of specimen damage. Typical 'low resolution' AES may be performed with a sample current of $\sim 10\,\mu A$ and a beam size of 100–200 μm giving a sample spot current density of 0.1 A cm^{-2}. While this may be reduced

148   *Electron spectroscopies*

substantially when required, very high spatial resolution systems may deliver 30 nA into a beam of less than 500 Å giving a 10^3 A cm^{-2}. At the lower current densities, electron stimulated desorption (see chapter 5) or dissociation are known to occur (typical current densities used in studies of these effects may be only 10^{-5} A cm^{-2} or less) and, while these processes are generally most efficient at relatively low electrons energies (~ 100 eV), typical AES conditions with incident beams of a few keV lead

Fig. 3.39. Micrographs of an area of a microcircuit test piece consisting of a 1 μm metalisation layer of Al on Si which has subsequently been etched through in places. (*a*) shows a conventional (secondary electron) scanning electron microscopy image, while (*b*) and (*c*) show Auger maps using the Si and Al Auger signals respectively. The Auger maps were recorded using an incident energy of 18 keV and a beam current of 2 nA; they consist of 128×128 pixel images with each pixel being obtained in 20 ms (i.e. a total collection time of about 5 min). The upper and lower halves of (*b*) show different stages in image enhancement which can be applied to the micrographs (after Prutton & Peacock, 1982).

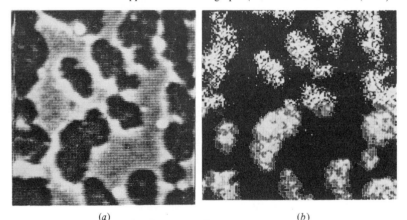

(*a*)   (*b*)

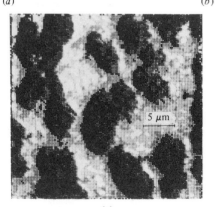

(*c*)

Table 3.2. *Some comparisons of XPS and AES*

	XPS	AES
Species identification and sensitivities	✓	✓
Absolute sensitivity – speed of analysis	—	✓
Spatial resolution – imaging	×	✓
Low damage	✓	—
Quantification	✓	—
Chemical information	✓	?

Ticks indicate a strength, the cross a clear weakness; a dash is 'neutral'.

to many secondary electrons in this low energy range. Some cases of electron assisted *adsorption* have also been reported, presumably associated with the electrons causing a dissociation or electronic excitation of species adsorbed onto the surface in some precursor state. At the highest current densities possible additional effects are as yet unknown. With highly focussed beams the total power is generally small and, as yet, there have been no reports of severe damage or disturbance of the system under study; nevertheless, much work remains to be performed.

3.3.4   *AES versus XPS – some comparisons*

As AES and XPS are by far the most widely used surface analytical techniques both within and outside the disciplines of 'clean' surface science, some comparisons of their relative merits seem appropriate. Table 3.2 summarises some of the major points, a tick being given for a 'good' technique and a cross for a 'poor' one; necessarily these judgements are not absolute but are defined relative to the whole spectrum of techniques used in surface science. A major virtue of both techniques is their relatively uniform sensitivity to all chemical species (other than H and He). The relative virtues of the techniques in several other aspects have much to do with their normal mode of operation. For example, it is easy to produce high intensity electron beams relative to X-ray sources so AES is typically faster and more sensitive than XPS because of these higher incident fluxes (and comparable excitation cross-sections). For the same reason AES is much more likely to incur sample surface damage; in the case of stimulated desorption, for example, photon and electron stimulated processes are intrinsically of comparable probability although the large secondary electron flux in the surface region may lead to more damage from incident electrons. Insofar as there is any significant difference in quantification the problems once again relate mainly to backscattering which can be important in AES but not in XPS; in this case, however, relative calibration is often possible and absolute quantification is not attempted.

As we have already mentioned, one of the most significant differences in the two techniques lies in their spatial resolution within the surface plane. While this can be exploited rather easily in AES it is difficult to develop such a facility in XPS. Typically the X-ray source is simply a target placed close to the sample with a broad emitted beam defined mainly by an aperture in front of the target. For systems utilising an X-ray monochromator a somewhat smaller beam is possible but spatial resolution remains poor relative to AES. The final item in table 3.2 is the ability to identify the chemical states of the surface species. As we have seen, this is widely used in XPS and while large effects are also expected in AES they have not been widely exploited. At least one reason for this is probably the tendency to use relatively low resolution, unretarded CMA electron energy analysers for AES but high resolution detectors for XPS. In doing so the higher speed of analysis is also exaggerated.

In summary, many of the differences in the application of AES and XPS relate more to instrumentation and traditional attitudes than to intrinsic differences. On the other hand, the spatial resolution capability of AES coupled with the availability of intense electron beams makes this technique superior for fast routine analysis, particularly of inhomogeneous surfaces. This speed and spatial resolution cannot be exploited, however, for surfaces which are readily damaged by electron beams. Finally, the more advanced state of knowledge and cataloguing of chemical effects in XPS currently favours this technique for the study of the chemical state of surface species.

3.4   Threshold techniques
3.4.1   *Appearance Potential Spectroscopy (APS)*

While in both AES and XPS the incident exciting electron or photon beam is used to produce a core state ionisation in a surface atom, the exact energy of this stimulating beam is relatively unimportant; it is simply chosen to be well above the threshold energy for the ionisation. In AES the incident beam energy may typically be three times the binding energy of ionised core levels and is chosen to optimise the ionisation cross-section. In XPS the same consideration of optimising cross-section is applied although for photoionisation this favours photon energies rather closer to threshold. Additionally, the energy above threshold defines the photoelectron energy, and this is kept low to fall in the energy range of short inelastic scattering mean-free-paths and thus good surface specificity.

By contrast, an alternative form of core level spectroscopy may be based on the detection of the onset of ionisation as the exciting beam energy passes through such a threshold (equal to the binding energy of the core level relative to the Fermi level of the solid). A technique of this kind involves sweeping the exciting beam energy and detecting the onset of ionisation either by the loss of flux from the incident beam or by the onset of emission of photons or Auger electrons associated with the refilling of the core hole thus created. These different modes of detection form a group of techniques under the general heading of APS. In principle, either incident electrons or photons could be used, but as the photon technique would call for a tunable light source in the 50–2000 eV energy range such experiments would only be possible using synchrotron radiation and have not been seriously pursued so far. All three modes of the incident electron technique have been used, however, the thresholds either being detected by a fall in the elastically scattered electron flux (Disappearance Potential Spectroscopy – DAPS), by the onset of photon emission (Soft X-ray Appearance Potential Spectroscopy – SXAPS) or by the onset of a new Auger electron emission channel (Auger Electron Appearance Potential Spectroscopy – AEAPS). Note that in all of these techniques the surface sensitivity is guaranteed by inelastic scattering of the *incident* electron beam because it is the threshold which is detected and so a small energy loss will reduce the incident beam energy at the onset to a value below this threshold. Thus the potentially long mean-free-path of the soft X-rays emitted does not destroy the surface sensitivity of the method. Moreover, because the technique detects the onset of a new emission channel it is not necessary to energy analyse the emitted photons or electrons; a sudden increase in either of these indicates the existence of a new threshold. Of course, if the total (background) emitted flux is large before a threshold this will affect the signal-to-background and thus the detectability of any new onset.

3.4.1.1 *Detection methods and relative merits*

DAPS involves the detection of thresholds without reference to the mode of decay of the core hole, and may therefore be regarded as the simplest of the three methods. However, the change in the elastically scattered yield from a surface when the primary energy passes through a threshold showing quite good cross-section is typically only 0.1% so, as in all three modes of detection, some electronic differentiation of the detected signal is needed to enhance the onset steps. While a conventional dispersive electron energy analyser such as a CMA may be used, the intrinsically small signals relative to the background favour a large angle detector and as we are only concerned with elastically scattered electrons (i.e. the most energetic electrons available) an RFA (such as a conventional LEED optics) is admirable (see fig. 3.40). By modulating the sample potential and thus the electron energy at the sample, but keeping the analyser potentials constant, the in-phase signal at the detector displays the required differentiation. A typical result is shown in fig. 3.41. Modulation amplitudes of less than 1 V are found to be satisfactory in this method and do not severely smear the detected signals. One potential virtue of DAPS is that the detection of the *elastically* scattered electrons ensures that inelastic scattering defines the surface specificity on both the inward *and* outward paths (as in LEED) thus giving greater surface sensitivity. In AEAPS it is not necessary to attempt to detect the Auger electrons specifically but only to detect the increased total electron emission resulting from their onset (cf. the discussion of SEXAFS in section 3.2.6.2). Indeed, by including the low energy secondary electrons in the detected signal one effectively uses the crystal surface as an 'electron multiplier' to detect many Auger electrons which do not escape the surface without loss. In its simplest form the secondary electron yield may be measured by detecting the sample current (equal to the incident current minus the total emission current). In this case the electron gun energy may be modulated and the in-phase alternating current arriving at the sample detected to provide electronic differentiations. Alternatively, the low energy secondary electrons alone may be detected at a suitably biased collector adjacent to the sample. This has the slight advantage that the elastically scattered component which *decreases* at threshold is not added to the *increasing* Auger electron and secondary yield; this effect is small, however, because only a small proportion of the incident flux does escape energy loss. AEAPS in

Fig. 3.40. LEED optics operated as a RFA for the detection of DAPS. While the incident primary energy, eV_E, is varied, the retarding grids are also varied to detect only pseudoelastically scattered electrons. Modulation of the sample potential allows the derivative of the DAPS spectrum to be measured as the in-phase modulated signal arriving at the collector.

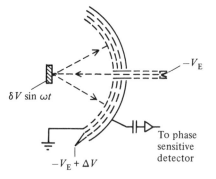

$\delta V \sin \omega t$

$-V_E$

$-V_E + \Delta V$

To phase
sensitive
detector

this simple form is perhaps a factor of 10 more sensitive than DAPS under similar conditions of incident current, etc.

By contrast SXAPS is substantially (a factor of $\sim 10^4$) less sensitive for similar conditions. A typical arrangement for this technique is shown in fig. 3.42 and is very simple. A filament emits the electrons which are accelerated to the sample by applying an attractive potential (plus a modulation component). This arrangement of voltages, together with a screening mesh in front of the detector, permits X-rays but not electrons to enter. The X-rays then photoemit electrons from the walls of the detector which are collected on a suitably biased thin wire placed inside the can. As may be seen from fig. 3.42, the basic instrumentation for the technique can be very simple and by the use of a naked filament rather than a refined electron gun substantially increased incident electron

Fig. 3.41. DAPS from a polycrystalline V surface taken in the derivative mode illustrated in fig. 3.40. Note that, as the electron reflectivity *decreases* when the energy passes through an ionisation threshold, the spectrum is plotted as the negative derivative signal to show these thresholds as peaks. The spectrum shows all the L-edges. The L_1-edge in particular is lifetime-broadened due to Coster–Kronig transitions (see section 3.3.3) (after Kirschner, 1977).

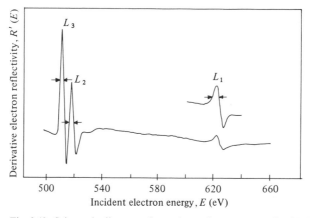

Fig. 3.42. Schematic diagram of experimental arrangement for SXAPS. The grounded screened 'box' acts as a photoelectron converter and these electrons are collected by the central wire or collector.

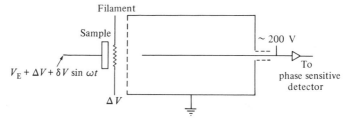

currents may be used ($\sim$ mA rather than $\sim \mu$A) to enhance the signal. Of course, without special care the detector efficiency of the soft X-rays may be poor (some experiments have even used the internal stainless steel walls of the vacuum chamber as the photoelectron converter). However, the principal problem with this technique is the very low efficiency of the soft X-ray emission process; we have already remarked that Auger electron emission is a far more probable mode of decay for a shallow core hole. At the energies of interest, typical fluorescent yields may be only 1% or less. Indeed, these yields are so low that the bremsstrahlung background may be 100 times larger than the characteristic line emission so that, not only is the signal small, but the background is large and leads to poor signal-to-noise ratios. Thus, while SXAPS was the first form of APS to be exploited in surface science and has the virtue of extreme simplicity in the basic equipment requirements, it may be the least satisfactory of the three techniques to follow.

3.4.1.2   *Sensitivities, cross-sections and line shapes*

In section 3.3 we remarked that one of the strengths of AES (in common with XPS) is the relative similarity of ionisation cross-sections, and thus sensitivity of detection, of a wide range of elements. In AES, however, we chose the incident energies to be well above threshold on the broad peak of cross-section versus energy. We also showed that near threshold there is a very substantial fall in ionisation cross-section. The cross-section very close to threshold also becomes very strongly species dependent in a fashion quite unlike the peak cross-section. This is because very close to threshold the initial state of a neutral atom and an incident electron of kinetic energy just greater than E_b (the binding energy of a level relative to the Fermi level) is transformed, after ionisation, to an ion and two electrons just above the Fermi level. The cross-section for this process depends on the density of final states, which is a self-convolution of the density of empty one-electron states above Fermi level. The cross-section therefore depends directly on the number of available empty states, and as the technique of APS relies on detecting the step in absorption or emission as threshold is passed, the sensitivity depends on the (self-convolution of the) density of states just above the Fermi level. This can vary radically from material to material. Consider, for example, Cu and Ni, two materials adjacent in the periodic table. Both have a valence band consisting of a relatively delocalised broad s–p character part, and a relatively localised and therefore narrow 3d character component. However, in Ni the Fermi level lies just below the upper edge of the d-band, while in Cu it lies above the d-band. The two situations are shown schematically in fig. 3.43. The different Fermi level location in these otherwise similar materials has the result that the density of empty states just above the Fermi level in Cu is very much lower than in Ni so that, while Ni has a strong, readily detected APS signal, Cu gives barely discernible thresholds. As a routine core level spectroscopy these variations in elemental sensitivity are a severe disadvantage. Moreover, in SXAPS this problem is further exaggerated by the significant changes in fluorescent yield associated with the same energy levels of different species. Because these fluorescent yields are small ($\ll 1$), however, AEASP is essentially unaffected by these changes in radiative to non-radiative branching ratio for the decay, and the yield sensitivity (as for DAPS) depends only on the intrinsic ionisation cross-section.

For those materials which do have reasonably high densities of states above the Fermi level, APS offers the ability to probe this density of empty states. These empty states are not accessible to most surface spectroscopies, although the fact that the final state involves two electrons in these states means that the APS signal is broadened to a self-convolution. In this regard it is worth noting that if APS were performed with incident *photons* the final state would involve only one electron above the Fermi level and would thus yield the density of empty states directly. Moreover, in the case of incident photons the dipole selection rule applies (i.e. that the final state must have $l+1$ or $l-1$ angular momentum character if the emission is from a state of angular momentum l) so that the density of states measured would be selective in angular momentum character. For incident electrons this selection rule only applies when well above threshold and very close to threshold, as in APS, it is not exact.

Fig. 3.43. Schematic diagram of one-electron density of states of filled and partly filled d-band transition metals such as Cu and Ni.

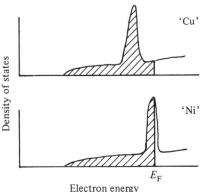

While chemical effects have not been explored in any depth in APS both energy shifts and shape changes are to be expected as the chemical environment of an atom is changed. Chemical shifts of the kind discussed in XPS are to be expected, but in addition changes in the density of empty states can also have a profound effect not present in XPS. For example, our discussion so far, in defining the threshold relative to the Fermi level, is strictly only valid for a metal in which empty states are available immediately above this level. If a chemical change causes a band gap to open up around the Fermi level then the threshold (which is strictly defined relative to the deepest empty states) will shift. Indeed, a band gap (or other strong feature in the density of states) appearing even some way above the Fermi level may lead to a significant shape change in the APS signal. An example of a chemical shift in APS, observed after oxidation, is shown in fig. 3.44.

In summary, therefore, the methods which comprise APS do provide much of the elemental and chemical information available from XPS or

Fig. 3.44. DAPS spectrum (in a doubly differentiated form) of partly oxidised Ti showing both L_2- and L_3-edge thresholds in the metal and a 'chemically' shifted component associated with the oxide (after Kirschner, 1977).

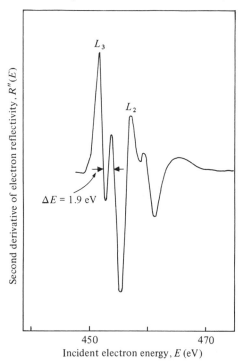

AES although much work remains to be done in characterising the breadth of applicability of the methods. Moreover, they do provide information on the density of *empty* states not available through XPS and AES and can, in some forms, be rather simple experimentally. On the other hand, they suffer from large variations in sensitivity to different elements and differing electronic environments which limits their value as general-purpose spectroscopies, and largely excludes their use on certain materials.

3.4.2 *Ionisation Loss Spectroscopy (ILS)*

The existence of a well-defined, discrete energy, photoelectron peak in XPS for each accessible core level in a species relies on the fact that in photoionisation a photon always gives up all of its energy to the system. Moreover, apart from the various many-body effects of shake-up, etc., this energy is essentially given to a single bound electron. In the process of electron ionisation the incident electron can give up any amount of energy up to its full kinetic energy. However, for the ionisation of any particular band state level, there is a well-defined minimum energy transfer which can result in the ionisation which is equal to the binding energy of the relevant level as in APS. Thus, if this binding energy is E_b and the incident energy is E_0 all interactions with the atom leading to ionisation will result in electrons being emitted with an energy less than $E_0 - E_b$. If we study the secondary electron emission spectrum in such a process, we therefore expect to see a step in the spectrum at $E_0 - E_b$ corresponding to the switching-off of this ionisation process. In this sense ILS is a threshold technique like APS; although the incident ionising electrons have an energy well above threshold, the ionisation is detected by a threshold process (i.e. a process in which the electron has lost the minimum possible energy to produce the ionisation). There are other similarities to APS. For example, the final state again consists of a core hole and two electrons in the continuum; in APS threshold, however, both of these electrons are in the deepest available empty states, while in ILS at 'threshold' one electron is in such a minimum energy state but the other lies well above the vacuum level. ILS sensitivities are therefore also dictated by the density of empty states just above the Fermi level. On the other hand, because one of the two electrons lies high in the continuum we normally expect the density of states to be smooth in this region and the line shape in ILS, which reflects the density of these final two-electron states, is essentially proportional to the one-electron density of states just above the Fermi level. In principle ILS should therefore yield clearer information on this empty density of states than APS. In practice the sharpness of the structure in the two techniques is determined not only by the presence or absence of a self-convolution process, but also by the intrinsic instrumental resolution of the mode of detection. In APS this resolution is limited by the intrinsic thermal energy spread of the electron beam and the modulation amplitude used in the electronic differentiation and might typically total 1 eV or less. In ILS both of these factors also contribute but in addition one has the effect of the energy resolution of the dispersive electron energy analyser used to detect the signal (an instrument not necessary in APS). At energies of several hundred eV it is not easy to obtain good resolution ($\lesssim 1$ eV) as well as good sensitivity through a large acceptance angle (see section 3.1.3). Typical ILS and DAPS spectra are compared in fig. 3.45 together with an X-ray absorption spectrum (which also monitors the one-electron density of empty states above the Fermi level). Evidently in these 'typical' data the ILS results do not show appreciably better resolution than the DAPS despite the self-convolution process involved in DAPS.

The sensitivity in ILS is also not particularly favourable despite the fact that the ionisation process occurs at incident energies well above threshold and so should have a substantially higher total ionisation cross-section relative to the near threshold value relevant to APS. This is because only threshold events corresponding to a particular combination of final state energies are detected in ILS, so that it is a differential, rather than total, cross-section which is measured. Indeed, it is found to be preferable to work with primary energies reasonably close to threshold rather than those corresponding to peak total ionisation cross-sections. Nevertheless, the energy must be high enough to ensure that the ionisation loss edge in the secondary electron peak is reasonably well removed from the very low energy peak where the signal-to-background would be poor. For this reason, and those of energy resolution, it is often preferable to use a dispersive analyser at a fixed energy and scan the incident beam energy to sweep the ionisation losses through the pass band of the detector, rather than varying the detection energy and keeping the incident energy fixed.

In general, therefore, APS and ILS are seen to have many common problems and deficiencies. Both sample the density of empty states close to threshold and therefore provide the same basic information (including chemical effects) and suffer from the same marked sensitivity variations depending on the size of this density of states at threshold. As such, they are unsatisfactory as routine analytic tools for composition analysis but do provide special information (in favourable systems) not readily obtainable by other methods.

3.4.3   *Structural effects in threshold spectroscopies*

In our discussion of XPS we described two sources of structural information which can arise due to coherent interference effects in elastically scattered photoelectrons in the vicinity of the emitter atom. One effect, the interference of the emitted electrons to produce a structure-dependent angular variation in the measured emission, is not generally relevant to the threshold spectroscopies (although such effects may occur in the methods involving electron detection). Similar diffraction phenomena must occur in the *incident* beam (as in AES) and will lead to variations in the local ionising electron flux at the emitter site

Fig. 3.45. Comparison of DAPS, ILS and X-ray absorption spectra involving the same $L_{2,3}$-subshells. The DAPS data are shown in a reintegrated form for comparison. Note that the fine structure seen in the X-ray absorption spectrum is lost in the lower resolution ILS technique, while it is smoothed out in the DAPS data due, in part, to the self-convolution process (after Kirschner, 1977).

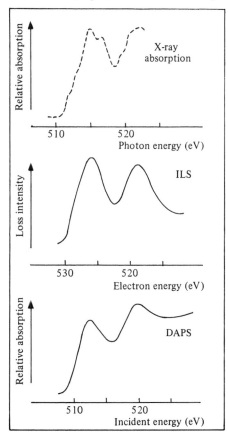

as a function of primary electron energy or direction. As yet, however, there have been no attempts to use this effect for structural studies. Of greater interest, however, is the effect of the scattering of the emitted electrons back onto the emitter to modulate the ionisation cross-section itself which in photon stimulation is referred to as EXAFS (see section 3.2.6.2). These oscillations relate to the decrease in the photoelectron wavelength in XPS as the photon energy is raised, thus permitting periodic matching of the nearest neighbour distances to a half-integral number of electron wavelengths (ignoring the scattering phase shifts). In EXAFS the final state of a single electron in the continuum greatly simplifies this picture relative to that in electron stimulated threshold spectroscopies where the final state involves two electrons in the continuum. Nevertheless such effects should also occur in all of these techniques.

Conceptually at least the situation is simplest in ILS when the primary electron energy is well above threshold. In this case the ILS line shape in the vicinity of the energy loss edge is dominated by the one-electron density of states because one of the two electrons (the detected one) has a high energy and essentially samples a smooth continuum of states. The near-edge line shape is dominated (as in EXAFS) by this density of states information, but at somewhat higher energies the EXAFS effect becomes the major one. Provided the incident electron energy is sufficiently high, the line shape in this region is dominated by a one-electron EXAFS effect because, although both final state electrons can be backscattered and thus influence the ionisation cross-section, the detected electron has such a high energy that the backscattering for this electron is negligible. In transmission experiments using thin films these effects have been observed and lead to a technique known as EXELFS (Extended Electron Loss Fine Structure) which is usually performed using conventional electron microscopes operating with typical electron energies of hundreds of keV. A spectrum of this kind, taken from graphite using 200 keV incident electrons and studying C K-edge loss and fine structure, is seen in fig. 3.46 together with the expanded EXELFS structure and its Fourier transform. Of course, as in EXAFS, phase shift corrections must be applied to interpret the data correctly and a proper calculation of these effects is potentially more difficult in the incident electron technique; this is because in photon stimulated EXAFS the dipole selection rule applies and the outgoing photoelectrons (which are backscattered) are described by a sum of only two partial waves of angular momentum $l-1$ and $l+1$ where l is the angular momentum of the initial bound state. In incident electron ionisation this selection rule no longer applies in general, although for the special case of incident energies far above threshold and measurements in the direct transmitted direction (corresponding to small momentum transfer in the loss) the dipole selection rule is valid. In this transmission form, however, EXELFS is necessarily a bulk technique. There seems to be no reason, however, why similar studies in reflection should not yield similar EXELFS variations dependent on the *surface* structure. In order to maintain surface specificity and good signals at the high incident energies the experiments would be conducted at grazing angles to the surface as in RHEED. As yet, however, no serious attempts to exploit this method have emerged.

In the case of APS a number of experiments have recently been performed to attempt to study and use EXAFS-like effects for structural investigations, but the data are intrinsically more complex than those

Fig. 3.46. ILS spectrum (*a*) in the vicinity of the carbon *K*-edge from graphite showing EXELFS. The primary electron energy was 200 keV. The extracted fine structure function $\chi(k)$, (*b*), and the Fourier transform of this (*c*) are also shown (after Kincaid, Meixner & Platzman, 1978).

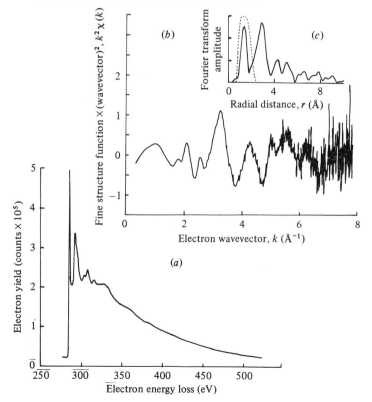

obtained from EXELFS. In particular the two-electron final state means that the line shape, which very close to threshold is dominated by the self-convolution of the density of empty states, is at higher energies dictated by the self-convolution of the EXELFS. Moreover, the fact that the incident electrons are always relatively close to threshold (at most a few hundred eV above threshold) ensures that the dipole selection rules are not applicable to the ionisation event and thus a proper theoretical description of the process and of the phase shift corrections is more difficult. At first glance these difficulties seem overwhelming. However, in conventional EXAFS the absorption coefficient fine structure function can generally be expressed as a sum of harmonic functions (equation (3.19))

$$\chi(k) = -k^{-1} \sum_i A_i(k) \sin\left[2kR_i + \phi_i(k)\right]$$

A self-convolution of such a function maintains the same basic periodicity, so that a Fourier transform of the Extended Appearance Potential Fine Structure (EAPFS) should reflect the basic real space distances R_i in essentially the same way as the transform of a simple (one-electron) EXAFS spectrum. Indeed, this is also true for derivatives of the EAPFS with respect to energy so that the normal electronic differentiation of APS may be used and this raw data may be subjected to the transform process. An example of raw and processed EAPFS data is shown in fig. 3.47 which certainly indicates that for this particular adsorption system (0 on Ni$\{100\}$) the results appear to be meaningful. Moreover, recent theoretical work (Laramore, 1981) indicates that this simple interpretive scheme may be justified. However, one problem which certainly does influence the APS fine structure in crystalline samples is the effect of incident electron diffraction which we mentioned earlier as a potential source of structural information. At low incident energies this is a very important effect which can obscure the otherwise interpretable EAPFS data. The extent of application of EAPFS is therefore not clear at the time of writing.

3.5   Ultraviolet Photoelectron Spectroscopy (UPS)
3.5.1   *Introduction*
While we have concentrated in the foregoing discussion on the use of electron spectroscopies to investigate the core levels of surface species both as a means of compositional analysis and for the determination of chemical state, many of these techniques can also provide information on the more weakly bound and less localised

valence electronic states. In most cases, however, these core level spectroscopies are not well suited to providing this information. For example, while Auger electron spectral line shapes may carry information on the local valence density of states, extracting this information from this two-electron process can be complicated. Moreover, the one-electron process involved in XPS leads to very high emission electron energies from the valence states which make high energy resolution difficult to obtain; additionally, the photon energy is far above the photoionisation threshold so that the cross-section is usually small. These and other difficulties favour the use of photoelectron spectroscopy at much lower photon energies.

UPS is typically performed using a He gas discharge line source which can be operated to maximise the output of either HeI (21.2 eV) or HeII (40.8 eV) radiation, although other inert gas species (providing somewhat lower energies) may be used and much work is now performed

Fig. 3.47. Derivative DAPS spectrum in the vicinity of the Ni $L_{2,3}$-edge from a thick oxidised film on Ni{100}, showing the extended fine structure (*a*). In (*b*) this fine structure is extracted while in (*c*) a Fourier transform is shown. The main peak in the transform corresponds to a Ni–O distance of 2.04 ± 0.05 Å if the outgoing wave is assumed to be predominantly d-like for the purposes of phase shift correction (after den Boer, 1980).

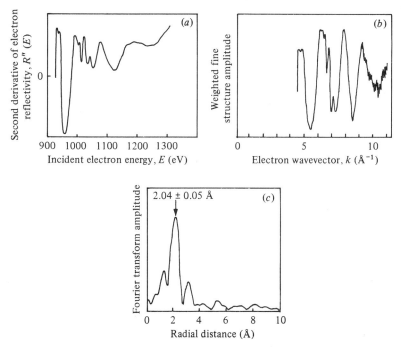

using the continuously tunable monochromated synchrotron radiation. Using these lower photon energies of normally less than about 40 eV and typically 21.2 eV, it is clear that only valence levels are accessible. These include the occupied band states of a clean solid surface as well as the bonding orbital states of adsorbed molecules. Particularly at the lower end of this photon energy range we see, from the energy dependence of the inelastic scattering mean-free-paths (fig. 3.2) that the sampling depth of UPS is potentially somewhat greater than that of most of the surface techniques discussed so far. Indeed, early work on UPS (at even lower energies) assumed that the technique probed bulk properties; later work concentrated on the surface sensitivity and cast doubt on the value of UPS to investigate bulk valence bands, while it now seems clear that the true state lies between these extremes. UPS is surface *sensitive*, but not necessarily as surface *specific* as some other surface analytic techniques. To a large extent these two different views of UPS have led to two relatively separate regions of application of the technique. The first of these is concerned with the investigation of the surface electronic structure – specifically the surface *band* structure of clean surfaces and ordered atomic adsorption layers on them. This work stems directly from earlier interests in investigating bulk solid band structures by photoemission. The second area is particularly concerned with the identification of molecular species on surfaces and the characterisation of their decomposition and reactions, using UPS to identify the characteristic electronic energies associated with *bonds* within the molecules. Both of these approaches have benefited considerably from the development during the last few years of Angle-Resolved UPS (ARUPS). While these two approaches exploit some common ideas there are sufficient differences in their basic approach and objectives to make it appropriate to discuss them separately.

3.5.2 *UPS in the elucidation of band structure*

The essential power of UPS as a technique for the investigation of the band structure of surfaces can best be appreciated by discussing the development of the technique to investigate *bulk* bands. It is therefore convenient to initiate this discussion on the assumption that UPS is actually a bulk technique; i.e. that the inelastic scattering mean-free-path is very long. Of course we know that this is not actually true but the description which emerges from this approach turns out to be rather close to reality. A vital consideration in this discussion is the role of momentum conservation. Photoemission, in common with other processes, must conserve energy and momentum and because the photon momentum is extremely small relative to that of a photoelectron having the energy dictated by energy conservation (the Einstein equation) the process is impossible for a free electron. An unbound electron can only absorb a part of the photon's energy, not all of it, resulting in Compton scattering. In the case of an electron bound in an atom, on the other hand, the nucleus can take up the momentum recoil of the photoelectron and photoemission is permitted, the emitted electron energy being essentially unaffected by this recoil. An electron in a valence (band) state of an infinite solid represents a rather special situation. In this case the electron is no longer bound to an individual atom and while it is itinerant in the solid only its *reduced* momentum or wavevector **k** is well defined. A change of electron momentum by any reciprocal lattice vector **G** of the solid is indistinguishable from no change. This means that energy and momentum can be conserved for photoemission of a valence electron in a solid by the whole solid taking up a recoil momentum of exactly one reciprocal lattice vector **G**. Thus in the *reduced zone scheme* the transition is a vertical one in energy involving no change in reduced **k**. This concept, and some of its consequences, can best be appreciated by recourse to a nearly-free-electron model of a solid. Fig. 3.48 illustrates a photostimulated transition of the type described in both the extended and reduced zone scheme. In the extended zone scheme the electron must suffer a Bragg reflection adding **G** to its wavevector to find an available state; in the reduced zone scheme the same transition is seen as vertical and **k**-conserving or 'direct'.

Assuming that the energy–wavevector relationship is essentially free-electron-like away from the zone boundaries the energy conservation condition

$$E(\mathbf{k} + \mathbf{G}) - E(\mathbf{k}) = h\nu \tag{3.28}$$

can be written as

$$\frac{\hbar^2}{2m} [(\mathbf{k} + \mathbf{G})^2 - \mathbf{k}^2] = h\nu$$

which reduces to

$$2\mathbf{k} \cdot \mathbf{G} = \frac{2m}{\hbar^2} h\nu - G^2 \tag{3.29}$$

As the right-hand side of this equation is a constant, for given photon energy and reciprocal lattice vector, this becomes the equation of a plane in **k**-space perpendicular to **G**. Evidently, for this to describe a photoemission process the initial **k** vector state must be occupied, i.e. it must lie within the Fermi surface, which for a simple metal is a sphere lying wholly within the first Brillouin zone. Fig. 3.49 illustrates the result of equation (3.29) when the plane does cut the Fermi sphere and a disc of states appears which can contribute to the photoemission. Notice that this disc corresponds to a fixed range of **k** between the maximum value at the Fermi surface $\mathbf{k}_F$ on the edge of the disc, and a minimum which corresponds to the point at the centre of the disc and depends on the relative values of hv and **G**. There is thus a minimum and maximum value of the energy of initial states sampled.

The experimental data of fig. 3.50 show that this conclusion is essentially correct. Here, UPS data are shown for the s–p-band of Ag at a number of different, rather low energies. Notice that the energy axis is drawn in terms of the initial state energy, $E - hv + \phi$ where E is the

Fig. 3.48. Photoemission in a nearly-free-electron band structure showing the momentum transfer **G** in the extended zone scheme which is seen as a 'direct' or 'vertical' transition in the reduced zone scheme.

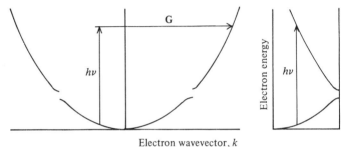

Electron wavevector, k

Extended zone   Reduced zone

Fig. 3.49. k-space diagram showing the plane defined by equation (3.29) intersecting the Fermi sphere of a simple metal. The hatched disc shows the surface of possible initial states for direct transitions involving photon adsorption energy of hv; $\mathbf{k}_{\mathrm{min}}$ and $\mathbf{k}_F$ (the Fermi surface value) are the minimum and maximum values of initial state momentum.

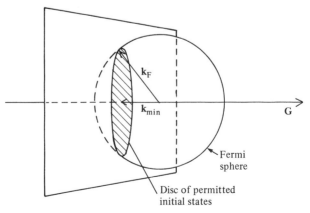

measured photoelectron kinetic energy and ϕ the work function. The
maximum energies all line up at the Fermi energy, while the main elastic
emission is confined to a small energy spread below this energy which
broadens as the photon energy increases. The lower energy limits of this
direct emission (indicated by the shaded regions) are found to agree
rather well with the predicted energies at the centre of the disc of
permitted states based on a free-electron model. Of course, all spectra
show a strong inelastic and secondary electron 'tail' as discussed earlier
for other electron spectroscopies.

While the requirement of **k**-conservation is a very important and
restrictive criterion, the simple considerations outlined above show that
in a three-dimensional band structure and with photoelectrons being

Fig. 3.50. Photoelectron energy spectra taken from an Ag sample at various
photon energies, showing the broadening range of accessible s–p initial states
(after Koyama & Smith, 1970).

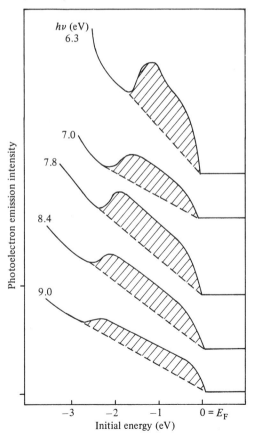

collected over all directions (or at least over a large angular range) the resulting photoelectron spectrum provides a rather averaged *joint density of states*. Thus, if we assume that the probability of photo-emission from all states is equally likely provided the transition is **k**-conserving, the spectrum represents a convolution of the density of initial states with the density of available final states for direct transitions. Far more specific information is available in the study of such photoemission processes if the detection is performed in an angle-resolved fashion. The same model which we have used above illustrates this feature rather readily. Thus we have considered photoemission from initial wavevector **k** to **k** + **G**. Fig. 3.51 illustrates these states on the 'plane and sphere' model for a nearly-free-electron metal. Notice that, while a range of initial values of **k** lie on the plane defining possible photoemission states, the emission direction defined by the final wavevector **k** + **G** uniquely defines the initial state involved. Thus the photoemission peak detected at a specific emission angle arises from a discrete initial energy and **k** vector. The observation of such a peak therefore allows the initial **k** state to be plotted on a band map. ARUPS of solid surface valence bands thus offers a method of plotting the initial state band structure directly. This is an extremely powerful concept. Conventional methods of band structure investigation provide far more averaged and less specific information.

An illustration of the application of this approach is shown in the data of fig. 3.52 in which ARUPS data are shown from a surface of InSe. The spectra, taken using a fixed photon energy, are shown for different polar angles of emission within a single high symmetry azimuth. The locations of the peaks in these spectra have then been used to infer the initial **k** and energy values of occupied electronic bands in the material as shown in

Fig. 3.51. As fig. 3.49 but including the final state **k** vector (**k** + **G**) and showing the disc of permitted final states associated with the reciprocal lattice vector **G**.

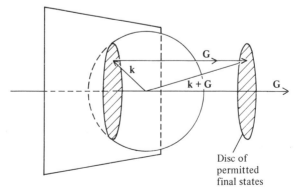

fig. 3.53. The simplicity of these results and the inferred band mapping are impressive but unfortunately are rather atypical. In general the determination of the solid surface band structure from the ARUPS spectra is far more complicated. These complications are not apparent in our treatment so far because we have used a number of simplifications. Most importantly, by developing our theory for a nearly-free-electron model, we have assumed that the final **k** value is related to the final energy by the free-electron relationship (equation (3.29)). Because this relationship is exactly the one relevant to the photoelectron which propagates outside the solid to the detector, the final **k** value of the

Fig. 3.52. ARUPS spectra from InSe in the ΓKM azimuth at the polar angles shown. The incident photon energy was 18 eV (after Larsen, Chiang & Smith, 1977).

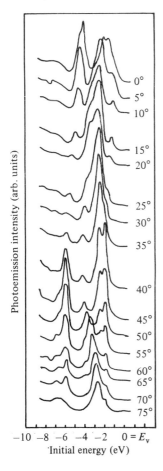

electron at the detector is identical to that of the photoelectron propagating through the solid. This is generally not true.

The essential limitation of our discussion so far is that we have considered only step one of the so-called *three-step model* of photoemission from solids first proposed and used by Bergland & Spicer (1964a, b). In this model we treat as separable the processes of photoemission, propagation or transport of the photoelectrons to the surface and transmission through the surface to the vacuum and the detector. As we have remarked, if we consider only a nearly-free-electron solid then the $E(\mathbf{k})$ relationship away from band gaps is essentially the same both inside and outside the solid and the transmission through the surface is a trivial problem. In such a situation the $\mathbf{k}$ value in the final state band is directly measured in the angle-resolved photoemission experiment and so both initial and final state bands can be plotted away from the zone boundaries.

Most solids, of course, do not conform to this simple nearly-free-electron picture and indeed it is their deviation from free-electron character which is precisely the point of interest and experimental investigation. In this case we must recognise (as in LEED) that transport of the photoelectron out from the surface which has only two-dimensional periodicity parallel to the surface is governed by the conservation of $\mathbf{k}_{\parallel}$, the component of $\mathbf{k}$ parallel to the surface. Thus the

Fig. 3.53. Initial state bands in InSe in the $\Gamma K M$ direction inferred from the ARUPS data of fig. 3.52 (points shown as circles) and similar data taken at a photon energy of 24 eV (squares) (after Larsen, Chiang & Smith, 1977).

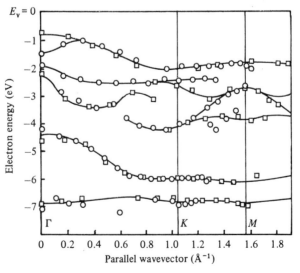

perpendicular component, $k_\perp$ is not conserved in transmitting the photoelectron to the outside world and the detector. The results we have shown in figs. 3.52 and 3.53, for photoemission from InSe, represent a special case. InSe is a layer compound whose structure consists of sandwiches in which two inner sheets of In atoms are located between two outer sheets of Se atoms. The bonding within the sandwiches is much stronger than between sandwiches and, indeed, it is rather easy to cleave this material between the sandwiches to present a clean surface of the orientation used in these experiments. This layer structure has itinerant electron states within the sandwiches but (as indicated by the bonding) little coupling between layers; the states are therefore highly localised relative to displacement perpendicular to the surface. This means, of course, that the states disperse little in $k_\perp$. The partial band structure of fig. 3.53 is therefore plotted by making use of the assumption that the bands are flat in $k_\perp$ so that only $k_\parallel$ is of interest, a quantity which is conserved and therefore well known.

For truly three-dimensional solids, however, the problem of $k_\perp$ determination remains a problem. Some success has been realised in the interpretation of angle-resolved photoemission data by assuming that the final states *are* free-electron in nature. The justification for this assumption is that, as the energy increases, the effect of the periodic crystal potential decreases. Thus, as the energy increases the bands must tend towards free-electron behaviour. It is therefore not totally inconsistent to assume free-electron final bands and use this assumption, coupled with the photoemission data, to infer the location of the initial state bands which, at much lower energy, are certainly not free-electronlike. The data of fig. 3.54 are taken from a study of angle-resolved photoemission along the surface normal of a Cu{110} crystal surface using a range of photon energies from 15 eV to 100 eV. Normal emission data are frequently used because the constant $k_\parallel$ means that the data relate to a high symmetry direction in k-space. The full lines correspond to calculated bands (Burdick, 1963) while the experimental points are shown as rectangles which also indicate the experimental errors. One interesting feature is the band labelled Σ_2 which is shown dashed; photoemission from this band should not be possible for normal emission for any polarisation of the incident light and the experimental points were obtained at collection angles slightly displaced from normal emission. Generally symmetry arguments can be powerful in identifying the nature of initial states and will be discussed in detail later.

The data of fig. 3.54 clearly present a strong case for the k-conservation or direct transition selection rule and for the use of free-electron final state bands; the agreement between theory and experiment is excellent. However, the situation is not always so favourable. The assumption of free-electron final state bands is certainly one which should be avoided if possible. One possible method of achieving this is by the use of a method originally proposed by Kane (1964) in which the

Fig. 3.54. Experimentally determined initial state occupied band structure points in the ΓKX direction in Cu obtained by plotting peaks in ARUPS data taken at normal emission from a Cu{110} surface using a range of photon energies. The associated photon energies are indicated on the top and bottom of the figure; note that the two scales are displaced due to the difference in binding energy at the top and bottom of the diagram. The size of the rectangular experimental data points indicates the experimental uncertainties. The full lines are the theoretical band structure of Burdick (1963). Transitions from the dashed Σ_2 band are symmetry forbidden at normal incidence and these data were obtained at $3°$ from the surface normal (after Thiry *et al.*, 1979).

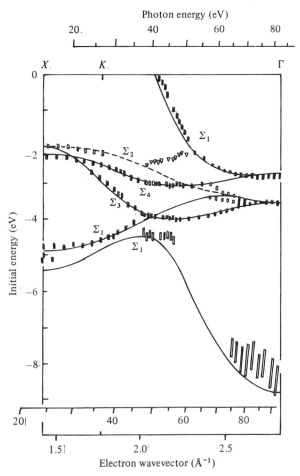

same transition is identified on two different surfaces of the same material. Fig. 3.55 illustrates the basic idea of this method. The angular distribution of the photoemission is investigated in the same three-dimensional plane from two different surfaces; if the same transition can be identified on each surface, the two values of $\mathbf{k}_{\parallel}$ found relative to the two surfaces uniquely define the value of $\mathbf{k}$ itself. Thus, both initial and final state energy bands may be plotted. This triangulation method relies on the ability to identify the same transition on the two surfaces which can only be easily achieved by assuming that peaks occurring at the same energy correspond to the same transition. This 'energy coincidence' condition is obviously not necessarily unique but, in practice, with checks to look for systematic behaviour in the same range of $\mathbf{k}$, can be rather effective. It provides a method of plotting bands which is far less dependent on simplifying assumptions and has recently been used with some success in another study of Cu (Pessa, Lindroos, Asonem & Smith, 1982).

One final aspect of the three-step model which we have so far neglected in our discussion is the role of the photoelectron transport to the surface (step two). It is this step, via inelastic electron scattering, which provides a degree of surface specificity to photoemission. One

Fig. 3.55. Schematic k-space diagram illustrating the method of initial state **k** determination by triangulation using ARUPS spectra taken at different polar angles θ from two different faces of the same material. The case illustrated corresponds to the use of {111} and {110} faces of an f.c.c. solid. Spectral features at the same initial state energy are identified in ARUPS data at angles θ_A and θ_B on the two faces; the associated **k** for this initial state is shown.

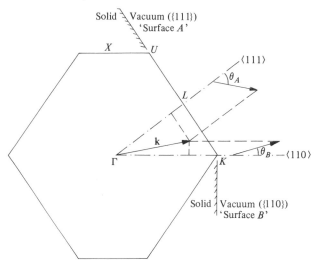

consequence of this for the study of 'bulk' band structure is that the loss of periodicity perpendicular to the surface leads to some relaxation of the strict reduced **k** conservation or direct transition rule for the photo-emission process. Some $\Delta k_\perp$ deviation must be permitted which increases as the surface specificity increases and the inelastic scattering mean-free-path decreases. One rather direct consequence of this is the phenomenon of 'band gap emission'; i.e. that photoemission may be observed at final state energies and **k** values corresponding to a band gap. Of course, these emissions are usually seen with lower intensity than adjacent 'allowed' direct transitions. In the terminology of the 'band structure approach' to LEED theory these transitions are to states with a complex **k**; an imaginary component of **k** leads to attenuation of the wave.

In order to interpret the intensities of photoemission features and to distinguish between 'band gap emission' and **k** conserving transitions we must pay some attention to the relative intensities of different **k** conserving transitions themselves. So far we have assumed that all such transitions are equally probable so that photoemission peak intensities should simply reflect the joint density of states in the relevant parts of the initial and final state bands. In reality this must be multiplied by the relevant matrix element for the transition. The photoemission matrix elements have already been discussed in section 3.2.4 in relation to XPS and we recall that it is usually reduced to a form $\langle f | \mathbf{A} \cdot \mathbf{p} | i \rangle$ with **A** the incident light vector potential and **p** the electron momentum operator. Strictly we should note that in using this form we have neglected the term $\mathbf{p} \cdot \mathbf{A}$ in the interaction Hamiltonian (see equation (3.10)) by assuming that the spatial dependence of the vector potential is not important and so the term vanishes. In fact it is generally believed that the vector potential *does* vary in the surface region but that it becomes unimportant if the photon energy greatly exceeds the plasmon energy of the solid surface. With typical plasmon energies falling in the range 10–20 eV it is clear that this approximation is therefore valid at typical XPS energies, but *not* at UPS energies. Some experimental and theoretical work has been devoted to investigations of the effect of the $\mathbf{p} \cdot \mathbf{A}$ term which is referred to as the 'surface photoeffect' (e.g. Kliewer, 1978) but it does not appear to give rise to any sharp structure in the photoemission spectrum which might be confused with other processes of more practical interest. Neglecting this effect, therefore, from future discussion, we note that within the three-step model it is relatively straightforward to evaluate the anticipated matrix elements from transitions between different parts of initial and final state bands; the calculation of the band structure necessarily involves the computation of the Bloch function (plane wave) composition of each point in the band and so the transition probability can be computed by direct evaluation of the momentum matrix elements.

One important consequence which arises from consideration of the momentum matrix elements even without the need to evaluate them numerically is the emergence of symmetry selection rules which can cause certain transitions, possible in terms of energy and **k** conservation, to be forbidden on symmetry grounds. These relatively simple and totally general rules are relevant not only in the study of photoemission from extended Bloch or band states of a solid surface, but also from molecular orbital states of adsorbed molecules. These rules have the strongest and therefore mose useful effect if we consider photoemission into a direction which lies on a mirror plane of the surface under study. In this case we note that the matrix element as a whole must also be symmetric while the initial and final state wavefunctions themselves must be either symmetric or antisymmetric relative to this plane. Evidently if a wavefunction is antisymmetric then its amplitude *in the mirror plane* is zero so that transitions to such a final state will not be observed if photoelectrons are collected on this plane. Consider, now, the symmetry of the remaining components of the matrix element. If the whole is to be symmetric but the final state is antisymmetric the product of the remaining terms must also be antisymmetric. Thus we see that transitions from a symmetric initial state will have zero intensity on the mirror plane if excited with an antisymmetric electromagnetic wave (i.e. with the **A** vector perpendicular to the mirror plane), while zero intensity will be observed for emission from an antisymmetric initial state when the **A** vector lies *in* the mirror plane.

These rules highlight the value of having a polarised light source in performing angle-resolved photoemission studies of surfaces. One commonly used source which is intrinsically plane-polarised is synchrotron radiation. It is also possible, however, to produce a significant degree of polarisation in the light from a standard HeI (21.2 eV) discharge source by successive grazing reflections from gold-coated surfaces. Fig. 3.56 shows schematically how an arrangement using three reflections, one having twice the grazing angle of the others, leads to an emergent beam coaxial with the incident beam. Rotation of the whole device rotates the polarisation vector. Using grazing angles of 15°, 30° and 15° in such a device at HeI energy leads to about 90% polarisation although only about 10% of the incident light emerges.

So far our discussion of the 'band structure' aspect of angle-resolved photoemission has concentrated on the 'bulk' band structure and invoked the limited penetration of the technique, due to inelastic scattering of the photoelectrons, only to rationalise the weakening of $k_\perp$ conservation in the real experiment. If this were the only way in which the surface entered the process we would be left with a valuable method of investigating bulk band structure (which is true) but with no information on the surface itself. In reality this cannot be so; if we use a technique which intrinsically restricts its sampling to electronic states in the top few atom layers, we must sample the electronic structure of this region and not of the bulk solid. However, calculations of the band structure of two-dimensional 'slabs' consisting of a single atom layer, or several atom layers which are two-dimensionally periodic, reveal two features. The first is that the electronic structure within an atomic layer only two or three layers from the surface is essentially identical to that in a layer inside an infinite three-dimensional solid. As a result ARUPS can provide valuable band structure information on the 'bulk' even though it is somewhat surface specific. In addition, however, these same calculations show that the top atomic layer or so possesses additional localised 'surface states'. These states are two-dimensional Bloch states but are damped or evanescent away from the surface. Because these states are localised in the surface, photoemission from them can be seen rather easily and it is this aspect of ARUPS from clean solid surfaces which provides the useful 'surface' information content. Recognition of these surface states in ARUPS spectra, in the absence of theoretical calculations, is usually obtained by the essential lack of dispersion of the states with $k_\perp$ due to their one-dimensional localisation (as for the states in layer compounds mentioned earlier), and by a commonly observed enhanced sensitivity to contamination. Evidently adsorbed species on the surface will generally have little effect on 'bulk states', but are likely to influence the surface electronic structure and so quench photoemission from these surface states of the clean surface.

An example of ARUPS spectra including emission from a surface state is shown in fig. 3.57(*a*). These spectra, collected at normal emission from a Cu{111} surface were taken using a range of different photon energies using incident light of so-called p-polarisation; i.e. with the **A** vector

Fig. 3.56. Reflection polariser based on two reflections involving a grazing angle θ and one of 2θ leading to coaxial incident and emergent beams. The degree of polarisation and total throughput depends on the photon energy, the angle θ and the coating of the reflecting surfaces.

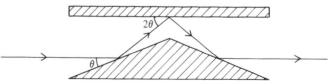

containing a component perpendicular to the surface (the alternative s-polarisation has the **A** vector parallel to the surface). Also shown in fig. 3.57 are the bulk Cu band structure in the ΓL **k**-space direction (equivalent to a $\langle 111 \rangle$ direction in real space) and a computation of the photoemission spectra to be expected from this band structure when the momentum matrix elements are included. The experimental spectra show three major features. One peak disperses strongly with photon energy from an initial state energy about 1 eV below the Fermi energy,

Fig. 3.57. (a) Normal emission ARUPS data at various photon energies for Cu{111} (Knapp, Himpsel & Eastman, 1979). (b) Calculated band structure of Cu in the ΓL direction. (c) Computed photoemission spectra using this band structure and the three-step model but including momentum matrix elements (Smith, Benbow & Hurych, 1980).

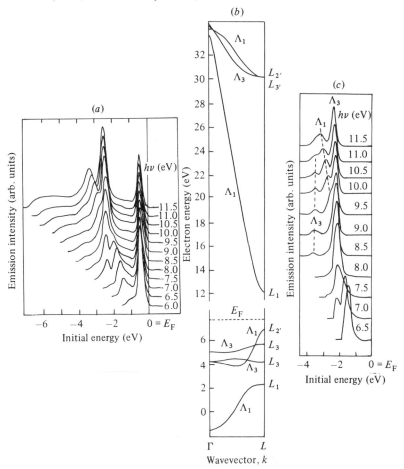

E_F ($hv \approx 6$ eV) to > 3 eV below E_F ($hv = 11.5$ eV) and is associated with emission from the initial state s–p-band labelled Λ_1. A second feature showing weak dispersion with a binding energy of ~ 2 eV is from the Λ_3 d-band of Cu. Notice that in the three-step model all emission is to the Λ_1 unoccupied final state band. A third feature which is strongly in evidence in the experimental data at a binding energy of ~ 0.4 eV is not reproduced in these theoretical calculations and clearly corresponds to emission from a state lying in a band gap of the bulk bank structure. This state is a surface state and has been found to be present in calculated surface band structures.

While the three-step model of photoemission is evidently an excellent basis for understanding ARUPS data from clean solid surfaces and provides a clear link between these data and the surface band structure which may be highly relevant to an understanding of other surface electronic properties, it is evident that it is not totally satisfactory for completely and quantitatively understanding the experimental data. In particular, we have already remarked on the phenomenon of 'band gap emission' resulting from a relaxation of $k_\perp$ conservation. Calculations based on a 'one-step' model of the process do appear to be able to rectify this deficiency although the direct relationship with band structure is lost. In this formulation the relationship between a LEED scattering experiment and the final state electron wavefield in photoemission is recognised explicitly. In LEED one has a problem of matching a known incident wave and outgoing scattered waves outside the crystal to the excited, multiple scattering, damped wavefield inside the crystal. In photoemission the source is inside, but the matching problem is similar and the methods for handling the multiple scattering and damping due to inelastic scattering are essentially the same. Formally the problem in photoemission can be expressed in terms of time-reversed LEED states. One interesting physical phenomenon which emerges from this treatment is the role of the hole lifetime in photoemission. As has been discussed in the context of LEED (chapter 2), the inelastic electron–electron (and electron–plasmon) scattering in the excited final state can lead to energy broadening. This broadening results from the coherent interference aspect of LEED, and should not lead to the same effect in photoemission. However, in the case of photoemission the excitation process leads to a hole state left behind in the solid and if this has a short lifetime the resulting photoemission energy spectrum will be broadened. For shallow hole states (i.e. less than deep X-ray levels) we have already remarked earlier in this chapter that the dominant mode of decay is by an Auger process. For a hole in the metallic valence band it is clear that, in the absence of strong matrix element (selection rule) effects this rate will be determined by the number of occupied states lying at shallower energies. Photoemission from states close to the Fermi level should therefore lead to sharp structure in the energy spectrum because the multiplicity of possible decay routes is small and the lifetime large. By contrast the lifetime of holes produced deeper in the valence band should be shorter so that the photoelectron energy distribution should show structure which is progressively broader as the initial state binding energy decreases. This basic effect is rather clearly seen in comparison with photoelectron spectra taken from Ni and Cu. These two materials have very similar band structure except that the 3d-band is displaced down to deeper energy in Cu to about 2 eV below E_F, while in Ni this band is just cut by the Fermi level (see fig. 3.43). This means that the density of states just below E_F is far higher in Ni than in Cu so that the hole lifetime for a state a few eV below E_F is much shorter in Ni than in Cu. Experiments do indeed show that, while the d-band structure of both materials shows up sharply in UPS spectra, the somewhat deeper s–p-band which can be clearly seen in Cu is largely obscured by broadening in Ni.

We should finally remark under the general topic of band structure investigations by ARUPS, that essentially the same considerations apply to the study of ordered overlayers of adsorbed species on the surface. Particularly in the case of atomic adsorbates the bonding valence electrons combine with the substrate valence electrons to produce a new surface band structure which can be investigated in the same way as for clean surfaces. Indeed, the same is true of the substrate bonding orbitals of ordered overlayers of adsorbed molecules. The relatively undisturbed internal molecular orbital states, on the other hand, can provide information on the adsorption of a far more direct kind.

3.5.3   *UPS in the study of adsorbed molecules*

The potential power of UPS in the study of adsorbed molecules on surfaces can most effectively be illustrated by one or two examples. Fig. 3.58 shows UPS spectra, taken in an angle-integrated mode using incident, unpolarised, HeI ($hv = 21.2$ eV) radiation. Panel (d) of this figure shows the photoemission spectrum from gas phase benzene (C_6H_6) molecules and the energy scale is shown in terms of the ionisation potential of the relevant initial states relative to the vacuum level. We see that the spectrum consists of a series of bands which at high detector resolution are seen to be composed of many fine lines associated with excited vibrational states. Each band is associated with a particular initial electronic state or molecular orbital state within the molecule (or occasionally comprises an overlap of two such bands). In panel (*a*) of fig. 3.58 is shown a UPS spectrum from a clean Ni{111} surface, and dashed, a spectrum from the same surface after exposure to 2.4 L of benzene at 300 K. Panel (*b*) then shows a 'difference spectrum' obtained from the difference of these two spectra, while panel (*c*) shows a further difference spectrum obtained by subtracting a clean surface spectrum from that

Fig. 3.58. (*a*) UPS data using a photon energy of 21.2 eV for a clean Ni{111} surface and the same surface with a chemisorbed layer of benzene. (*b*) shows the difference spectrum for these data while (*c*) shows a difference spectrum after condensing a layer of benzene on the surface. (*d*) shows the gas phase UPS spectrum of benzene (Turner, Baker, Baker & Brundle, 1970) displaced by an arbitrary energy to align peaks (after Demuth & Eastman, 1974).

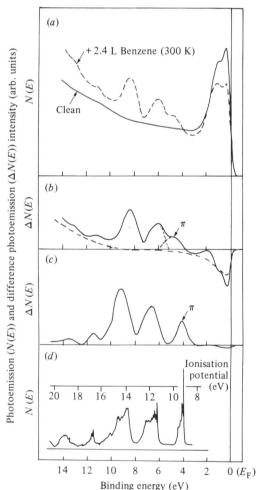

obtained when benzene was condensed on the surface at 150 K under a benzene pressure of 2×10^{-7} torr. Notice that, apart from a negative excursion corresponding to initial states just below the Fermi level (due to suppression of the Ni 3d-band emission), the difference spectra are dominated by the 'extra states' produced by the adsorbate. A comparison of the condensed benzene spectrum with that from the gas phase molecule shows that, apart from some broadening or smoothing, the relative positions of all the peaks are the same. In order to align these two spectra, as has been done in the figure, the ionisation energies of the gas phase spectrum have been shifted not only by the difference between Fermi level and vacuum level (the work function) but by an additional relaxation shift, $E_r = 1.4$ eV. The origin of this shift is the same as in XPS (see section 3.2.3); it is the additional intermolecular or solid state relaxation effect. The broadening relative to the gas phase spectrum is likely to result from a number of sources including instrumental effects (typically UPS in surface science is performed with instrumental resolutions of a few tenths of an eV) and shake-up effects discussed in the context of XPS (although the small available energy excludes most recognisable shake-up structure such as plasmons). In addition some weak overlap of orbitals may cause more conventional 'solid state broadening'. However, the striking feature of panels (c) and (d) is their similarity, indicating that the UPS 'fingerprint' spectrum of the molecule is preserved and, in fact, that the molecular orbitals are relatively undisturbed.

Comparison of the spectrum from condensed benzene and from chemisorbed benzene now shows two further changes. The first is a small additional relaxation shift associated with the more metallic environment, but we note that one band in the spectrum, that of lowest binding energy labelled π, is seen to shift not up in energy, but down, i.e. this one band no longer retains its relative position in the molecular fingerprint. This shift is associated with bonding to the substrate and so identifies the particular molecular orbital involved in the molecule–substrate bonding; i.e. the one orbital which is significantly disturbed relative to the free molecule by hybridisation with the surface.

In summary, therefore, we see that in this case the UPS fingerprint allows us both to identify the presence of benzene on the surface, and to identify the orbital which bonds it to the surface. Indeed, in this case at least, this identification also allows us to infer the orientation of the molecule on the surface. This is because the bonding orbital is a π-state in which the electrons are localised out of the plane of the molecule; if this orbital is the one hybridising with the surface it implies that the benzene molecule lies down on the surface. This basic distinction in localisation of σ- and π-orbital states is illustrated in the diagram of some of the benzene orbitals in fig. 3.59. Evidently benzene is not the simplest of examples but in general a σ-orbital is symmetric about the main plane (or axis) of the molecule, while π-orbitals are antisymmetric about this plane. This antisymmetry implies a node in the wavefunction in this plane and thus antinodes out of the plane. By contrast the σ-states have antinodes in the plane and so possess maximum electron densities in the plane. Bonding to σ-orbitals would therefore imply an end on configuration for benzene which would, of course, reduce the symmetry of the molecule and lead to complications due to loss of degeneracies. Note that, as seen from the energies in fig. 3.59, a second π-orbital exists whose calculated energy lies in the vicinity of the second band. Gas phase work suggests it may actually appear in the third band, but in any case possible disturbance of this π-component is masked by the σ-bands.

A further example of the value of UPS fingerprints of adsorbed molecules, taken from the same piece of work as fig. 3.58, is shown in fig. 3.60. This diagram shows only the difference spectra obtained on the same Ni{111} surface, firstly (a) on exposing to 1.2 L of acetylene (C_2H_2) at ~ 100 K, and secondly (b) after the clean surface has suffered a similar exposure to ethylene (C_2H_4) at ~ 100 K, while the lowest curve (c) shows the result of warming this latter surface above ~ 230 K. These data show

Fig. 3.59. Occupied molecular orbitals of benzene believed to contribute the spectra in fig. 3.58 (Jorgensen & Salem, 1973). The orbitals are identified by their full symmetry notation and are also labelled as σ or π. Computed gas phase ionisation energies are also shown as a guide to the approximate energies and ordering.

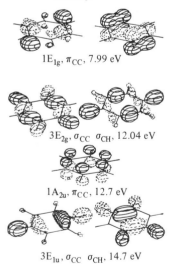
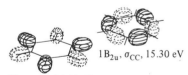

1E$_{1g}$, π$_{CC}$, 7.99 eV

1B$_{2u}$, σ$_{CC}$, 15.30 eV

2B$_{1u}$, σ$_{CH}$, 16.43 eV

3E$_{2g}$, σ$_{CC}$ σ$_{CH}$, 12.04 eV

1A$_{2u}$, π$_{CC}$, 12.7 eV

3A$_{1g}$, σ$_{CH}$, 18.22 eV

3E$_{1u}$, σ$_{CC}$ σ$_{CH}$, 14.7 eV

2E$_{2g}$, σ$_{CC}$ σ$_{CH}$, 21.14 eV

two features: firstly, the upper curves, by comparison with gas phase spectra, indicate that the adsorbed species are, indeed, acetylene and ethylene, both showing bonding shifts of the most weakly bound π-orbitals. The second point is that on warming the adsorbed ethylene to around 230 K, the spectrum changes to one typical of acetylene. Thus the fingerprinting capability of UPS is used to follow a primitive catalytic process of dehydrogenation in this simple system.

These two examples illustrate rather vividly the potential power of UPS as a method of studying the adsorption, dissociation and reaction of molecules on surfaces. Nevertheless, some limitations should also be stressed. In particular we note that using HeI incident radiation, as is most common, the range of kinetic energies of photoelectrons emitted

Fig. 3.60. Difference UPS spectra after adsorption on a Ni{111} surface of acetylene chemisorbed at ~ 100 K or ~ 300 K (*a*), for ethylene chemisorbed at ~ 100 K (*b*), and after warming this chemisorbed ethylene layer to ~ 230 K or by adsorbing ethylene at ~ 300 K (*c*) (after Demuth & Eastman, 1974).

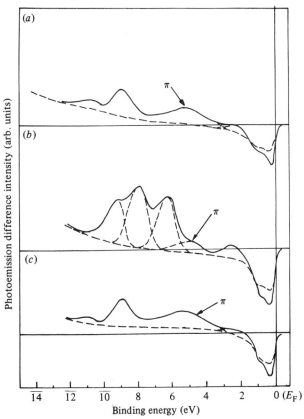

($hv - \phi$ where ϕ is the work function) is only ~ 16 eV, and, as there is a steeply rising background at low kinetic energies due to inelastic and secondary electrons, only about 10 eV of spectrum contains useful information. Furthermore, we note that molecular orbital photoemission peaks appear to have widths of typically ~ 1 eV. In such a situation, 'fingerprints' can only be useful in being recognisable and unique if the number of possible orbitals being involved is small, i.e. if the adsorbed molecules are small, or there are no more than about two different species, or both. A further difficulty which can occur, particularly for reactive species when it may be difficult to produce a physisorbed or condensed layer, is that, in comparisons of the gas phase spectrum and that from the chemisorbed species, ambiguities may occur in determining the two parameters of relaxation shift and bonding shift. Such ambiguities may lead to spurious assignment of the bonding orbital. An example of this problem occurred in early studies of CO adsorption on transition metals. In the gas phase spectrum of CO three occupied orbitals are accessible to UPS with HeI radiation; these are shown in fig. 3.61 together with their observed bonding energies. The shallowest orbital is the 5σ bonding orbital, derived from 2p states on the C and O, the next deepest is the 1π, also derived from 2p states, while the lowest lying level is the 2s-derived antibonding 4σ-orbital. Typical UPS spectra of adsorbed CO on a range of transition metal surfaces, however, shows only two bands (see fig. 3.62). Evidently either the 5σ suffers a bonding shift to overlap the 1π and form P_2 on the figure, or the 1π

Fig. 3.61. Occupied molecular orbitals of CO seen in UPS spectra (Jorgensen & Salem, 1973). Experimental gas phase ionisation potentials are also shown (Turner *et al.*, 1970).

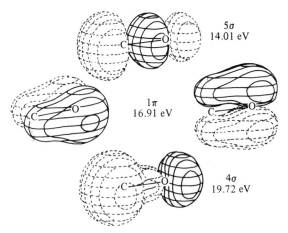

suffers a bonding shift to overlap the 4σ in P_1. While this latter possibility was originally the proposed solution, it is now recognised that it is the 5σ- and 1π-bands which overlap and that bonding is through hybridisation of the 5σ-orbital, the molecule standing up on the surface with its C end closest to the substrate. This assignment and the determination of the molecule orientation has proved possible through the application of ARUPS.

The application of ARUPS to the study of adsorbed molecules has been concentrated on two basic approaches; the use of point group symmetries and selection rules, and the comparison of measured and calculated angular distributions of emission from particular orbital bonds. Both approaches can be conveniently illustrated by reference to

Fig. 3.62. UPS data from CO adsorbed on a range of transition metal surfaces using HeII (40.8 eV) incident light. Note the presence of two adsorbate bands P_1 and P_2 at similar energies on all surfaces. These bands are similar to those seen in UPS from a metal carbonyl molecule. For comparison a low resolution gas phase spectrum (arbitrarily shifted) showing three bands is included (after Gustafsson & Plummer, 1978).

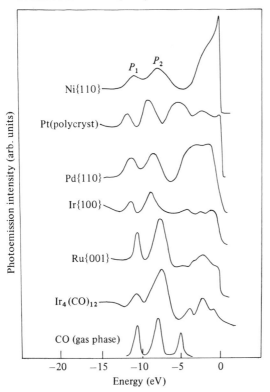

studies of CO adsorption on transition metal surfaces; this is the system most heavily studied so far and this is the one for which the methods are best developed.

The use of symmetry selection rules follows the approach already outlined for determining initial band symmetries in studies of photo-emission from a substrate band structure. We can apply similar methods to the CO molecule providing that the adsorption does not destroy its symmetry properties. Strictly, this requires that the principal axis of the molecule lies on a sufficiently high symmetry axis of the surface (along a surface normal) but in practice the degree of 'symmetry breaking' which has a consequence for the experiments depends on the degree of interaction of the molecule and the surface and the importance of electron scattering by the substrate. We shall return to this question later, but for the time being assume that we can consider the symmetry consequences by reference to an 'oriented molecule'; i.e. to a free CO molecule which has an orientation assumed to be fixed in space by the surface. We note that, as seen in fig. 3.61, the occupied molecular orbital states of CO, which has perfect axial symmetry (point group C_∞) have either σ (symmetric) or π (antisymmetric) symmetry relative to the molecular axis. Consider the geometry of fig. 3.63 in which we define z as along the molecular axis, assume the plane polarised incident light is in the zy plane characterised by an angle between the **A** vector and the z axis of θ_A and consider photoemission into a direction defined by the wave vector **k** and characterised by polar and azimuthal angles θ and ϕ.

Fig. 3.63. Schematic diagram of photoemission from a CO molecule oriented along the z axis and defining angles used in the text.

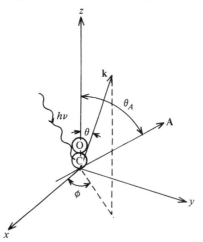

Turning first to σ-states, we see that these are symmetric under reflection in any plane containing the z axis while if $\theta_A = 90°$ the dipole component of the matrix element is antisymmetric about the xz plane. In this plane, therefore, the matrix element into symmetric final states is zero so that no σ emission is seen in this mirror plane, i.e. for any value of θ if $\phi = 0°$. If we now consider π-states we must restrict our simple considerations to emission along the z axis ($\theta = 0°$). Now we see that if we chose $\theta_A = 0°$, the dipole part of the matrix element is symmetric with regard to the emission direction, while the π initial state is antisymmetric so that no emission is seen from such an initial state in this direction. Evidently these considerations provide a simple method of distinguishing emission from initial σ- and π-states for this simple axial molecule.

These general arguments, involving no computation, allow one to predict only those matrix elements which are identically zero and do not provide a basis for quantitative comparison of theoretical and experimental angular dependences of photoemission from particular initial state bands. With the aid of suitable computation, however, these data do provide a basis for determining the orientation of molecules on surfaces with some precision. We have already seen, in the section on XPS, that angle-resolved photoemission from core states of adsorbed molecules can provide a basis for the determination of the local adsorbate–substrate structure (section 3.2.6.1). In that case the mechanism producing the angular effects of interest was the coherent interference of the directly emitted wave with components elastically scattered from the surrounding substrate atoms. The effect was therefore entirely associated with final state scattering in the substrate and the use of emission from (localised) core states ensured that this was the only source of structural information. In our present discussion we will concentrate on *initial state effects* and actually neglect those final state scattering processes involving the substrate. Initial state effects give rise to quite strong angular variations in the emission and when computed in this way do appear to interpret successfully data from adsorbed CO. Nevertheless, it is clear that the neglect of substrate effects in the final state scattering is not strictly justified; it is possible that the effect is small compared with initial state processes for adsorbed molecules, but the apparent success for CO which we will describe does not necessarily mean that the problem can always be neglected.

Returning to the 'oriented molecule' approach, therefore, we should remark that the initial state angular effects arise from the spatial distribution of the initial states; in effect photoemission from bonding orbitals involves *coherent* emission from several atomic centres within the molecule so that interference between them leads to angular distributions which reflect the relative locations of these centres in space relative to the emitted electron wavelengths. Of course, the relative phase and amplitudes of these coherent sources in the description depends on the details of the orbitals themselves.

For a cylindrically symmetric molecule such as CO, Davenport (1976, 1978) has shown that the angular dependence of the emission from any state, equal to the matrix element squared, may be written as

$$\frac{\mathrm{d}\sigma}{\mathrm{d}\Omega} = A(\theta) \cos^2 \theta_A + [B(\theta) + C(\theta) \cos 2\phi] \sin^2 \theta_A$$
$$+ D(\theta) \sin \phi \sin \theta_A \cos \theta_A \qquad (3.30)$$

where the functions of θ, A, B, C and D must be determined by calculation. As the final states for such a molecule, like the initial states, must have either σ or π symmetry, these three terms in equation (3.30) arise from the square of the amplitudes of the emission into these two channels independently, while the third term is the interference term between the two channels. Much the same structure is seen in the emission intensity from a core level of angular momentum l into the two final state channels of angular momentum $l-1$ and $l+1$. Some rather simple remarks can be made about the values of the coefficients A to D on the basis of symmetry. In particular, we note that for $\theta = 0$ (emission along the axis) there can be no azimuthal (ϕ) dependence, so $C(0) = D(0) = 0$. Also, using the earlier discussion, if the initial state has π symmetry, then with $\theta_A = 0$ the matrix element vanishes along the axis so $A(0) = 0$. Finally, for an initial σ-state with $\theta_A = 90°$, the matrix element vanishes in the $\phi = 0$ plane so $B(\theta) = -C(\theta)$; i.e. for an initial σ-state

$$\frac{\mathrm{d}\sigma}{\mathrm{d}\Omega} = A(\theta) \cos^2 \theta_A + E(\theta) \sin^2 \phi \sin^2 \theta_A$$
$$+ D(\theta) \sin \phi \sin \theta_A \cos \theta_A \qquad (3.31)$$

The results of some numerical calculations for CO are shown in fig. 3.64. These calculations included multiple scattering in the final state *within the molecule*, but take no account of the surface. The photon energy was taken to be 41 eV (i.e. approximately the HeII resonance line) and calculations are shown for the 4σ- and 1π-levels and for several different polarisations. The unpolarised summation used for fig. 3.64(c) takes no account of the change in the direction of the **A** vector due to optical refraction at the surface. The data of fig. 3.64 show several features. Firstly, it is clear that the angular emission patterns for the two initial states are very different and could clearly be distinguished by angular photoemission measurements if the calculations are sufficiently reliable. Panels (*a*) and (*d*), corresponding to an **A** vector along the *z* axis, have an appealing similarity to the actual σ and π wavefunctions themselves, the σ emission being concentrated along the molecular axis while the π emission is a minimum in this direction. This similarity cannot be generally true, however, and different photon energies, corresponding to different final state energies, can lead to quite different patterns. A second feature of these data is that both panels relating to 45° incidence of the light lead to angular distributions significantly skewed towards the y axis. In these cases the incident **A** vector breaks the symmetry of the photoemission process from the highly symmetric molecule. The main skew is towards the direction of the **A** vector but this is not a general result – skew away from the **A** vector is also possible as in the small lobe in panel (*c*). Finally, it is clear from the emission patterns from the 4σ-

Fig. 3.64. Calculated differential photoionisation cross-section for photo-emission from a CO molecule aligned along the z axis with the C end down. (*a*)–(*c*) correspond to emission from the 4σ-state at a photon energy of 41 eV, (*d*)–(*f*) correspond to emission from the 1π-state at a photon energy of 21 eV. (*a*) shows the result of an **A** vector along the (molecular) z axis, (*b*) shows the result of an equal (random) mixture of x and y polarisations and (*c*) shows the effect of 45° incidence of unpolarised light in the yz plane. For the 1π-state (*d*) has the **A** vector along the z axis, (*e*) has this vector along the y axis and (*f*) again shows the result of 45° incidence but with polarised light and now takes account of the reflectivity of a Ni substrate in determining the z and y components of the **A** vector (after Plummer, 1977).

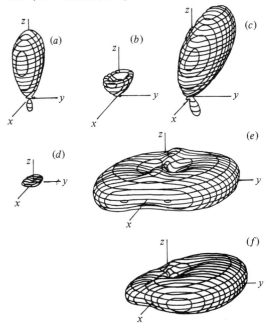

level at this energy, that experimental studies of this angular distribution should provide information of a reasonably precise nature on the molecular orientation on the surface.

While this effect has been investigated, the method is not without its difficulty. A common restriction of angle-resolving UPS spectrometers lies not only in the use of unpolarised light but in the fact that the electron energy analyser can only be moved in the plane of incidence. As may be seen in fig. 3.64(c), 45° incidence relative to the CO molecular axis leads to a polar angular dependence in this (yz) plane which peaks not along the molecular (z) axis, but about 20° away from it. Calculations show that the exact location of this broad peak is rather insensitive to the angle between the CO axis and the incident light direction; for 45° incidence on a surface, CO axis tilts of up to 20° in the yz plane lead to only small changes in the location of this peak relative to the surface normal. This problem is eased, however, if measurements are made in the xz plane; i.e. perpendicular to the plane of incidence. As may be seen in the results of the theoretical calculations, the incident light **A** vector does not break the symmetry in this direction and the observed polar angular dependence peaks along the surface normal if the molecule is perpendicular to the surface. Of course, if the molecule is tilted to the surface normal, the component of the incident **A** vector perpendicular *to the surface* is no longer along the molecular (z) axis, but this 'symmetry breaking' is precisely the one carrying the information of interest, namely the angle between the surface normal and the molecular axis.

Research indicates that such studies of the polar angular dependence of the CO 4σ photoemission from unpolarised HeII incident light with collection in the plane perpendicular to the incidence plane is capable of identifying tilt in the adsorbed CO species. Fig. 3.65 shows a schematic diagram of the structure proposed for the system studied which is a (2×1) adsorption structure of CO on Pt$\{110\}$. The tilt angle of 26° shown in this diagram is the value derived from the ARUPS experiment. Notice that the adsorption structure involves a monolayer coverage of CO and would be (1×1) but for the zig-zag tilt of the molecules proposed along the $\langle 110 \rangle$ direction rows. Indeed, this introduces glide symmetry lines into the structure which leads to characteristic missing LEED beams and identification of the space group as $p1g1$ (see chapter 2). It is the presence of this glide symmetry which led to the suggestion of tilted CO in this structure. The essentially 'atop' or one-fold coordination of the CO to the top layer Pt atoms used in the diagram is deduced not from the ARUPS, but from vibrational spectroscopy data (see chapter 9). Fig. 3.66 shows the results of the ARUPS experiments in the form of 4σ emission polar angle dependence for photon incident planes along the azimuths shown and collection perpendicular to that plane. The results of calculations, assuming that the CO molecules are tilted in the ⟨211⟩ azimuths and that all symmetrically equivalent directions are equally probable (leading to equally populated domains) are shown in fig. 3.67 for the two principal azimuths and for a range of tilt angles. These calculations show that tilt of ~20° should be readily discerned in the data and, indeed, the essential correspondence of the experimental data and the calculations for 25° tilt support the 26° tilt conclusion arrived at on the basis of a more quantitative comparison. We should perhaps remark that the same experiments (Hofmann *et al.*, 1982) also studied CO adsorbed on Pt{111} and found *no* evidence of tilt in agreement with

Fig. 3.65. Plan and sectional view of the structure proposed for the Pt{110} (2 × 1)*p*1*g*1–CO phase by Hofmann, Bare, Richardson & King (1982).

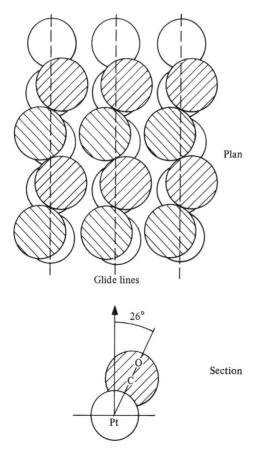

Plan

Glide lines

26°

O
C

Pt

Section

studies by other techniques. In this case the measured polar angle dependence peaked along the surface normal.

Despite our earlier reservations about the possible effects of the substrate other than a simple orientation of the molecule, these data are encouraging. Of course, the 4σ-level of CO is the most suitable for study in this molecule in that its emission suffers no energy overlap with other bands (as do the 1π and 5σ) and is not involved in bonding to the surface (as is the 5σ). It should therefore be easily identified and suffer little interaction with the surface. The potential role of the substrate in final state scattering, however, remains. Indeed, cluster calculations for the hypothetical linear molecule NiCO rather than CO show that the 5σ- and 1π-orbital photoemission angular dependences are significantly

Fig. 3.66. Polar angular dependence of experimentally observed ARUPS 4σ peak using 40.8 eV radiation in three different azimuths from the Pt{110} $(2 \times 1)p1g1$–CO structure. Different symbols correspond to different experimental runs (after Hofmann *et al.*, 1982).

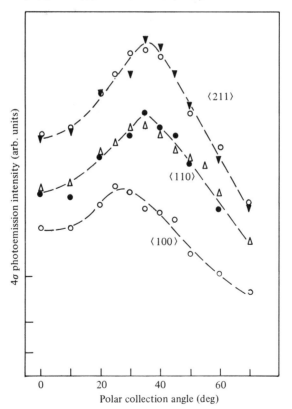

altered, although the qualitative form of the data for the 4σ-state is relatively unchanged.

One further approach to the use of UPS in the study of molecular orientation on surfaces which has also been applied to the adsorption of CO is the study of so-called 'shape resonances'. If one studies the photoionisation cross-section of the CO 4σ-level as a function of photon energy in the range 20–40 eV a rather prominent peak is seen (around 32 eV for gas phase CO) which is attributed to a final state scattering 'resonance' within the molecule itself. Detailed calculations by Davenport (1976) have revealed that this effect, which is also found for the rather similar molecules N_2 and NO, is specifically associated with transitions from σ initial states to a σ final state. While detailed calculations of the exact energy at which this effect occurs are susceptible, as are the angular distributions, to proper inclusion of the role of the substrate, this symmetry result should be general. This means that not only should the existence of the resonance in the correct range of final state energies label the associated state as of σ symmetry, but also it should only be observed when there is a component of the incident **A** vector along the molecular axis to couple the σ initial and final states. The resonance should therefore not be observed when **A** is perpendicular to the molecular axis. This measurement, which necessitates synchrotron radiation (polarised and tunable), has been applied to

Fig. 3.67. Computed polar angle dependence of the CO 4σ photoemission in two different azimuths for the Pt{110}(2 × 1)p1g1–CO structure assuming various tilt angles in the ⟨211⟩ azimuths. These results of Hofmann *et al.* (1982) are based on the calculations of Davenport (1976, 1978).

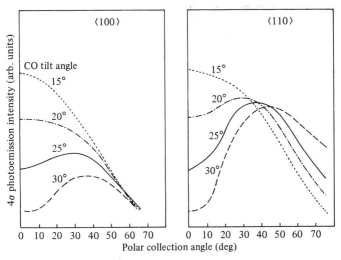

confirm the vertical orientation of CO on Ni{100} (Allyn, Gustafsson & Plummer, 1977) for example.

Further reading

A number of reviews of specific areas of electron spectroscopy have already been mentioned in the individual sections of this chapter. A few collections of papers and reviews are particularly relevant, however, and may be highlighted. *Electron Spectroscopy for Surface Analysis*, edited by Ibach (1977a), contains useful reviews of electron spectrometer design, core level spectroscopies and photoemission. Two excellent collections of reviews on photoemission (both XPS and UPS) are edited by Feuerbacher, Fitton & Willis (1978) and Cardona & Ley (1978).

4   Incident ion techniques

4.1   Introduction

At low incident kinetic energies (at most a few tens of eV) the interaction of incident ions with a surface is dominated by charge transfer to neutralise the ion. This produces electron emission characteristic of the electronic structure of the surface, and therefore forms a valence level spectroscopy known as INS. This will be discussed in detail in the next section.

By contrast, a number of techniques in surface studies utilise the kinetic energy transfer of more energetic incident ions to provide information on the surface. Most of these techniques use incident inert gas ions He^+, Ne^+ or Ar^+ in the energy range from a few hundred eV to a few keV although some use is also made of far more energetic ($\gtrsim 1$ MeV) He^+ and H^+ ions. While these incident ions may also suffer charge transfer at the surface, and can produce electronic excitations both in the form of core level ionisation and plasmon excitation, most techniques concentrate on the kinetic energy transfers between the incident ion and the atoms which comprise the surface. Despite the fact that these atoms are bound to a solid the kinetics of the initial primary ion–surface atom collision are almost exactly described by a simple free atom two-body collision. The duration of the collision is short, the interaction energy large and the local binding forces small. It is therefore easy to demonstrate, simply on the basis of energy and momentum conservation, that if an incident ion of energy E_0 and mass M_1 strikes a surface atom of mass M_2 and is therefore scattered through an angle θ_1 (in the laboratory frame of reference, see fig. 4.1), then the scattered ion has an energy, E_1, given by

$$\frac{E_1}{E_0} = \frac{1}{(1+A)^2} [\cos \theta_1 \pm (A^2 - \sin^2 \theta_1)^{\frac{1}{2}}]^2 \tag{4.1}$$

where $A = M_2/M_1$ and the positive sign in the formula is for $A > 1$, both signs for $A < 1$. Similarly the atom which is struck gains energy and assuming it was initially at rest it recoils with an energy E_2 at an angle θ_2 relative to the incident ion trajectory such that

$$\frac{E_2}{E_0} = \frac{4A}{(1+A)^2} \cos^2 \theta_2 \tag{4.2}$$

Energy conservation requires that $E_0 = E_1 + E_2$ and so provides a unique relationship between θ_1 and θ_2; the specific values of these depend on the exact incident ion trajectory and will be discussed in section 4.3. We note, however, that the scattered ions have an energy which, for a specific emergent (scattered) angle is defined only by the ratio of the masses of scatterer and scattered particle. Studies of the scattered ions and their energies therefore provide a potential means of composition analysis which will be discussed in the section 4.3.

Unfortunately the recoiling surface atom will usually recoil *into* the surface and so cannot provide this same simple information. On the other hand, the recoiling particle, together with the scattered ion if it is scattered into the surface, causes a collision cascade in the surface region which can evidently produce damage to the order of the surface but may also lead to fragments (individual atoms or clusters of atoms) leaving the surface. This 'sputtering' is valuable in two ways; firstly the sputtered fragments may be mass analysed to provide compositional information on the surface, and secondly the sputtering leads to erosion of the surface which may be used to clean the surface or to 'peel off' successive layers of the solid to reveal the subsurface layers. The analysis of the sputtered fragments is usually made on the charged fragments only and leads to the technique of SIMS described in section 4.5. Before this, however, the sputtering process is described briefly in section 4.4 as it is not only the primary process in SIMS but is also important in 'depth profiling', a method widely used in AES, XPS, SIMS and other techniques enabling these surface probes to be used to study subsurface composition with quite fine depth resolution.

4.2   Charge exchange between ions and surfaces

All of the techniques described in this chapter concern the incidence of ion species on a surface, and in most cases involve the detection of ions emerging from the surface. These ions are almost never

Fig. 4.1. Schematic representation of the scattering of an ion from a surface atom.

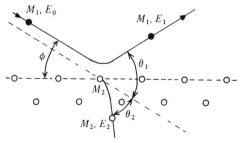

in charge equilibrium with the surface and the mechanisms of charge exchange between ion and surface, and their efficiency, can influence the results of all the techniques in a significant way. In the remainder of this section we will discuss spectroscopies which make explicit use of this charge exchange to investigate the electronic structure of surfaces. First, however, we will outline the possible mechanisms of charge exchange in general, and highlight some features of these mechanisms which will be recalled in the discussion of specific techniques.

Fig. 4.2 illustrates schematically the four processes with which we will be concerned. In each energy level diagram a metallic surface is shown on the left with a conduction band filled to the Fermi level, while the localised well of a closely approaching ion is shown on the right. Fig. 4.2(*a*) illustrates the processes of resonant charge exchange. In this case the hole state on the ion forms a broadened energy level (see below) which straddles the Fermi level of the surface. An electron in the metal surface can therefore tunnel across to the ion without energy change leading to *resonance neutralisation*. If the relevant state on the ion does straddle the Fermi level, however, it is also possible for an electron occupying this state on the incident species, above the Fermi level of the surface, to tunnel back across to the metal leading to *resonance ionisation*. The broadening which permits this two-way exchange occurs, at least in the static case, because as the ion and surface approach, their valence levels overlap and hybridise forming a 'surface molecule'. Given sufficient time, and a broadened level which does overlap the Fermi level, an equilibrium would result between ionisation and neutralisation in which the incident species would have a well-defined fractional average

Fig. 4.2. Schematic diagram showing resonant (*a*) and quasi-resonant (*d*) charge exchange, Auger neutralisation (*b*) and Auger de-excitation (*c*) as discussed in the text. In (*a*) 1 shows the direction of electron transfer for neutralisation, 2 for ionisation.

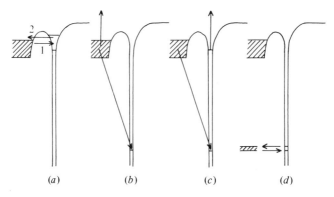

(*a*)   (*b*)   (*c*)   (*d*)

charge. If these species were then removed instantaneously from the surface, their fractional charge would appear as a well-defined fraction of ionic and neutral species. Incident ions likely to undergo these effects must have their ionisation potentials close to the work function of the metal; alkali metal ions typically satisfy this criterion. Of course, other ions, having ionisation potentials which are in the middle of the surface conduction band, can also undergo resonance neutralisation, but in this case resonance ionisation is very improbable and the process is essentially a one-way exchange. Fig. 4.2(*d*) shows a related process in which a deeper lying occupied state of the surface lies energetically close to the hole state of the ion. In this case the ultimate charge equilibrium, following resonance neutralisation, would involve filling the deeper lying hole with an electron from the conduction band, but in the meantime two-way resonant exchange can occur. This situation can lead to interesting and rather spectacular effects when the ion interacts only briefly with the surface as in ion scattering experiments with ions of ~ 1 keV energy. The process is known as quasi-resonant exchange and is discussed further in section 4.3.1. Notice that in the case shown in fig. 4.2(*d*), the hole state on the ion lies energetically far below the conduction band of the surface so that the equilibrium state, if there is time for it to be achieved, would clearly make the incident particle neutral.

The more usual case for an incident ion of this type, when the surface does not have a close-lying occupied state, is illustrated in fig. 4.2(*b*). In this case an electron from the conduction band of the surface tunnels into the well of the ion but then falls down to the deeper lying hole, giving up its excess energy to another conduction band electron which is emitted from the solid and can be collected outside. This Auger electron carries information on the density of states of the surface and its detection forms the basis of INS which is believed to be dominated by this process of Auger neutralisation. The final, rather special, process illustrated in fig. 4.2(*c*) is of Auger de-excitation in which the incident particle is not an ion but an excited neutral atom. The process involves the filling of the deep-lying hole by a conduction band electron and the ejection of the electron bound in the excited state of the atom to carry away the excess energy. Its relevance stems from the energy level diagrams of He and Ne in particular, which have metastable excited states lying energetically close to the Fermi level of typical metals. It is therefore possible that an incident He$^+$ ion, for example, could be resonance neutralised as in fig. 4.2(*a*) to produce an excited He* atom, which might then de-excite to the ground state by this process. Similarly, if He* is brought up to the surface deliberately, this process competes with the de-excitation mechanism involving resonance *ionisation* followed by Auger neutralisation. So far, the evidence of ion scattering experiments and INS involving both incident He$^+$ and He* is that Auger de-excitation is only an important process in the case of incident He* and then usually only with adsorbate covered surfaces.

One further general remark about ion–surface charge exchange concerns the *rate* of this process. We have already mentioned that one view of the level broadening on the incident species is that it relates to the formation of a surface molecule. More generally, this energy broadening, Γ, can be related to the charge exchange transition rate, R_n, by the uncertainty principle

$$\Gamma = \hbar R_n \qquad (4.3)$$

Particularly for large ion–surface separations, the broadening and transition rate will be governed by the overlap of the wavefunction tails which can then be reasonably approximated by an exponential function in the distance from the surface s

$$R_n = A \exp(-as) \qquad (4.4)$$

where A and a are constants of the ion–surface system. One interesting consequence of this for INS, which utilises very low energy ions, is that given sufficient time much of the neutralisation can occur well outside the surface, leading to information which is highly surface specific; indeed, it is determined by the electronic structure just *outside* the surface. For more energetic ions, as in LEIS, the consequences of equation (4.4) will be developed in section 4.3.1. However, for energetic ions, we might also notice that exchange processes (such as resonant exchange) which might otherwise be energetically forbidden, can be permitted if the interaction time, Δt, is sufficiently short to lead to substantial uncertainty principle broadening. This possibility of special behaviour for fast ions rather than very slow ones is an example of so-called non-adiabatic effects in charge exchange. The special consequences of quasi-resonant exchange also fall into this category. A very crude guide to the interaction time can be obtained by noting that a 1 keV He$^+$ ion travels 2.2 Å in 10^{-15} s; and an interaction time of this duration, using the uncertainty principle as given in equation (4.3), leads to a broadening of ~ 1 eV.

4.2.1 *Ion Neutralisation Spectroscopy (INS)*

The neutralisation of positive ions at metal surfaces was first studied in the late 1920s, notably by Oliphant (1929), who observed that when positive ions of He$^+$ were incident on a Mo target, the ions were neutralised and electrons given off. Oliphant also observed that some of the neutral He atoms impinging on his target caused the emission of electrons. He rightly concluded that these were in fact excited or metastable He* atoms that were being de-excited at the Mo surface.

When a slow ion with a large neutralisation energy, or ionisation potential, is incident on a metal surface, the ion is neutralised by a two-electron Auger-type process (Hagstrum, 1954, 1961). This is the process illustrated in fig. 4.2(*b*), and in more detail in fig. 4.3. When the incoming ion is just outside the metal surface, two electrons in the filled valence band of the metal interact, exchanging energy and momentum. One electron, the neutralising electron (moving down in fig. 4.3), tunnels through the potential barrier into the potential well presented by the ion, and drops to the vacant atomic ground level which lies at an energy E_i below the vacuum level. The energy released in this transition is taken up by the second interacting electron which may now have sufficient energy to escape from the metal, if properly directed. This will happen if the component of momentum normal to the surface is sufficiently large. These Auger-type transitions can, of course, take place anywhere within the filled valence band so that the ejected electrons have a range of energies rather than one specific energy. Outside the metal surface the

Fig. 4.3. Energy level diagram for an ion just outside a metal surface. The electron transition pairs 1, 2 and 1^1, 2^1 illustrate the Auger-type transitions of the ion neutralisation process. E_i is the ionisation energy of the ion, ϕ metal work function, E_1, E_2 electron energies in the metal.

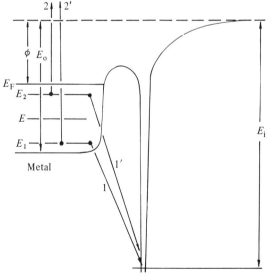

electron energy distribution can be measured quite straightforwardly. As we have already mentioned, the neutralisation of ions at a metal surface is a true surface process rather than one which occurs at some depth within the metal; the ejected electrons arise predominantly from just outside the metal surface or, at most, from the first layer of metal atoms (Heine, 1966; Wenass & Howsmon, 1968). For the slow-moving incident ions, the metal electrons involved in the neutralisation process originate from the region in which the metal electron wavefunction tail, outside the metal, overlaps with the wavefunction of the ion.

The presence of adsorbed atoms on the metal surface causes changes in the electronic states of the surface region; these changes have a profound effect on the ejected electron energy spectra. An example of the type of change observed is provided by fig. 4.4, which shows a number of ejected electron energy distributions taken for He$^+$ ions of 5 eV incident kinetic energy. These data were obtained for atomically clean Ni$\{100\}$, Cu$\{100\}$ and Ge$\{111\}$ (fig. 4.4(a)) and for the Ni$\{100\}$ surface covered with adsorbed layers of O, S and Se in turn (fig. 4.4(b)). The energy distributions show that the ion neutralisation process is sensitive to both the nature of the solid and the character of the chemisorbed material. It is from these energy distributions that spectroscopic information about the surface and surface adlayers is extracted. The means by which this end is achieved and the finer details of the neutralisation process will now be described.

The details of the electron ejection process which are set out in fig. 4.5, are expressed in terms of the escape probabilities and density of states in the metal. Referring back to fig. 4.3, it is fairly easy to work out the limits of the energy distribution, that is, the minimum and maximum energies of the ejected electrons; these limits can be determined by simple energy conservation principles. Thus the maximum energy, $E_{K_{max}}$, that the ejected electrons may attain, is given by

$$E_{K_{max}} = E_i - 2\phi \tag{4.5}$$

where E_i is the incident ion's effective ionisation potential and ϕ is the work function of the metal. The effective ionisation potential of an atom near a metal surface is less than the unperturbed value by an amount equal to the classical image potential of the ion outside the metal surface. The minimum energy $E_{K_{min}}$ of the ejected electrons is given by:

$$E_{K_{min}} = E_i - 2E_F - 2\phi \quad \text{if } E_i - 2E_F > 2\phi \tag{4.6}$$

or

$$E_{K_{min}} = 0 \quad \text{if } E_i - 2E_F < 2\phi \tag{4.7}$$

From these equations it is clear that we must expect the position of the ejected electron energy spectra to shift along the energy axis with

Fig. 4.4. (*a*) Electron kinetic energy distributions obtained using He⁺ ions of 5 eV kinetic energy incident on Cu, Ni and Ge surfaces. (*b*) Electron kinetic energy distributions obtained using 5 eV kinetic energy He⁺ ions on Ni{100} surfaces covered with O, S or Se (after Hagstrum, 1966).

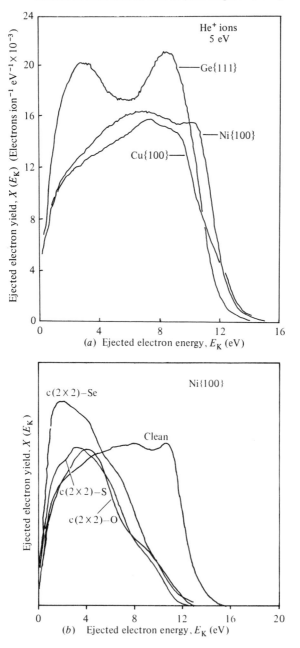

(*a*) Ejected electron energy, E_K (eV)

(*b*) Ejected electron energy, E_K (eV)

changes in work function due to chemisorbed material on the metal surface. More generally, if we arbitrarily assign energies E_1 and E_2 to the electrons participating in the neutralisation process, the ejected electrons will have energies given by

$$E_K = E_i - (E_1 + \phi) - (E_2 + \phi) \qquad (4.8)$$

In order to deal with the Auger neutralisation process, certain simplifying assumptions are made. First, that the transition probability for the ejected or neutralising electrons is constant, and independent of band energy, and also of the symmetry character of the valence band electrons. Secondly, that the transition probabilities for both the ejected and neutralising electrons are equal. Thirdly, that the final states density is constant, and energy broadening inherent in the transition is

Fig. 4.5. Energy level diagram showing functions of energy related to the ion neutralisation process. The functions U, F, P and X are here those appropriate to the $\{111\}$ face of Cu. S is the distance from the surface; S_t the distance at which electronic transitions occur; $E_{K_{max}}$ is maximum energy of ejected electrons.

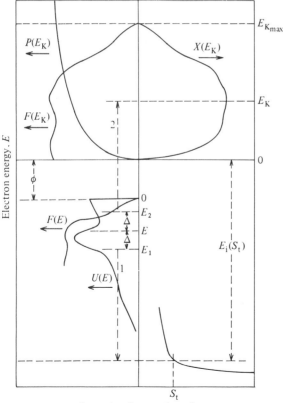

Separation from surface, S

neglected. Following these rather drastic assumptions, the probability that an electron in the range $E + dE$ will participate in the Auger process depends solely on the density of initial states in the valence band $N_v(E)$; this is equal to the function $U(E)$ in fig. 4.5.

It is the function $U(E)$ which is used to find the internal energy distribution $F(E)$, of the electrons excited in the ion neutralisation process. Ejected electrons lying in the energy range dE_K at an energy E_K may be obtained from any neutralisation process, for example, for which the initial states of the two participating electrons are symmetrically situated on either side of the energy E lying halfway between E_K and the ground state of the atom at $-E_i$. Values of E_K and E which will satisfy this condition satisfy the relation

$$E_K = E_i - 2(E + \phi) \tag{4.9}$$

Equation (4.9) is obtained from equation (4.8) by putting $E_1 = E_2 = E$. For initial states at $E_1 = E + \Delta E$ and $E_2 = E - \Delta E$ which are symmetric with respect to E, the probability of the specific neutralisation process involving these states must be proportional to $N_v(E + \Delta E) \times N_v(E - \Delta E)$; that is the product of the state densities at the initial energies. The total probability of producing an excited electron in the interval dE_K at E_K is the integral of the above product over the energy increment ΔE. Using the more general band function $U(E)$ we may write this probability as

$$F(E) = \int_{-E}^{+E} U(E + \Delta E) U(E - \Delta E) \, d\Delta E \tag{4.10}$$

The internal energy distribution function $F(E_K)$ is obtained from $F(E)$ merely by changing the energy variable in accordance with equation (4.9), followed by normalising above the Fermi level to an area equivalent to one electron per incident ion.

The function $F(E)$ is the pair density function for all electron pairs in the initial band which can produce an excited Auger electron at $E_K = E_i - 2(E + \phi)$. This pair density function is also often referred to as the self-convolution, fold, or convolution square of the function $U(E)$. Once the internal distribution of excited electrons $F(E_K)$ is known, the distribution of externally observed Auger electrons $X(E_K)$ can be obtained from the escape probability $P(E_K)$ for electrons crossing the surface barrier. This is done using the expression

$$X(E_K) = F(E_K) P(E_K) \tag{4.11}$$

which merely multiplies the internal energy distribution by the appropriate escape probability in order to arrive at the external distribution.

The total yield γ of all ejected electrons is the integral of $X(E_K)$ over all E_K. All of the functions described thus far are depicted in fig. 4.5 which shows, in fact, the functions which apply to the {111} surface of a Cu single crystal and are taken from the data of Hagstrum. A more detailed description, and a more thorough justification of the procedures and approximations used above can be found in the literature (Hagstrum, 1954).

From the experimental point of view, it is the function $X(E_K)$ which is obtained as the measured kinetic energy distribution. From $X(E_K)$ we must obtain the function $U(E)$. To do this we must endeavour to use incident ions of as low an energy as possible, since the energy spectra of higher energy ions is significantly broadened from that of an ideal, zero energy, ion. Generally the ideal, zero energy, ion can only be approximated to by extrapolating the spectra of two sets of different, low energy ions.

The procedure which has to be followed to obtain spectroscopic information from ion neutralisation data is as follows:

(1) The ejected electron energy distributions $X(E_K)$ are obtained at two different, low, incident ion energies.

(2) The distributions are then used to extrapolate to the distribution which would be obtained for ideal incident ions of zero energy.

(3) The resulting ideal energy distribution is then divided by the function $P(E_K)$ of fig. 4.5 to obtain $F(E_K)$. Note, $P(E_K)$ is a function of the solid used rather than of the incident ion.

(4) The function $F(E)$, obtained from $F(E_K)$, by change of variable, is unfolded or deconvoluted to produce $U(E)$, the transition density function.

When foreign atoms are adsorbed on the clean surface of a metal, changes ensue in the electronic states of the surface region. Notably, changes occur in both the density of states and wavefunction magnitude in the vicinity of the adsorbed atom. When the atom is adsorbed a large number of electronic states of the metal–atom system pass through the atom position. The discrete states of the free atom, corresponding to the lowest energy electron configuration, are replaced by a broad energy region in which the wavefunction magnitude at the atom is larger than it would be had it not been adsorbed on the metal surface. The situation is shown schematically in fig. 4.6 where the adsorbate resonance has a width ΔE_A at a band energy E_A. This is a virtual bound state because electrons in it are in quite intimate contact with the continuum of filled electronic states of the metal. The point of maximum wavefunction magnitude has been shifted by an amount D_A from the energy of the free atom ground state, which lies below the vacuum level by an amount equal to the free space ionisation potential. Several resonances can occur if electrons go into non-equivalent orbitals in the adsorbed atom.

The effect on the ion neutralisation process, of the virtual bound state of the adsorbed atom, for the case of an electronegative atom of relatively large free space ionisation potential, can be visualised as follows. The virtual bound state lies completely below the Fermi level of the metal whilst the probing ion, say He$^+$, presents a second potential well just outside the metal. At the adsorbed atom, the wavefunction magnitude is greater over the energy range of the virtual bound state, thus the magnitude of the wavefunction tail at the positive ion position will also be enhanced. The ion neutralisation process may therefore be expected to be more probable in the presence of an adsorbed atom. This does not necessarily imply that the overall yield of ejected electrons will be any greater.

Fig. 4.6. Electron energy diagram showing energy levels of a solid and adsorbate atom for two positions of the adsorbate atom (*a*) adsorbed, (*b*) desorbed. $\psi_A{}^2$ is the wavefunction magnitude for the surface electronic orbital of the adsorbed atom.

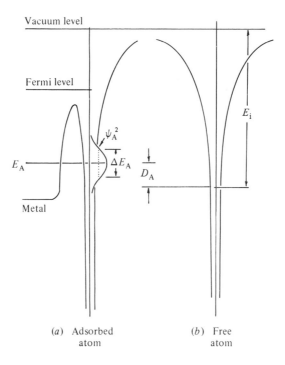

(*a*) Adsorbed
atom

(*b*) Free
atom

4.2.2　*INS with metastable intermediates*

INS can, in principle, be carried out using an alternative route whereby the ion is generated right at the metal surface from an incident, excited, neutral atom. An advantage of this approach is that the ion generated has merely thermal energy, typically $1/40$ eV at room temperature.

If an atom with an excited metastable level is incident on a metal surface it can be ionised by the excited electron tunnelling into the metal. Clearly this is only possible if the metastable level lies above the Fermi sea of the metal. In situations where there is no suitable vacant energy level in the metal, the metastable atom may interact directly with the surface in an Auger de-excitation process. This type of interaction is a one-electron process and yields the density of states directly, without the necessity for data deconvolution.

He and Ar both possess convenient metastable levels lying above the top of the Fermi sea, for most metals. In the case of He there are two metastable levels, a singlet level 2^1S_0 at 20.614 eV above the ground state and a triplet level 2^3S_1, at 19.818 eV above the ground state. For these metastable levels to be useful it is necessary that they possess a sufficiently long lifetime, a condition which is satisfied by both atoms. He metastables can be de-excited by collision or by radiative decay. Provided the pressures are correctly chosen, the metastable atoms will collide with the target metal surface, or the walls of the experimental chamber, more often than with other He atoms. The radiative lifetime for the singlet metastable state is of the order of 10^{-2} s for the allowed two-photon process. Decay by this mode yields radiation at 585 Å, which is itself capable of ejecting photoelectrons from the metal surface.

The lifetime for the triplet state is much larger (10^3 s) since the transition to the singlet ground state is not allowed. The first radiative state of He is the 2^1P_1 level at 21.2 eV. One would expect to produce resonant-photon radiation when raising the exciting potential above the 2^1P_1 threshold level. Consequently, operating conditions must be chosen to ensure that the metastable excitation cross-section is quite large relative to the higher excited states.

Fig. 4.7 shows schematically, in more detail than before, the ion formation and neutralisation process using the He metastable as an intermediate. With the exception of the first step, where the metastable atom is converted into an ion, the process is exactly as described in the preceding paragraphs for an incident low energy ion. It was pointed out previously that using a metastable intermediate produces at the metal surface an ion of thermal energy only, say, $1/40$ eV at room temperature. However, in reality this is not quite true since it neglects the effect of the classical image potential of the ion in the metal. The image potential V_{im} is given by

$$V_{im} = -\frac{e^2}{4x} = \frac{3.6}{x} \qquad (4.12)$$

the final form giving the potential in volts with the distance x separating the ion from the surface in angstrom units, e is the electronic charge. The magnitude of the image potential depends, of course, on the distance from the metal surface at which ionisation and neutralisation occur. The evidence suggests that the effect of the image potential is to give an ion of no more than 1 eV at the metal surface. Thus, by means of the metastable intermediate a beam of very low energy ions can be produced right at the metal surface. The effects of energy broadening are thus greatly minimised.

4.2.3   *Experimental arrangements for INS*

In order to use the technique of INS, the usual requirements of UHV apply. The incident beam of He$^+$ ions or He metastables does not

Fig. 4.7. Energy level diagram for ion formation and consequent neutralisation using incident metastable He atoms. The electron pairs 1 and 2 illustrate Auger de-excitation; the electron pair 3 is involved in the resonance ionisation and Auger neutralisation process. E_x is the metastable excitation energy.

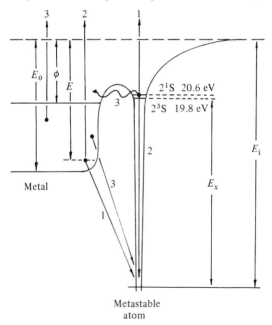

prejudice this requirement since, except at very low temperatures, they do not remain bound to the target surface. Nevertheless, it is necessary to ensure that the He source is free from impurities which would gradually build up on the surface. Further, it is necessary to have sufficient pumping speed to maintain the target chamber in the 10^{-9}–10^{-10} torr range, even when the He beam is incident.

The principle exponent of INS has been H. D. Hagstrum of Bell Laboratories and it is to one of his experimental arrangements that one must turn for an elegant example of the technique in practice. A schematic diagram of the type of apparatus used by Hagstrum is displayed in fig. 4.8. The apparatus is enclosed in a stainless steel vacuum envelope in the form of a three-dimensional cross of tubing 16.5 cm in diameter. There are four horizontal flanged ports, one port on top, and another on the bottom. The top port carries a target turning mechanism which can present the target face to any one of the horizontal ports by rotation about the axis perpendicular to the plane of the paper. The

Fig. 4.8. A schematic diagram of apparatus used in measuring electron energy distributions in ion neutralisation experiments (after Hagstrum, 1966).

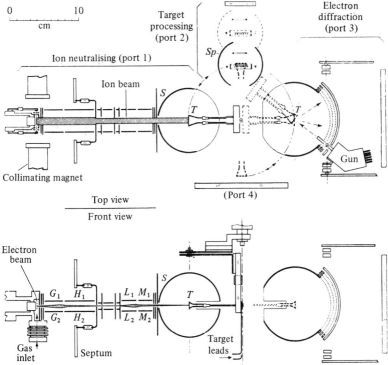
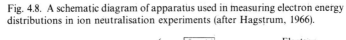

bottom port connects to the high speed ion pumps and gas inlet system. This particular system is fitted with additional experimental arrangements to allow LEED measurements of the target surface as an auxiliary measurement.

Port 1 carries the ion neutralisation apparatus. In it, ions are formed by electron impact in an electron beam and then focussed by two lens systems (G, H and L, M) onto the face of the target, T. Ejected electrons are collected at S and the kinetic energy distribution is determined as the slope of the retarding potential curve of electrons collected at S.

Port 2 carries the target processing system. Here the target may be enclosed in a rectractable sphere Sp. Whilst inside this sphere, the target may be sputtered and bombarded by ions of neon or argon at a pressure of 10^{-2}–10^{-3} torr.

Port 3 carries the LEED system by means of which the surface structure of the target may be examined. Port 4 is a viewing post.

The basic experimental requirements of the apparatus are as follows: (1) The incident ions must be slow to reduce the effects of energy broadening. Usually two distributions at incident ion kinetic energies of, say, 5 and 10 eV are obtained for extrapolation to a distribution having relatively small energy broadening. (2) The resolving power of the apparatus must be sufficiently high in the energy distribution measurements, since the retarding potential curve of electron current to the electrode S is differentiated to produce the electron kinetic energy distribution. Resolving power depends on the relative sizes of the target T and the electron collector S. The degradation in resolution inherent in the measurements amounts to convoluting the distribution by an instrumental broadening function, whose width at half maximum is of the order $\frac{1}{2}(d_T/d_S)^2 E$ where d_T is the target diameter and d_S is the sphere diameter. This instrumental broadening can be reduced to about 0.1 eV without great difficulty. (3) The data should have as little noise as possible, particularly low frequency noise, since digital deconvolution is required. Generally, it is most satisfactory to make a number of runs, and store and average them in a multichannel scalar. (4) Finally, the surface conditions at the sample must be maintained such that the background pressure of active gases is below 10^{-10} torr.

All of the above conditions and restrictions of course apply to any attempt to obtain ion neutralisation spectra using metastable intermediates. The experimental arrangements are of necessity somewhat different since the incident beam is of excited or metastable atoms rather than ions. Indeed, ion production is a thoroughly undesirable side effect of the production of metastable atoms, since both metastables and ions are formed at the same time by electron bombardment with electrons of appropriately low energy. In the case of He metastables, the excitation voltage which provides a balance between yield of metastable He atoms and yield of resonance photons, is generally around 25–30 V, although much higher potentials are commonly used.

A straightforward metastable source contains a filament situated outside a cylindrical electron collector grid, held at $+25$ V with respect to the filament. The filament provides the bombarding electrons, the whole operating in an ambient pressure of 10^{-2} torr of He. Positive ions are filtered out by charged plates. For more complex studies of the ejection process it is usually necessary to provide as intense a metastable source as possible. This can really only be done by two methods, each method employed separately or simultaneously.

The first of these methods forms the He atoms into a beam as efficiently as possible before attempting excitation by electron bombardment. The usual means of forming a beam is to use a multichannel array or supersonic nozzle of some sort. A typical example of the multichannel array is the Bendix glass array having, for example, a transparency of 50% and containing capillaries 2×10^{-4} cm in diameter and 6.3×10^{-2} cm long, with a length to diameter ratio of 3.2×10^2.

Further improvements in the metastable source are obtained by utilising coaxial electron impact rather than transverse electron impact. Generally this is achieved by having a cylindrical electron source around the axis of the He atom, and magnetic focussing to constrain the electrons to travel essentially within the He beam, until their final collection at a biassed grid (Rundel, Dunning & Stebbings, 1974). An alternative approach, uses an electron lens system to focus the electron beam along the He beam with an external magnetic field to cause the electrons to follow a spiral path. This increased path length greatly improves the efficiency of metastable production. Since the metastable is an uncharged species it cannot be focussed, only collimated once formed, thus using a preformed beam has great advantages. The metastable flux obtainable with this sort of arrangement is of the order of 5×10^{13} metastables s^{-1} steradian^{-1}, see, for instance, Brutschy & Haberland (1977) and Johnson & Delchar (1977).

The experimental set-up for obtaining ion neutralisation spectra using metastable intermediates is shown in fig. 4.9. Here a metastable source fitted with an electron gun and surrounding electromagnet is situated at *A* and *B*. The resulting beam of metastables, ions, and photons passes through deflection plates *D* which deflect the ions out of the beam leaving only metastables and photons to be collimated by the aperture *E*. The beam thus formed impinges on the target *G* after passing through the centre of the three grid energy analyser *F*. Electrons ejected by the neutralisation process are energy analysed by the retarding potential method.

4.2.4   *Experimental results from neutralisation at metal surfaces*

The theory of the Auger-type neutralisation which converts incident ions into neutral species, or incident metastable species into neutral species, is straightforward and simple to understand. Furthermore, some of the predictions of the theory are essentially simple to check. In this section are set out some of the background data preliminary to examining the results of INS for clean and adsorbate covered surface.

Electron kinetic energy distributions from a Cu{111} surface for the three ions He^+, Ne^+ and Ar^+, each of 5 eV incident energy, are shown in fig. 4.10. Each of these ions has a different ionisation potential. Nevertheless, the same structural features appear at comparable points in the curves relative to the maximum energy of each curve, and this is direct evidence that these features arise from structure in the initial state of the process, namely, the valence band of the solid. Similar curves can be obtained using metastable intermediates and fig. 4.11 shows the electron energy spectra obtained from He metastables incident on the {111} and {100} planes of W. Once again there is evidence of structure in the spectra. Note, however, that the shift in the energy spectra along the energy axis, which one would expect from the work function difference between the two surfaces, is not especially evident (MacLennan & Delchar, 1969).

Fig. 4.9. Schematic diagram of the experimental arrangement used in measuring electron energy distributions by metastable de-excitation: *A*, hypodermic needle He source; *B*, electron gun; *C*, earthed plate; *D*, ion deflector plates; *E*, beam defining aperture; *F*, hemispherical electron collectors; *G*, target crystal; *H*, solenoid (after Johnson & Delchar, 1977).

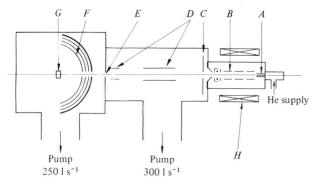

Fig. 4.10. Electron energy distributions for the three ions He$^+$, Ne$^+$ and Ar$^+$ each of 5 eV energy incident on a Cu{111} surface (after Hagstrum, 1966).

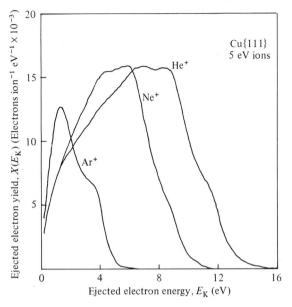

Fig. 4.11. Electron energy distributions obtained from He metastable atoms incident on the {111} and {110} planes of W (after Delchar, MacLennon & Landers, 1969).

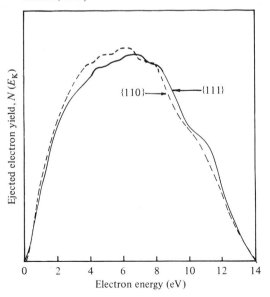

That the ion neutralisation process is very sensitive to surface contamination has already been demonstrated by fig. 4.4. Similar sensitivity to adsorbates is obtained using metastable intermediates and these are displayed in fig. 4.12 for CO, and H adsorbed on W surfaces. Some differences are apparent with these combinations, most notably that whilst a CO chemisorbed layer has a pronounced effect on the ejected electron energy distribution, the H does not.

More detailed studies, using He metastable intermediates, have enabled a direct comparison to be made between the data obtained from metastable de-excitation and that obtained from UPS from the same surface under the same conditions. Data obtained during the adsorption of O and CO on Mo{110} showed that the mechanism for metastable de-excitation was probably resonance ionisation followed by Auger neutralisation (Boiziau, Garot, Nuvolone & Roussel, 1980).

A study of CO adsorption on the Pd{111} surface has been made by Conrad *et al.* (1979) using metastable de-excitation together with UPS measurements. This study showed that, for the clean Pd surface, the de-excitation of the metastable He atoms occurs by resonance ionisation

Fig. 4.12. Electron energy distributions for He metastable atoms incident on CO and H covered W surfaces (after Delchar *et al.*, 1969).

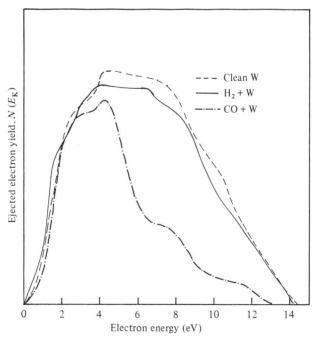

followed by Auger neutralisation (fig. 4.13(a)), so that deconvolution of the data is necessary before it becomes akin to that obtained using UPS, in particular before the peak due to the d-band electrons becomes evident.

For the CO covered surface, on the other hand, the spectra obtained were shifted along the energy axis by 0.8 eV when the beam was changed from essentially all singlet state He to all triplet state (by quenching the singlet state with radiation at 2 μm wavelength). Clearly, for this CO covered surface, Auger de-excitation (also referred to as surface Penning ionisation) must be the mechanism operating. The kinetic energy of the emitted electrons is then simply determined by the excitation energy of the metastable atom (19.8 or 20.6 eV), by the ionisation energy of the target and by the interaction potentials between the excited and ground-state noble gas atom and the target. In this instance the interaction potentials for the excited and ground-state atoms are similar and rather flat, so that we can say that the difference between the two potentials is just the excitation potential. If the ionisation energy is not referred to the vacuum level, but instead to the Fermi level, and if we assume a work function of 5.0 eV, then the peaks arrowed in fig. 4.13(d) correspond almost exactly with those observed by UPS.

A point of great significance which emerges from this result is that, for the CO covered surface, only adsorbate-derived levels appear and no features arising from the metallic d-states are seen. This result indicates that the metal is completely 'shielded' by the adsorbate, an idea first put forward by Delchar, MacLennan & Landers (1969) for CO on W surfaces. The metastable atom may be considered to probe only those states whose wavefunctions overlap sufficiently with the He metastable orbitals during the collision.

The importance, and indeed the usefulness, of this effect can be seen in the work of Bozso *et al.* (1983), who have used the screening effect, and the particular surface selectivity of the technique, to demonstrate the occupation of the $2\pi^*$-orbital of CO and NO adsorbed on Ni{111}, for which there had hitherto been no unambiguous, direct evidence, although it had long been postulated.

4.2.5 *Information from INS of metals*

Up to the present time, INS has been systematically applied to four solids, Cu, Ni, Si and Ge. Under conditions in which the sample surface was known to be clean, the results have been compared with those obtained by UPS. In particular, these comparisons have been made for Cu and Ni surfaces and the differences are rather interesting.

216   *Incident ion techniques*

In fig. 4.14 are shown a comparison of the INS data for the Ni{100} c(2 × 2)–Se surface with data obtained by photoemission spectroscopy. Once again there are differences evident between the two. These differences stem from the fact that INS results represent the d-bands of

Fig. 4.13. Electron energy distributions from a clean Pd{111} surface (*a*) and (*b*) and a CO covered Pd{111} surface (*c*) and (*d*). (*a*) and (*c*) are using photon excitation ($hv = 21.2$ eV), (*b*) and (*c*) using He* 2^1S excitation ($E^* = 20.6$ eV). E_b is the electron binding energy with respect to E_F and E_K is the kinetic energy of emitted electrons (after Conrad *et al.*, 1979).

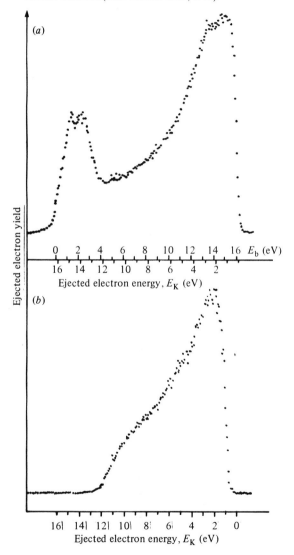

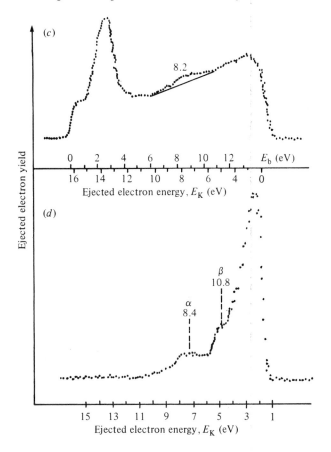

Fig. 4.14. Comparison of INS and UPS data for the Ni{100}c(2 × 2)–Se surface. Normal incidence ions (*U*), normal incidence light (*L*1), 45° incidence light (*L*2) (after Hagstrum & Becker, 1972).

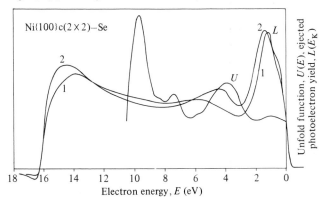

the surface atoms, which are 'different' in character from the d-bands of the bulk atoms, whilst photoemission spectroscopy represents the d-bands of the bulk. Indeed, this is a manifestation of the surface selective nature of the INS technique which is to be found in more recent work of Hagstrum & Becker (1972) where they have combined INS with the techniques of UPS, AES, and LEED, within a single vacuum chamber. This combination allows a surface to be characterised with a high degree of precision; that is, the crystallography, chemical impurity content and energy level structure can all be measured. In particular, a very direct comparison can be made between the information yielded by the techniques of INS and UPS.

The early results from this sophisticated experimental arrangement provide an interesting comparison of the relative sensitivity of angle-integrated UPS and INS to electrons in the surface monolayer. There is a contrast between the band energies as seen by UPS at normal incidence and at $45°$ incidence, on the one hand, with INS on the other for the Ni$\{100\}$ surface covered by a c(2×2)–Se layer. Notice that the INS measurements 'see' levels not 'seen' by the UPS technique using either polarisation. That there should be differences between the two spectroscopies is not surprising in view of the clear-cut differences between the electron ejection processes upon which they depend.

As an example of the detailed structural information which can be obtained by the ion neutralisation route one can take the energy spectra of fig. 4.4(b) and subject them to the detailed processing and unfolding outlined in preceding paragraphs to yield the curves for $U(E)$ of fig. 4.15. The levels marked p in the figure correspond to the atomic p-orbitals in free O, S and Se. The dashed lines are molecular orbital energies in the free molecules H$_2$X where X is O, S or Se. Three types of molecular orbital spectrum are to be found amongst the six curves for adsorbed species in fig. 4.15. The last two are the most complex spectra, with peaks near the orbitals indicated for the free H$_2$X molecule. These peaks were attributed to bridge-type bonding. Relatively small negative charging of the S, Se end of the surface molecule is indicated by the fact that the orbital peak lies near the atomic p-orbital energy of the free H$_2$S or H$_2$Se.

When the structure is changed from c(2×2) to p(2×2) by removal of half the number of adsorbed molecules, we see that the molecular orbital spectra change completely to curves in which there is a single peak below the Ni d-band peak, suggesting a change in the local bonding symmetry. For O adsorbed on Ni$\{100\}$ both the c(2×2) and p(2×2) structures show a single peak shifted by a much larger amount towards the Fermi level than is the case for either S or Se. This result was interpreted on the basis of a reconstructed surface in which the adsorbed atoms are incorporated into the top layer of substrate atoms, where relatively large charge does not result in a large work function change. In fact, the chemisorption of O on Ni{100} has subsequently been found, by a variety of techniques, to be complex but it now seems agreed that the local coordination structure of the c(2 × 2) and p(2 × 2) phases is the same. It is possible that the INS data are affected by a coexistent oxide phase. In any case, the changes seen in INS in these examples clearly illustrate the sensitivity of the technique, and its potential utility in investigating adsorbate bonding.

Fig. 4.15. Curves for $U(E)$ for clean Ni{100} and the same surface covered with O, S and Se (after Hagstrum & Becker, 1971).

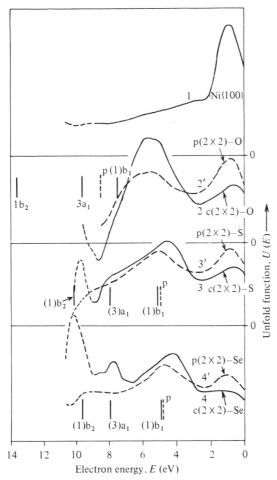

4.3   Ion scattering techniques

Ion scattering techniques, in which the scattered primary ions are studied, may be divided into two main categories; those using low energy ions (typically less than 10 keV) which are usually referred to as Low Energy Ion Scattering (LEIS) or Ion Scattering Spectroscopy (ISS) and those using high energy ions (typically greater than 100 keV up to ~ 2 MeV) variously referred to as Medium or High Energy Ion Scattering (MEIS) or (HEIS). For convenience we will adopt the acronyms LEIS and HEIS for the two general areas, and will discuss the low energy techniques first to develop some of the common ideas, and then highlight in the HEIS discussion the features which differ in the two energy ranges.

4.3.1   *LEIS: basic principles*

LEIS is primarily concerned with the exploitation of equation (4.1) to determine the composition of surfaces although, as we shall see, considerable structural information is also obtained. Experimentally, therefore, an approximately monoenergetic beam of ions, typically He^+ or Ne^+ in the energy range ~ 0.5–3.0 keV (although in some cases up to ~ 10 keV) is directed at the surface in some well-defined direction and the energy of the primary scattered ions is measured at a well-defined emission (scattering) direction. In this way, E_0, M_1 and θ_1 in equation

Fig. 4.16. 'Typical' 1 keV He^+ ion scattering spectrum of a contaminated alloy surface using an incident current of 3.5×10^{-7} A and a scattering angle of 90° (after Taglauer & Heiland, 1976).

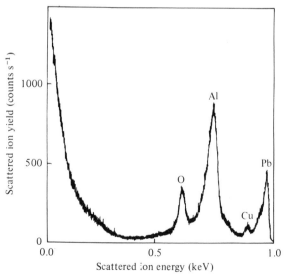

(4.1) are fixed so that the energies of the scattered particles, E_1, give the mass of the scatterers M_2. A typical LEIS spectrum is shown in fig. 4.16, the different surface species corresponding to the scattered peak values of E_1 being labelled. The ability to resolve different mass numbers on the surface is related to the energy widths of the scattered peaks in the spectra which in turn depend on the energy and angular resolution of the experimental arrangement. A large angle collection (or poorly collimated incident beam) leads to poor definition of θ, and thus a broadening of the energy peaks. For a sufficiently small angle detector, on the other hand, the mass resolving power $(M_2/\Delta M_2)$ is dictated by the energy resolving power $(E_1/\Delta E_1)$ of the experiment by the relationship

$$\left(\frac{M_2}{\Delta M_2}\right) = \left(\frac{E_1}{\Delta E_1}\right)\frac{2A}{A+1}\left[\frac{A+\sin^2\theta_1 - \cos\theta_1(A^2-\sin^2\theta_1)^{\frac{1}{2}}}{A^2-\sin^2\theta_1 + \cos\theta_1(A^2-\sin^2\theta_1)^{\frac{1}{2}}}\right]$$

(4.13)

which is shown graphically for several values of A and an energy resolving power of 100 in fig. 4.17 as a function of the scattering angle θ_1. Note that at low scattering angles the resolution is poor because the energy loss is low so that all scattered peaks are 'bunched up' at the high energy end of the spectrum close to E_0. The best mass resolution is obtained with a small value of A. On the other hand, we note that as A decreases, the range of possible scattering angles also decreases. For

Fig. 4.17. Mass resolution as a function of laboratory scattering angle, θ_1, for various values of $A = M_2/M_1$ as given by equation (4.13).

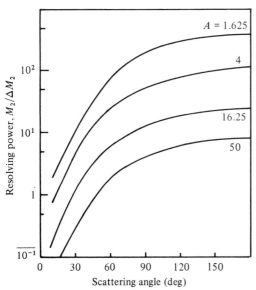

example, in the particularly simple case when $\theta_1 = 90°$, equation (4.1) reduces to

$$\frac{E_1}{E_0} = \frac{A-1}{A+1} \tag{4.14}$$

with no solution possible for $A < 1$. Thus if, at this scattering angle, we chose a heavy incident ion to optimise resolution for heavier target atoms ($A > 1$) we exclude from the analysis any lighter surface species. For smaller values of θ_1 the minimum value of A detectable falls below unity.

While the energetics of the scattering process in terms of the relationship between (E_1/E_0), A and θ_1 are independent of the nature of the ion–atom interaction potential, $V(r)$, the scattering cross-sections are dependent on the potential. Indeed, for any potential there is a unique relationship between the scattering angle and the impact parameter, b, for any particular incident ion trajectory (b being defined as the perpendicular distance between this trajectory and a line, parallel to it, but passing through the scatterer's centre). In the centre of mass frame for the ion–atom scattering event this scattering angle is given by

$$\theta_{cm} = \pi - 2 \int_{r_{min}}^{\infty} \frac{b \, dr}{r^2 \left(1 - \frac{b^2}{r^2} + \frac{V(r)}{E}\right)^{\frac{1}{2}}} \tag{4.15}$$

This allows us to calculate the total scattering cross-section for all angles greater than θ_{cm} which is given by the area πb^2 and the differential cross-section into the angle θ_{cm} (with some spread $d\theta$) which is related to the differential $\sigma = 2\pi b \, db$. The choice of scattering potential $V(r)$ is not well defined although in the energy range of LEIS it is usually taken to be a purely repulsive potential between nuclei with some account of electron screening. Such a potential is that involving the Molière approximation to the Thomas–Fermi screening function between atoms of atomic numbers Z_1 and Z_2

$$V_m(r) = \frac{Z_1 Z_2 e^2}{r} \phi\left(\frac{r}{a}\right) \tag{4.16}$$

where in this case the screening function is given by

$$\phi\left(\frac{r}{a}\right) = 0.35y + 0.55y^4 + 0.10y^{20} \tag{4.17}$$

and

$$y = \exp\left(-0.3r/a\right) \tag{4.18}$$

and a is a characteristic screening length such as the Firsov value

$$a = 0.885\,34\, a_B (Z_1^{\frac{1}{2}} + Z_2^{\frac{1}{2}})^{-\frac{2}{3}} \tag{4.19}$$

where a_B is the Bohr radius. However, neither this form for the screening nor the potential itself are universally regarded as accurate and this problem of proper choice of potential is one of the difficulties of quantification in LEIS. For example, another potential considered comparable in acceptability is the Born–Mayer potential

$$V_{BM}(r) = A\,e^{-Br} \tag{4.20}$$

suitable values for A and B for different combinations of ion and scatterer atomic number Z_1 and Z_2 having been given by Abrahamson (1969); this can frequently lead to differential cross-sections differing by factors of two or more from those calculated using the Molière potential.

Using any potential of this kind, however, we can compute the relationship of scattering angle θ_1 and impact parameter b. Fig. 4.18 shows a set of ion trajectories drawn from such a set of data using a Molière potential to describe the interaction of a 1 keV He$^+$ ion scattering from an O atom at the origin of the diagram. For simplicity each trajectory has been drawn as two straight lines joined at a single scattering node; strictly some curvature is involved, particularly close to this node, but this detail does not affect the overall picture. A conspicuous feature of this diagram is the existence of a 'shadow cone' behind the O scatterer which is not 'seen' by He$^+$ ions incident along the given direction. Thus if another scatterer lies within this cone it will be shadowed and cannot contribute to the scattering process. For the conditions for which fig. 4.18 is drawn we see that at a typical interatomic spacing in a solid of ~ 2 Å behind the scatterer, the width of the shadow cone is ~ 1.5 Å; i.e. it is comparable with typical interatomic spacings.

Fig. 4.18. Scattering trajectories for 1 keV He$^+$ ion scattering from an O atom, located at the origin, assuming a Thomas–Fermi–Molière potential. Note the existence of the 'shadow cone'.

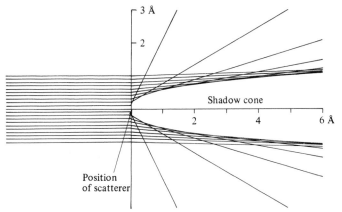

This means that in an ion scattering study of a surface under these conditions the top atom layer will typically shadow much of the second atom layer and all of any deeper atom layers. Fig. 4.19 shows 1 keV He$^+$ shadow cones around the top layer Ni atoms of a Ni{100} surface which in this azimuthal section ($\langle 100 \rangle$) entirely shadow all lower layers. We see, therefore, that elastic shadowing is a major source of surface specificity in LEIS and can restrict the signal to the top atom layer alone. In this regard the technique may be regarded as more surface specific than most electron spectroscopies which sample successive layers in an exponentially reducing factor with a mean-free-path of typically two atom layer spacings. Some indication of the scaling of the size of the shadow cone, and of the scattering cross-sections, as a function of incident ion energy and species may be deduced by calculating the impact parameter for a particular small scattering angle for a range of conditions. The results of such calculations, again using a Molière potential are shown in fig. 4.20; all impact parameters are for 5° scattering and are calculated for incident He$^+$, Ne$^+$ and Ar$^+$ ions scattered from O or Cu atoms in the energy range 0.2–10.0 keV. Clearly the very low energy incident ions yield far more surface specific scattering data than those at higher energies, although the heavier incident ions will provide more surface specific data than the light ions; on the other hand, the high energy heavy ions will produce far more surface damage than the low energy light incident ions of comparable shadow cone and cross-section size.

So far we have assumed that the interaction of the incident ions with the surface can be described in terms of a single scattering event of the ions with the individual surface atoms and results such as the spectrum of fig. 4.16 seem to support the notion that this simple description is adequate. In reality, of course, the ion–atom interaction potential, while heavily screened and therefore basically a short-range force, has

Fig. 4.19. 1 keV He$^+$ ion shadow cones from top layer atoms of a Ni{100} surface projected onto the substrate atoms.

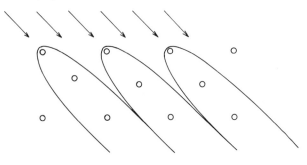

adequate range to ensure that ions do feel some repulsive force from a number of surface atoms. While an ion trajectory 'aimed' very close to one surface atom may suffer only small deflections by adjacent atoms, ions on intermediate trajectories may well suffer comparable scattering by several atoms. Indeed, at very grazing incidence angles it is clear that single scattering is impossible (each surface atom would lie in the shadow cone of adjacent atoms) but the ions can 'skim' off the repulsive potentials of several atoms. To appreciate some of the effects of mutliple scattering it is simplest to perform a one-dimensional calculation on a linear regular 'chain' of atoms; this covers many events leading to emission within the plane of incidence (which is usually studied) and is of particular relevance for scattering along a low index azimuth of a single crystal surface. The calculations, which include scattering from all the atoms in the chain (typically of less than 10 atoms), allow ions to be 'fired' along a regular array of parallel trajectories to simulate a beam in this incidence direction. The emitted ions for each trajectory may then be

Fig. 4.20. Energy dependence of 5° impact parameters for scattering of He$^+$, Ne$^+$ and Ar$^+$ ions off O and Cu atoms, assuming a Thomas–Fermi–Molière potential.

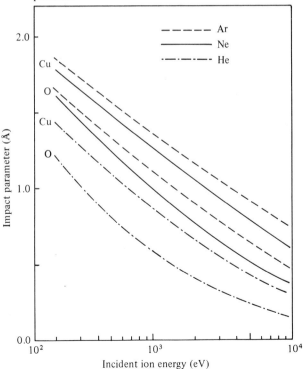

plotted on an energy–scattering angle graph. Examples of results of this kind are shown in fig. 4.21, which shows calculations for 1 keV Ar$^+$ ions from a chain of Cu atoms spaced 2.55 Å apart (corresponding to the $\langle 110 \rangle$ azimuth on a Cu$\{100\}$ surface) with grazing incidence angles (i.e. angles between the surface plane and the ion direction) of 25° and 30°. The results show the characteristic scattering 'loops' and, as each point corresponds to a trajectory on a regular mesh, the density of points at any position on a loop indicates the relative probability of this event. These calculations show two main features: the existence of minimum and maximum possible scattering angles, and the occurrence of two different energies of emission for the same scattering angle. The minimum scattering angle is related to a minimum value of the grazing emergence angle due to multiple ('skimming') scattering on the outward path. The maximum possible scattering angle is a feature only of grazing incidence and strong scattering which ensures that no atom can be hit properly 'head-on' to produce large angle scattering. Investigations of

Fig. 4.21. Scattered ion energy versus scattering angle 'loops' for 1 keV Ar$^+$ ions scattered from a regular linear chain of Cu atoms of spacing 2.55 Å. Calculated results are for grazing incidence angles of 25° and 30°.

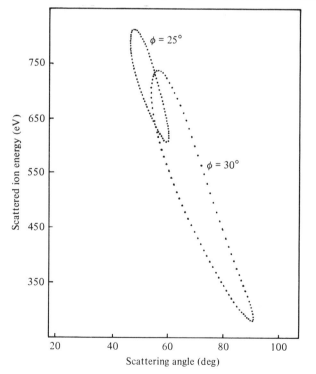

the trajectories through the scattering process show that the loop structure is attributable to two basically different kinds of scattering; the lower energy part of the loop corresponds to scattering events mainly off one surface atom (pseudobinary collisions) while the upper loop relates to ions following a double collision course with two atoms contributing significantly to the scattering. Such an event is shown schematically in fig. 4.22. It is found (cf. equation (4.1)) that two separate scattering events lead to less total energy loss than one scattering through the same total angle. While these general forms of behaviour are characteristic of all LEIS conditions, the conditions used for the calculations in fig. 4.21 are very extreme. The importance of double collisions and the narrow width (in scattering angle) of the scattering loops is a feature of very strong scattering (which is exaggerated by heavy ions at low incident energies) and of grazing incidence angles. Thus the same calculation with all conditions identical apart from the substitution of He^+ for Ar^+ leads to a scattering loop with no maximum scattering angle, a smaller minimum scattering angle, and a noticeably smaller density of points on the upper (double scattering) part of the loop relative to the lower part. Thus, for such conditions, the minimum scattering angle persists but otherwise the behaviour is dominated by essentially single scattering events. However, with Ne^+ and Ar^+ scattering, even at higher energies, the extra double scattering peak is seen in an energy spectrum. Moreover, fig. 4.21 indicates that close to the ends of the loops the density of points in some scattering angle width $d\theta$, is high and enhanced signals are seen close to the maximum and minimum scattering angle; thermal vibrations, however, do reduce this effect in destroying the exact periodicity of the chains and thus producing scatter on the calculated scattering loop diagrams. Of course, multiple scattering is not, in reality, restricted to scattering within the plane of incidence and in calculations from real three-dimensional systems, some very strong zig-zag scattering events, along surface 'channels' of atoms, are also found.

However, while such calculations do show that 1 keV He^+ ion scattering, for example, should show far less multiple scattering

Fig. 4.22. Schematic illustration of in-plane double scattering of an incident ion from two adjacent atoms (scattering angles θ_1 and θ_2) in a surface.

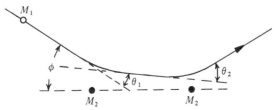

structure, than indicated, for example, in fig. 4.21, the absence of such effects from spectra for many He$^+$ and even low energy Ne$^+$ experiments is also attributable to another effect; namely the role of neutralisation. When an inert gas ion approaches a surface, charge transfer from the valence electrons of the solid can occur leading to neutralisation of the ion and thus its loss from the scattered ion signal. The mechanisms of this process have already been discussed in detail in section 4.2. Following Hagstrum (1961) it is usual to assume that the neutralisation rate has an exponential dependence on the distance of the ion from the surface s, so that the rate is given by

$$R_n = A \exp(-as) \tag{4.4}$$

The probability of neutralisation in an interval of time dt is $R\,dt = (R/v_\perp)\,ds$ where $v_\perp$ is the component of the ion velocity perpendicular to the surface. Taking the product of the probabilities of ion escape along a trajectory from far away from the surface to some minimum spacing s_{min} leads to the escape probability

$$P_{ion} = \exp\left[\int_{s_{min}}^{\infty} \left(\frac{A}{v_\perp}\right) \exp(-as)\,ds\right] \tag{4.21}$$

and for the case of the rather energetic ions of LEIS (relative to those in, for example, INS) it is usual to approximate s_{min} to zero so

$$P_{ion} = \exp(-A/av_\perp) = \exp(-v_0/v_\perp) \tag{4.22}$$

where v_0 is some characteristic velocity which depends on the ion species and target. While the exact values of the parameters involved are not well known this simple model appears to describe adequately many (but not all) of the observed effects; typical values of P_{ion} for 1 keV He$^+$ scattering from a metal surface after taking account of both inward and outward trajectories may be 10^{-2} or less. As a result any scattering process which keeps the ions in the immediate vicinity of the surface for an extended period (such as double scattering) greatly increases the probability of neutralisation. Moreover, these high neutralisation probabilities contribute substantially to the surface specificity of the technique; ions penetrating below the top atom layer suffer a much higher neutralisation probability. The inverse dependence on velocity and thus energy in equation (4.22) therefore increases the surface specificity at low energies as does elastic shadowing. On the other hand, it causes a loss of scattered ion signal at low energies in contrast to the increasing elastic cross-section σ. Thus, in many systems, the total detected percentage ion yield, given by

$$\frac{N_{det}}{N_{inc}} = \sigma P_{ion} \tag{4.23}$$

(where N_{inc} is the number of incident ions) shows first an increase and then a decrease as the energy is increased (see, e.g. the case of Ag in fig. 4.23). However, this simple picture is apparently not valid with certain combinations of incident ion and target atom. Fig. 4.23 also shows similar yield versus energy curves for He^+ ion scattering from In and Sn in particular which show strongly oscillating structure. This effect is attributable to the effects of quasi-resonant charge exchange (fig. 4.2(*d*)) when the neutralising species has a filled electronic level close in energy to the empty (ionised) level of the incident ion. In these cases 4d-levels lie close to the binding energy of the He 1s (ionised) state. Under these circumstances the total neutralisation probability (or of ion escape, P_{ion}) is an oscillatory function of the time which the ion spends in the vicinity of the surface which, of course, depends on the velocity and thus energy of the incident ion. Specifically this time is proportional to $(1/v)$ and thus to $[1/(E_0)^{\frac{1}{2}}]$; the widening of the periodicity in energy as the energy is increased as seen in fig. 4.23 is therefore to be expected. Fig. 4.23 also shows that the effect damps out for surfaces whose 4d-levels are further displaced in energy. The effect is attributed to a coherent interference of the ionisation and neutralisation processes between the two narrow energy levels involved. These periodic neutralisation effects are observed for quite a number of ion–atom combinations where the energy levels are appropriate; for example, Ga, Ge, As, In, Sn, Sb, Tl, Pb and Bi show the

Fig. 4.23. (*a*) He$^+$ scattered ion yields for 90° scattering as a function of incident ion energy for a series of materials having successively deeper 4d-states. The energy levels (relative to the vacuum) are shown in (*b*) and compared with the 1s-state of He. Apart from the Ag 4d-level, valence states have been omitted.

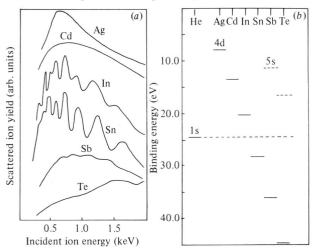

effect with incident He$^+$ ions, and Ga also behaves similarly with incident Ne$^+$. Quite a good quantitative, quantum mechanical understanding of these effects now exists (see, e.g. Tully, 1977). Evidently these strong oscillations in ion yield make quantitative assessments of the surface composition from LEIS very difficult. On the other hand, the oscillatory behaviour does act as a 'fingerprint' of a particular atomic species on the surface and it has been suggested as a means of distinguishing between species having mass numbers too close to be resolved by the scattered particle energy. Moreover, experiments involving a study of a particular atomic species on a surface in different chemical environments (in different compounds and at various coverages on different surfaces) indicate that subtle differences in the periodic structure occur which can be attributed to the different chemical environment. As the effect is an electronic one this might be expected but it does indicate that in these special circumstances LEIS may provide chemical information. On the other hand, the fact that the periodic structure is substantially the same in these different environments (and, indeed, in free atoms) highlights the fact that the neutralisation, at least for these materials, is a rather highly localised ion–atom, rather than ion–surface, interaction.

While we have already remarked on the role of neutralisation in suppressing surface double scattering features from many LEIS spectra, the process is even more important in extracting ions which have undergone more complex multiple scattering or inelastic collisions deeper in the subsurface region. But for the removal of these ions the scattered energy spectra would not show narrow peaks as in fig. 4.16, but peaks with broad low energy shoulders. This effect may be appreciated by comparing the ion scattering spectrum with that obtained when the neutral particles are also detected. Such a pair of spectra for 5 keV Ne$^+$ scattering from an Au surface are shown in fig. 4.24; although the spectrum labelled ions plus neutrals was taken using a lower resolution (time-of-flight) spectrometer than that used for the ions alone, there is a conspicuous loss of the inelastic tail in the ion spectrum. Neutralisation can therefore simplify the interpretation of LEIS spectra and enhance the surface specificity of the technique.

Some further insight into the role of neutralisation in suppressing multiple scattering effects in He$^+$ ion scattering, in particular, can be obtained by comparison with ion scattering involving species not susceptible to Auger neutralisation. We have already remarked that alkali metal atoms have ionisation potentials similar to the work function of many metals so that charge exchange is likely to occur by a resonance process, and moreover the metal–particle charge equilibrium state at close separation (if achieved) is unlikely to involve total charge neutrality of the scattering species. Fig. 4.25 shows the results of an interesting comparison between 600 eV He^+ and Li^+ ion scattering from a Ni{110} surface in which the specular scattered yield (for 60° scattering angle) of each species is shown as a function of the azimuthal scattering plane. Notice that the Li^+ yield varies by a factor of ~ 30 as the crystal is rotated about its surface normal. This variation is reproduced rather well by three-dimensional multiple scattering calculations which take no account of neutralisation. By contrast, the He^+ yields are almost totally independent of azimuth, although the theoretical computations predict similar anisotropy due to multiple scattering. It is clear that essentially all multiple scattering events are removed from the He^+ scattered yield due to the enhanced neutralisation experienced in these trajectories which necessitate the ion spending a longer time close to the surface. The Li^+ results, of course, could be reconciled with a total absence of charge exchange but independent studies indicate that as much as 40% of the yield may be neutral species. Clearly, however, the extent of charge exchange is essentially independent of scattering trajectory as there is no attenuation of multiple scattering. This is probably because the Li species achieve temporary charge equilibrium with the surface during scattering, so that the charge state remains unchanged if the ion spends longer close to the surface.

While the suppression of multiple scattering by neutralisation potentially simplifies the use of LEIS for compositon analysis, neutralisation also has a pronounced effect on the single scattering yield,

Fig. 4.24. Comparison of 5 keV Ne^+ scattering spectra from Au through an angle of 90° detecting both ions and neutrals (*a*) and ions only (*b*). (*a*) was taken using a time-of-flight energy analyser, (*b*) using an electrostatic deflection analyser (after Buck *et al.*, 1978).

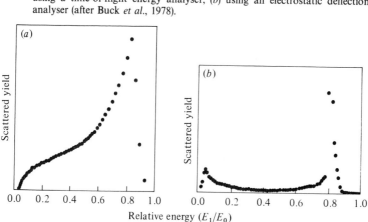

Relative energy (E_1/E_0)

making absolute composition determinations very difficult. Relative measurements, however, using calibration experiments, are possible and for low energy He^+ scattering from most systems the apparent dominance of local ion–atom neutralisation effects, and hence the lack of 'chemical' sensitivity, is a major virtue.

4.3.2   *Structural effects in LEIS*

We have already seen that LEIS is a very surface specific probe due both to neutralisation effects and to elastic shadowing. This has led to the use of LEIS to obtain simple structural information. For example, it is relatively difficult with electron spectroscopic techniques such as AES or LEED to distinguish between adsorption of species A *on* the surface of B or *under* the surface layer; most adsorbates are believed to be

Fig. 4.25. Azimuthal dependence of Li^+ and He^+ 600 eV ion scattering intensities from a Ni{110} surface with specular reflection and a 60° scattering angle. The experimental data are compared with multiple scattering (no neutralisation) and single scattering (ion fraction $P_{ion} = 0.11$) calculations (after Taglauer, Englert, Heiland & Jackson, 1980).

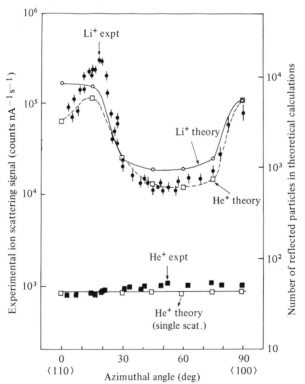

on the surface although there certainly appear to be instances of underlayer adsorption (e.g. N on Ti(0001), Shih, Jona, Jepson & Marcus, 1976). If LEIS is used in conditions under which only the top atom layer is sampled, this question is easily resolved. Moreover, in the case of small molecule adsorption some similarly simple questions may be resolved. For example, if CO is adsorbed on a surface, does the CO molecule stand on end on the surface (and, if so, which end is bonded to the surface?) or does it lie down or dissociate? These three cases should give quite different relative ion scattering signals from the C and O atoms on a surface. Of course, in assessing the relative strengths of these signals some corrections must be applied to account for the different LEIS sensitivities due to differences in elastic scattering cross-sections and neutralisation.

In the case of adsorption involving only a fraction of a monolayer coverage, the substrate atoms will generally be only partly shadowed by the adsorbate atoms; however, assuming that the adsorbate atoms adopt well-defined adsorption sites relative to the substrate atoms the amount of substrate shadowing should be dependent on the polar and azimuthal incidence angles of the ions and the width of the shadow cone. By varying these conditions the shadow cones may be swept through the substrate atoms, thus providing data on the adsorbate–substrate registry. Moreover, on an atomically rough surface in which the adsorbate atoms may be accommodated into quite deep adsorption sites on the surface, it is even possible that directional shadowing of the adsorbate species by the substrate atoms could be detected. For example, in the study of O adsorption on an f.c.c. $\{110\}$ surface to produce a (2×1) overlayer structure there are three basically reasonable structural models depicted in fig. 4.26. These are either a top layer reconstruction with O atoms replacing half of the top layer substrate atoms (fig. 4.26(a)) or adsorption of the O into the channels in the surface, either in sites directly above the next layer substrate atoms (fig. 4.26(b)) or bridging these atoms (fig. 4.26(c)). Some consideration may also be given to bridge sites on top of the top (ridge) layer substrate atoms. In many cases the azimuthal anisotropies to be expected for the O and substrate scattered signals are quite different. In particular, if the O atoms lie above the top substrate layer in any of the structures we expect little azimuthal anisotropy in the O signal. If the O lies in the hollow channels (models (b) or (c) of fig 4.26) below the top substrate layer, we expect strong shadowing of the O in the $\langle 100 \rangle$ directions (particularly for model (c)) but not in the $\langle 110 \rangle$ directions. In the case of the Ag$\{110\}(2 \times 1)$–O structure this latter behaviour is observed favouring site (c) (Heiland, Iberl, Taglauer & Menzel, 1975). In the case of the Ni$\{110\}(2 \times 1)$–O structure (Heiland &

Taglauer, 1972) little azimuthal anisotropy is seen for the O signal while the nickel signal is reduced in the $\langle 100 \rangle$ directions favouring the reconstruction model (*a*). These simple deductions are difficult to quantify, however, as they rely on the exact width of the shadow cone which can change significantly depending on the model potential used. Moreover, recent work on 1 keV He^+ ion scattering from O and C adsorbed in known structures (Godfrey & Woodruff, 1979; Woodruff & Godfrey, 1980) indicates that the shadowing is not controlled by elastic scattering effects alone. Instead it is proposed that the neutralisation must be described by a local ion–atom interaction and not simply by an ion–surface interaction as implied by equation (4.4). By assuming a similar dependence of the neutralisation rate R_n on the distance r from any particular surface atom

$$R_n = A \exp(-ar) \tag{4.24}$$

one obtains neutralisation rates which are trajectory dependent in the surface region and thus lead to azimuthal anisotropies even in the absence of elastic shadowing. In essence this leads to a picture of shadowing as arising from a 'hard' elastic shadow cone surrounded by a 'soft neutralisation cloud' of exponential character. One implication of this interpretation is that this broad 'soft' shadowing may make LEIS rather insensitive to the details of adsorbate–substrate registry at least for high symmetry substrate surfaces, although it probably does not affect the value of the technique in making some qualitative structural assignments.

An alternative method of obtaining structural information in LEIS relies on the use of conditions under which neutralisation is rather less

Fig. 4.26. Models of a (2×1) adsorption structure on an f.c.c. $\{110\}$ surface. Adsorbate atoms are shown as black spots, substrate atoms as full (top layer) and dashed (second layer) circles.

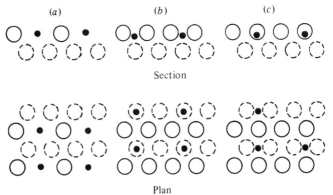

dominant so that multiple scattering effects can be observed. Alkali ion scattering clearly is ideal for this purpose (cf. fig. 4.25) but has not yet been explored widely. However, significant effects can also be obtained, for example, using 5 keV Ne$^+$ ions at grazing angles of less than about 20°. Structural studies may then be performed in one of two ways. By choosing small scattering angles (e.g. near specular reflection) a strong double scattering peak may be observed and the energy of this feature and its dependence on incident azimuth may be used to deduce atom spacings ('chain periodicity') along these directions. For example, a doubling of periodicity due to surface reconstruction (e.g. fig. 4.26(*a*)) can be identified. Alternatively, by observing the scattered particles emitted along the surface normal (at a relatively large scattering angle), and once again studying azimuthal effects due to differences in atom spacing, the self-shadowing and 'focussing' along atom rows may be studied. This phenomenon of 'focussing' may be readily appreciated from the trajectory diagram of fig. 4.18. In this single scattering model we see that, while the flux of ions arriving within the shadow cone is zero, the flux just outside the edge of the cone is substantially enhanced relative to the flux well outside the cone. This is particularly true well behind the first scatterer where the enhancement of the flux comes from ions scattered through relatively small angles by the first scatterer. Thus if a second scatterer lies in such a location just outside the shadow cone of the first scatterer an enhanced scattered signal will be observed (providing neutralisation does not suppress it) which is pseudosingle scattering rather than true double scattering and which, as seen in fig. 4.21, can lead to substantially different ion energies. Thus for a particular atom spacing we see that strong shadowing should occur at very grazing angles while as the incidence angle is increased away from grazing the scattered signal should peak strongly at some critical angle before falling to a simple single scattering value. Of course, as we have already remarked, the shadow cone model relates to a single scatterer and is not strictly appropriate for a discussion of 'chain' scattering. Nevertheless the basic features of this picture are found to apply to more qualitative treatments of chain scattering. Some effects of this kind are to be seen in the results shown in fig. 4.27. Note in particular that at low grazing angles the scattered signal from a Cu{110} surface for 5 keV Ne$^+$ is weak along the closest packed directions $\langle 100 \rangle$, $\langle 110 \rangle$ and $\langle 211 \rangle$ and also that for $\phi \sim 16°$ an enhanced signal is seen along $\langle 100 \rangle$ attributable to this focussing. Much of the detail of these azimuthal dependences can be reproduced by simple multiple elastic scattering calculations. Thus changes in such results on adsorption can be used to deduce the nature of the structural changes occurring. Moreover, studies of clean known surface structures allow some of the critical shadow cone dimensions to be checked experimentally, thus allowing the choice of model potential to be less arbitrary. Studies using these relatively energetic and heavy ions are not without their difficulties, however; in particular, far more surface damage is incurred and many such studies of adsorption have involved 'dynamic adsorption' studies in which the surface is constantly recharged with adsorbate and in which the surface temperature may be elevated to allow damage to heal out quickly. In this way a steady state may be studied; the surface structure, on the other hand, may differ from that investigated statically at lower temperatures by other techniques.

4.3.3   *Instrumentation, problems and prospects: LEIS*

The basic instrumental requirements for LEIS are a mono-energetic parallel beam of positively charged ions, and an energy analyser of small acceptance angle (to define the scattering angle). Although most work to date has used inert gas ions, there is growing interest in alkali ions. The ion gun generally consists of three parts: an

Fig. 4.27. (*a*) 5 keV Ne$^+$ scattering yields along the surface normal of a Cu{110} surface as a function of incidence azimuth for different grazing incidence angles. (*b*) shows a plan view of the surface (open circles are second layer atoms) with some of the directions of closest 'chain spacings' labelled (after de Wit, Bronckers & Fluit, 1979).

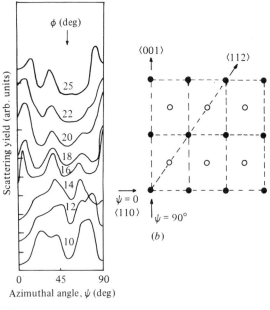

(*a*)

ion source, extraction, acceleration and focussing stages and a mass filter. The inert gas ion source either consists of a hot filament to produce electrons which are accelerated into a gas ionisation region (basically a Bayard–Alpert ion gauge) or a plasma source. Evidently the former is simpler although it may produce a broader energy spread in the ion beam due to the potential gradient in the ionisation region. In either case some potential gradient is necessary to ensure efficient extraction of the ions so that energy spreads of 5–10 eV are typical without special precautions (and evidently limit the final resolution of the system – see equation (4.13)). Moreover, in both cases the inert gas pressure in the ionisation source must be relatively high (at least $\sim 10^{-7}$ torr and often much higher) so that some differential pumping between ion source and spectrometer chamber is usually necessary. However, this need is not always acute as quite high pressures of inert gas can often be tolerated in the spectrometer due to their exceeding low reactivity with the sample. On the other hand, gas purity is important; a 1% impurity level in a 10^{-7} torr total pressure implies 10^{-9} torr of impurity which, if reactive, could severely limit the useful time scale of an experiment. While the extraction, acceleration and focussing of the ion beam is then generally performed electrostatically, magnetic mass filtering is usual (although quadrupole mass filters have been used). Again the need for mass filtering relates to gas purity. However, because many reactive impurity species (e.g. CO) have high ionisation cross-sections relative to those of the inert gases, the ion beam may have a reactive impurity content enhanced by an order of magnitude or more relative to the residual neutral gases without mass filtering.

Energy analysis of the scattered ions is readily performed with analysers identical to those described in section 2.1 for use in electron spectroscopy; only the polarities need to be reversed. For systems permitting variable scattering angles in particular (of especial value for structural studies), movable CHAs are generally used, very like the instrumentation applied to angle-resolved photoemission. Some work has also been performed using CMAs, but for ion guns off the axis of the analyser it is necessary to obscure some of the entrance aperture to ensure only a single scattering angle is accepted. Only two significant departures from this scheme have been used. The first is the use of time-of-flight analysers which may be used for the study of scattered neutrals as well as ions; the time-of-flight is independent of charge, so energy analysis is possible for both species with such an analyser and studies using them can provide direct information on neutralisation processes. However, as we have seen, neutralisation in LEIS can be a major simplifying feature so that this mode of study is of less general interest.

Moreover, while neutrals are effectively *analysed* by their time-of-flight, their eventual detection is difficult at the lowest energies. Around 3 keV and higher, for example, channel plate multipliers and similar devices prove quite efficient, but become rather inefficient for 1 keV He neutrals, for example. Finally, one piece of LEIS instrumentation has been developed which uses a CMA with an axial ion gun and is of particular interest for routine surface composition analysis. In this arrangement the whole of the CMA subtends a constant 138° scattering angle relative to the incident beam. Thus while this large scattering angle leads to low cross-sections, the large acceptance angle is a unique feature for LEIS and leads to considerable efficiency. Insofar as LEIS is suitable for routine analysis of surface composition, this arrangement seems ideal.

One further general feature of LEIS information worthy of note is that the detection systems are invariably digital rather than analogue (and for this reason necessitate dispersive energy analysers). This arises because of the low efficiency of the process coupled with the potentially damaging influence of the incident beam. In a typical scattering experiment into a well-defined angle a scattered peak contains $\sim 10^{-6}$ of the incidence flux after neutralisation and elastic cross-sections have been accounted for. On the other hand, each incident ion can produce substantial surface damage; for example, a 1 keV He^+ ion beam can sputter, on average, about 0.2 surface Ni atoms per incident ion from a Ni surface, while similar Ne^+ or Ar^+ incident beams sputter an order of magnitude more efficiently. Thus it is important to use the minimum incident flux rates for an analysis to keep damage rates low, implying the use of digital detection methods. In practice a typical incident current of a few nA may be used to give up to 10^4 ions per second into a scattered peak to give 1% statistics using ~ 1 second per energy point. As the incident beam typically has a width of several mm experimental times ~ 1 hour or more are possible without severe damage using incident He^+ ions, although the problem is far more severe for Ne^+ or Ar^+ ions.

In summary, therefore, LEIS can provide compositional information on surfaces in a particularly simple and extremely surface specific way. Absolute quantitative analysis is difficult, on the other hand, due to the difficulties of quantifying both elastic and neutralisation effects; some calibration studies have indicated that in a number of systems, LEIS yields can be used to quantify relative coverages, but it is probable that in the area of routine analysis the greatest strength of LEIS lies in its strong surface specificity when used to complement an electron spectroscopy such as AES.

The strong surface specificity is also a key to the potential value of LEIS in structural studies and some particularly basic information such as molecular orientation can be obtained rather easily. More detailed structural investigations are possible both in the strictly single scattering regime and in double or pseudosingle scattering regimes, but much work remains to be done to define the potential of these methods.

4.3.4 *High Energy Ion Scattering (HEIS)*

Some of the most significant differences between HEIS and LEIS may be deduced from the trends within LEIS across its energy range. Thus in moving to He^+ and H^+ ions in the hundreds of keV to MeV range scattering cross-sections become small and shadow cones narrow, while neutralisation ceases to play an important role. As the strong neutralisation and broad shadow cones are the key factors producing surface specificity in LEIS, this would suggest HEIS is essentially a bulk technique, and indeed most work performed with the method is aimed at studies of subsurface rather than surface regions. Because of the essential absence of neutralisation effects, incidence along an arbitrary direction leads to a scattered ion energy spectrum which does not contain sharp peaks as in LEIS (e.g. fig. 4.16) but simply steps whose high energy edges are governed by equation (4.1) and result from scattering from surface atoms. The lower energy component corresponds to ions scattered from the same atomic species but deeper in the surface; these ions have lost additional energy to electronic excitations. In subsurface studies, however, the occurrence of the energy losses can be put to good use. Because the elastic scattering cross-sections of the atoms are small, energy losses due to multiple atom scattering are rare and the electronic energy losses, typically of only a few eV, cause no deflection of the ions. Moreover, the rate of this energy loss is essentially constant so that different loss energies are related linearly to depth below the surface. Thus, for any particular scattering species we can relate the size of the scattered signal at the binary collision energy (equation (4.1)) to the composition at the surface and the size of the signal at energies lower than this to the composition at a depth below the surface proportional to the energy loss. A rather elegant example of this idea is shown in fig. 4.28, in which a section of the scattered ion energy spectrum from 250 keV He^+ ions, scattered through $127°$ from an Al–Au multiple sandwich sample, is shown. The sample consists of successive layers of ~ 10 Å of Au and 100 Å of Al. The spectrum shows the Au scattered signal at the highest energy from the surface and periodic peaks behind associated with the deeper lying layers corresponding to periodic amounts of energy loss. In this case the scattered He^+ ions were measured with an electrostatic deflection analyser giving good energy resolution ($\Delta E/E = 7 \times 10^{-3}$) leading to a depth resolution of ~ 10 Å. More commonly such experiments are performed with MeV incident ion energies when the scattered ions are energy analysed using the nuclear techniques of solid state detectors; they provide substantially coarser resolution in energy and thus in depth (~ 100–300 Å). Moreover, this method only works well for the study of heavy (high atomic number, Z) elements in light (low Z) matrices because scattering cross-sections are proportional to Z^2 and for light elements the low energy scattered ions are superimposed on a large background of scattering from heavier atoms. While the lack of neutralisation removes surface specificity from such studies it does greatly simplify quantification. Moreover, at these high incident ion energies the role of screening on the scattering potential is also very weak and a simple Coulomb interaction may be used. As a result these Rutherford scattering experiments can be used to give absolute surface and subsurface compositions.

As we have already remarked, the high ion energies also ensure that shadow cones are narrow and so do not generally lead to surface specificity. However, by proper choice of incidence angle relative to the

Fig. 4.28. 250 keV He$^+$ ion scattering spectrum, scattering angle 127°, from an Al–Au multiple sandwich sample (after Feurstein *et al.*, 1975).

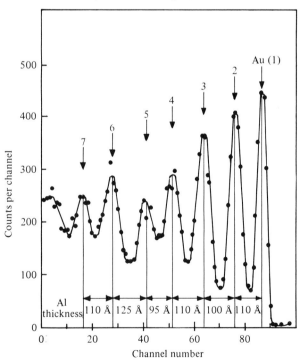

crystallographic structure these shadow cones can be used to exclude the subsurface scattering. Such a situation is shown schematically in fig. 4.29. For any low index single crystal surface (in this case a cubic {100} surface) normal incidence ions will only see the top one or two atom layers and the shadow cones, aligned along principal crystal axes, will exclude all deeper layer scattering. In such a situation the nature of the scattered ion spectrum contains only the surface (no electronic loss) scattering and, as in LEIS, becomes dominated by peaks at energies given by equation (4.1); an example is shown in fig. 4.30 which shows scattered ion spectra from 2 MeV He$^+$ ions incident on a clean W{100} surface. For incidence along a 'random' ($\langle$Random$\rangle$) direction the considerable substrate scattering contribution is seen but with incidence along the surface normal ($\langle 100 \rangle$) a single 'surface peak' is seen. This behaviour is characteristic of HEIS with incidence along a 'channelling' direction of the solid. Of course, the narrow shadow cones ensure that this surface scattered spectrum is found only over a narrow range of incidence angles but this narrowness may also be utilised in the determination of surface structures. This may also be seen from the schematic diagram in fig. 4.29; while the surface normal for a low index surface is always a channelling direction, other non-normal incidence directions also satisfy this condition and the diagram shows one such direction ($\langle 110 \rangle$ for an f.c.c. solid or $\langle 111 \rangle$ for a b.c.c. solid). In the schematic section of fig. 4.29 the top atom layer has been expanded by an amount d_z which has led to these atoms shifting out of the non-normal incidence channelling atom 'strings'; the result is that the size of the scattered 'surface peak' will be enhanced. This means that any distortion

Fig. 4.29. Schematic diagram of a section of a cubic {100} surface with a top layer spacing expansion of d_z showing high energy scattering shadow cones in two bulk channelling directions.

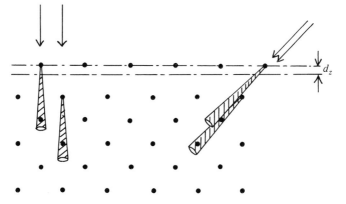

in the structure of a solid in the surface region should lead to enhancements in the anticipated surface peak size in certain channelling directions so that surface structure may be investigated. Because the technique of HEIS is rather quantitative, with well-known, Rutherford backscattering cross-sections, the number of 'atoms per string' in the surface peak may be determined absolutely, although, as this quantity will vary for different bulk channelling directions if the surface is distorted, this absolute calibration is not strictly necessary. For example, in fig. 4.29 the normal incidence data should yield one atom per string while the off-normal incidence conditions will be enhanced above this value. More exactly, however, it is necessary to take account of thermal vibrations which on average make even an ideal bulk structure give rise to a surface peak larger than one atom per string, because atoms close to the first shadowing atom in the string spend some time outside the shadow cone. For a Coulomb potential the shadow cone radius R_s at a distance d behind the first scatterer is given by

$$R_s = 2(Z_1 Z_2 e^2 d/E)^{\frac{1}{2}} \tag{4.25}$$

Fig. 4.30. 2 MeV He$^+$ ion scattering spectra from W{100} with the incident beam along the surface normal (⟨100⟩) and in a 'random' (⟨Random⟩) non-channelling direction (after Feldman *et al.*, 1977).

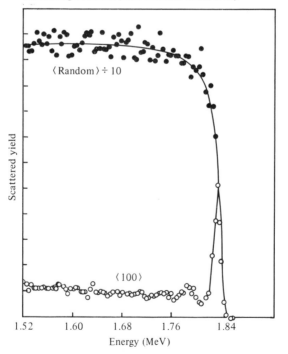

where Z_1 and Z_2 are the atomic numbers of ion and scatterer and E is the ion energy. Evidently the size of the surface peak for an ideal string is determined by the ratio ρ/R where ρ is the root mean square thermal vibrational amplitude which for a given temperature (fixed ρ) depends on (E/d). Some results for the size of the surface peak for 2 MeV He^+ scattering from a $W\{100\}$ surface are shown in fig. 4.31 for a number of different energies, mainly for normal incidence ($\langle 100 \rangle$) but also along two non-normal bulk channelling directions ($\langle 110 \rangle$ and $\langle 111 \rangle$). The full line on the graph shows the expected behaviour for an ideal structure and the fact that the non-normal incidence results (normalised to the same value of E/d) lie within the scatter of normal incidence data indicates that no significant change in the top atom layer spacing is present. The authors of this work estimate an upper limit in the layer spacing change of 6% to be compatible with their data.

An alternative method of using HEIS for surface structure determinations which does not rely on absolute determinations of the size of the surface peak is the method of 'double alignment' shown schematically in fig. 4.32. In this method the ions are again incident along a channelling direction but the scattered yield is then measured at a range of angles close to a 'blocking' direction as shown; this is essentially equivalent to shadowing except that the alignment of a scatterer (the effective emitter of scattered ions) with another surface atom leads to the blocking of the

Fig. 4.31. Experimentally determined number of atoms/string in the HEIS He^+ surface peak from a $W\{100\}$ surface at different energies and for different channelling directions. The full line shows the values expected from theory for an ideal bulk-terminated structure (after Feldman *et al.*, 1977).

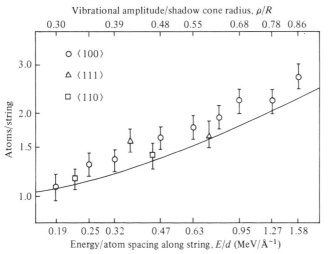

scattered signal. In this way the relative positions of the surface atoms may be obtained. As an example, data for the determination of the surface layer spacing of a Pt{111} surface are shown in fig. 4.33. The size of the surface peak, and of the 'bulk' (energy loss) signal are shown as a function of measured scattering angle. The bulk signal shows strong blocking along a bulk crystallographic direction while the minimum in the surface peak is slightly displaced to larger scattering angles, indicating a surface layer expansion of $1.5 \pm 1\%$.

This example serves to illustrate the rather high precision which appears to be possible using the HEIS technique. The very narrow shadow cones lead to a very fine probe of the registry of this top layer with the substrate in simple structural problems of this kind. Moreover, the interpretation is rather simple compared, for example, with the rather opaque quantum mechanical multiple scattering effects in LEED. Nevertheless, in more complex structural problems the interpretation of the data relies on guessed structures for modelling as in LEED. Perhaps the most complex system studied so far by HEIS, and in particular, by the double alignment method, is the surface structures of Si. In the case of Si{100}(2 × 1) at least, the data do appear to offer a rather clear preference between proposed structures (Tromp, Smeenk & Saris, 1981). By contrast to LEIS, however, HEIS appears to be of limited value in investigating adsorbate structures. Typical low Z adsorbate atoms, which produce strong substrate shadowing in LEIS, have little effect in HEIS. This is because for high energies the scattering cross-sections are always small (making narrow shadow cones) but scale as Z^2. For an

Fig. 4.32. Sectional view of a surface showing the high energy blocking cones for incidence along the surface normal and collection along either a bulk or a surface blocking direction.

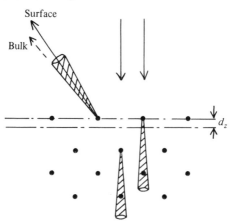

incident channelling direction, the low inelastic background of the
typical high Z substrate does allow the weak and lower energy low Z
adsorbate scattered signal to be measured. However, the very weak
shadowing and blocking imposed by the light adsorbate seems to
prevent structural studies. So far it has only been possible to investigate

Fig. 4.33. Double alignment blocking patterns for the surface peak and for the
bulk scattering signal for 173 keV H^+ scattering from a Pt(111) surface. The
scattering plane was ($\bar{1}$10) and the incident beam direction was along the [11$\bar{6}$]
channelling directions. The displacement between the angles of the minima
corresponds to a relaxation of the surface layer spacing of $1.5 \pm 1\%$ and is used
to construct the full (theoretical) lines (after van de Veen *et al.*, 1979).

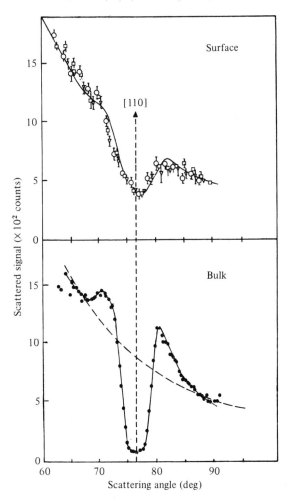

the influence of the light adsorbate on the location of the heavier substrate atom positions.

Experimentally the technique is not without its difficulties. The rather high energy ion beams must be produced in a particle accelerator and, as the vacua of these instruments are usually poor, differential pumping is necessary to maintain good UHV in the spectrometer chamber. Moreover, the strong alignment effects mean that specimen manipulators (goniometers) must be rather more precise than those commonly used in other surface techniques. Nevertheless a number of laboratories are now developing the instrumentation to explore and exploit HEIS in surface studies.

4.4   Sputtering and depth profiling

We have already remarked that when an energetic ion strikes a surface it dissipates at least some of its kinetic energy into the surface. In the simplest case when the ion is scattered back out of the surface after a binary collision with a surface atom, this surface atom then recoils into the solid with an energy given by equation (4.2). This energy, which may be a substantial fraction of the incident ion's energy, greatly exceeds the local binding energy of atoms within the solid and the recoil can thus lead to 'knock-on' collisions with many atoms, and to substantial local damage. In the case in which the incident ion is scattered *into* the solid, this effect can be even more substantial. A graphic illustration of this effect is shown in fig. 4.34 which shows the results of calculations on the outcome of several 4 keV Ar^+ ions fired into a target of randomly distributed Cu atoms. The diagram shows the incident projectile trajectories (*a*), the recoil atom trajectories (*b*) resulting from 10 incident particles, while in (*c*) the trajectories of sputtered particles from 50 incident ions is shown. The sputtered particles are those fragments caused to leave the surface as a result of the energy dissipation; fig. 4.34(*c*) shows rather clearly that most of these sputtered particles originate from the top one or two atom layers of the surface.

The existence of these sputtered particles giving rise to removal of atoms (or groups of atoms) from the surface leads to two important applications in surface science. The first is that the sputtered particles, being fragments of the surface, provide information on the chemical composition of the surface, so that if they are mass analysed they lead to a method (albeit intrinsically destructive) of surface composition analysis; this technique, SIMS, is discussed in detail in the next section. The second application is in the use of the destructive aspect of sputtering to remove atom layers from the surface. While such a process is invaluable in preparing clean crystal surfaces by removing the surface layers rich in contaminants and adsorbates, it is also widely used in the process of 'depth profiling'. As we have seen, surface analytical techniques are characterised by the ability, through different mechanisms, to sample only the top one or two atom layers of the surface. If such a technique is used in conjunction with a method of 'peeling off' atom layers from the surface, one obtains a method of bulk analysis having high spatial resolution perpendicular to the surface. This capability is invaluable in the study of corrosion layers, thin film electronic devices, coatings, etc. It is this combination which has led to the most widespread and technologically valuable application of techniques such as AES, XPS and SIMS.

The use of ion sputtering in surface science does involve certain difficulties, however. For example, what is the rate of surface erosion and how does it depend on the target material? Moreover, the process does not simply lead to the regular sequential removal of atomic layers, but substantial damage occurs in the subsurface region including atomic mixing, while the surface morphology may be changed as, for example, the second atomic layer is eroded while the first layer is still partially complete. These effects lead to problems in the calibration and depth resolution of depth profiling. In addition, in a multicomponent system, preferential sputtering of one species may occur relative to another, leading to a change in surface composition which is an artifact of the

Fig. 4.34. Results of Monte Carlo calculations (Ishitani & Shimizu, 1974) for 4 keV Ar^+ ions incident on a target of randomly distributed Cu atoms. (a) shows 10 incident particle trajectories, (b) the resulting recoil distribution and (c) the sputtering events resulting from 50 incident ions.

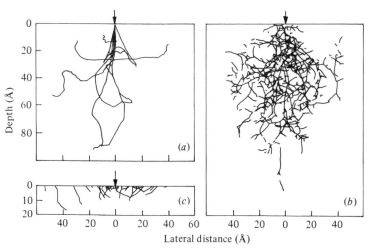

sputtering process. Some of these basic problems and the underlying processes will be discussed below.

As fig. 4.34 illustrates rather clearly, the result of the incident ion hitting the surface is a 'collision cascade' and sputtering occurs because the surface intersects this cascade. Note that one consequence of this is that the sputtered particles generally have only small kinetic energies (with a distribution peaking at less than 10 eV). Of course, under certain conditions, particularly for light atom targets and grazing ion incidence, direct recoil of surface atoms out of the surface is possible leading to rather energetic sputtered particles, but this behaviour is not a dominant process in sputtering. Because the effect is dominated by the slower 'randomised' cascade recoil atoms, it is possible to develop a semi-empirical analytic theory for this essentially statistical process. Such a theory is simplest for a target consisting of a random array of identical atoms. Experimentally, too, the study of clean elemental targets is simplest to characterise and polycrystalline samples are usually taken to approximate to the random model. For this situation Sigmund (1969) has shown that a multiple collision theory yields the result that the sputtering yield S (the number of sputtered atoms per incident ion) as a function of energy E is given by

$$S(E) = \Lambda F(x, E) \tag{4.26}$$

where $F(x, E)$ is a deposited energy function depending on the distance x into the surface. Thus sputtering depends on the amount of energy deposited at the surface ($x = 0$). The parameter, Λ, which is dependent on the target, is given by

$$\Lambda = \frac{0.042}{N U_0} \tag{4.27}$$

where N is the atomic number density of the target (in units of Å^{-3}) and U_0 is the surface binding energy, usually taken to be the sublimation energy of the material. Furthermore, the deposited energy function can be written as

$$F(0, E) = \alpha N S_n(E) \tag{4.28}$$

where $S_n(E)$ is the nuclear stopping power, usually taken for the energy range of interest to us as being dominated by elastic stopping (loss of energy through scattering) rather than electronic stopping, while α is a parameter dependent on the ratio of target atom mass M_2 and incident ion mass M_1. Final evaluation of these functions requires the use of a specific form for the elastic scattering cross-sections used to determine the elastic stopping. Sigmund has used a simple parameterisation of these and, at low energies (i.e. ion energies $\lesssim 1$ keV), the result is particularly simple and leads to the expression

$$S(E) = \frac{3\alpha}{4\pi^2} \left(\frac{T_{max}}{U_0} \right) \qquad (4.29)$$

where T_{max} is the maximum recoil energy of the target atoms (given by equation (4.2) with $\cos^2 \theta_2 = 1$) and the function α is shown in fig. (4.35). At higher energies his expression is rather more complex. However, this theory permits comparison of the effect of changing incident ion species and energy, and target material, on the sputtering yield. For example, the energy dependence of the sputtering of Cu by Ar^+ ions is shown in fig. 4.36; the figure shows the prediction of the low energy formula (dashed line) and high energy formula (full line) and a compilation of many experimental results from polycrystalline targets. Note in particular that the linear dependence of S on E implied by equation (4.29) is only valid at very low energies and in general the energy dependence is much weaker. Indeed, at rather high energies the sputtering yield decreases again; however, in surface studies (other than HEIS – see previous section) ion energies rarely exceed 10 keV.

Fig. 4.35. The factor α (equation (4.29)) as a function of mass ratio for low energies (dominated by elastic stepping) (after Sigmund, 1969).

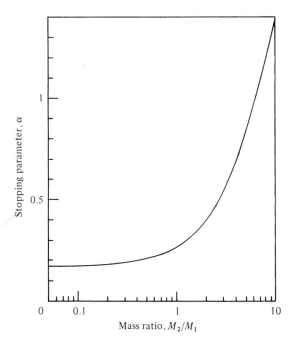

The variation of sputtering yield with target material is most conspicuously influenced by the surface binding energy although other factors such as the mass ratio M_2/M_1 and the elastic scattering cross-sections are also involved. If the high energy formula is written in the form

$$S(E) = \beta/U_0 \qquad (4.30)$$

to separate out the binding energy dependence we see that β is a rather smooth function of the target atom atomic number Z. Fig. 4.37 shows β calculated from Sigmund's high energy formula as a function of Z for incident Ne^+, Ar^+ and Xe^+ ions of 1 keV and 5 keV energy. Evidently the dependence of β on Z is not strong except for light elements and while the heavier incident ions do produce increased sputtering, this effect decreases as the incident ion becomes substantially heavier than the target atoms, particularly for the lower energy ions. This is also shown in the insensitivity of α to M_2/M_1 when this ratio is less than about one as seen in fig. 4.35. On the other hand, actual sputtering yields from different materials do vary substantially; fig. 4.38 shows measured sputtering yields from various targets for 400 eV incident Xe^+ ions. Theoretical values, including estimated values of U_0, are also shown and are mainly (although not always) in good agreement. The surface binding energy is therefore (as may be expected) an important parameter determining relative sputtering yields from different targets.

Sigmund's theory therefore gives an adequate description of sputtering yields from polycrystalline elemental targets and the influence of target species, ion species and energy. Strictly all this theory relates to

Fig. 4.36. Compilation of experimental sputtering yields for Cu by Ar^+ ions at various energies compared with theory using the low energy formula (dashed line) and the more general high energy formula (full line) (after Sigmund, 1969).

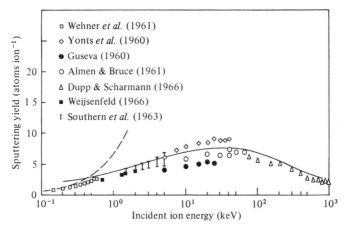

□ Wehner *et al.* (1961)
◇ Yonts *et al.* (1960)
● Guseva (1960)
○ Almen & Bruce (1961)
△ Dupp & Scharmann (1966)
■ Weijsenfeld (1966)
I Southern *et al.* (1963)

Fig. 4.37. Calculated β values (equation (4.30)) to give the sputtering yields of Ne$^+$, Ar$^+$ and Xe$^+$ ions of 1 keV and 5 keV as a function of target species atomic number (after Andersen, 1979).

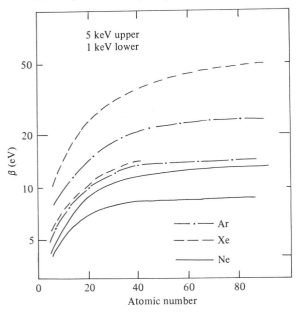

Fig. 4.38. Calculated and experimental sputtering yields for 400 eV Xe$^+$ ions for different species (after Sigmund, 1969).

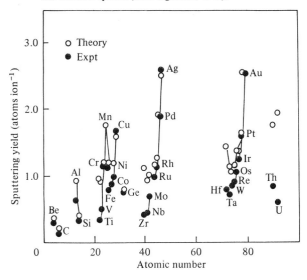

normally incident ions but the major effects of changing the angle of incidence can be understood rather easily. In particular, as the incident direction is varied away from the surface normal by some angle θ the penetration of the ion into the surface will decrease by $\cos \theta$ (for a constant range) and the whole scattering cascade will therefore become more concentrated into the surface region. We might therefore expect the sputtering yield to increase by a factor $1/\cos \theta$. Fig. 4.39 shows the angular dependence of sputtering yields for various materials and an approximate $1/\cos \theta$ behaviour is followed up to $\theta \sim 60°$. At very grazing angles, however, the sputtering yield falls off steeply. This is due to the fact that the incident ions are all scattered off the surface and do not penetrate as discussed in the previous section. Moreover, because they are scattered through small angles, the recoil energy of the surface atoms is small, leading to little energy deposition to produce the scattering

Fig. 4.39. Incident angle dependence of sputtering yields for various materials for 105 keV Ar^+ ions (after Oechsner, 1973).

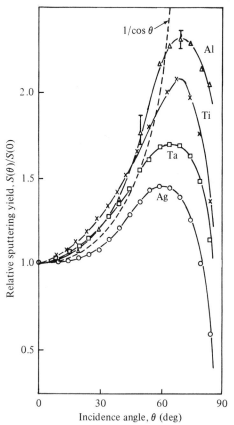

cascade and thus sputtering. Maximum sputtering is therefore achieved at intermediate angles where, in addition to the $1/\cos \theta$ factor, some direct surface atom recoils are out of the surface leading to an additional contribution to the sputtering.

Of course, all of our considerations so far have assumed that the target is essentially amorphous whereas in general we are concerned with crystalline materials. While we have seen that the models appear to be satisfactory for real polycrystalline samples, indicating that *on average* crystallinity is relatively unimportant in sputtering, studies on single crystal surfaces do show crystallographic effects in the sputtering yield. In particular the total sputtering yield is found to be dependent on the incident ion direction relative to the crystal axes. Fig. 4.40 shows an example of this effect from the work of Onderdelinden (1968) in which sputtering yields from a Cu{100} surface due to incident Ar$^+$ ions at various energies are shown as a function of incident beam angle relative to the surface normal in a $\langle 110 \rangle$ azimuth. The results show minima in the yield at angles of approximately 0°, 19° and 35° corresponding to incident directions of the type $\langle 100 \rangle$, $\langle 411 \rangle$ and $\langle 211 \rangle$. These minima along the low index directions can be understood in terms of channelling. We have already discussed this effect briefly in the context of HEIS but simply noted in that case that for very high energy incident ions the shadow cones were very narrow and so ions incident along low index directions could penetrate deep into the crystal along 'open channels' without interaction. In fact, the process of channelling is not so passive; the atom strings along these low index directions are seen as producing

Fig. 4.40. Sputtering yield from a Cu{100} surface for Ar$^+$ ions of various energies as a function of incidence angle in a $\langle 110 \rangle$ azimuth (after Onderdelinden, 1968).

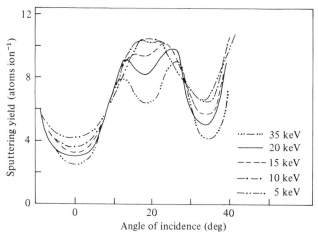

linear potential walls so that if the ions graze these 'walls' they are 'channelled' along the open holes between the strings in a positive fashion. However, if the lateral component of the energy $\psi^2 E$ (where ψ is the angle 'off-channel') exceeds some critical value they penetrate the wall, are scattered more strongly and thus are 'dechannelled'. This channelling can be seen as removing a portion of the incident ion flux sufficiently deep into a crystal to prevent it from contributing to the surface sputtering and the data of fig. 4.40 can be understood if only the non-channelled component of the incident beam is assumed to contribute to sputtering as though it were along a random direction and is thus equivalent to polycrystalline surface sputtering. Notice that channelling is more conspicuous at high energies because the constant value of $\psi^2 E$ for channelling requires that the effect becomes sharper in angle as the energy increases. Indeed, at the lowest energies shown in fig. 4.40 the narrow $\langle 411 \rangle$ channel is lost and at very low energies the very large elastic scattering cross-sections should minimise the effect. This is also seen in data for sputtering yields versus energy for Ar^+ sputtering of different low index faces of Cu along the surface normal as shown in fig. 4.41. In each case the surface normal corresponds to a channelling direction and particularly at the high energies we see a substantial reduction in sputtering yield relative to the polycrystalline material (cf. fig. 4.36). The theoretical curves follow the methods just described but for two different values of x_0, a parameter which is essentially the range of the non-channelled component of the incident ion flux. While the effect is not pronounced at high energies and is minimal at energies of only a few

Fig. 4.41. Sputtering yield for Ar^+ ions normally incident on Cu surfaces as a function of energy. The lines are from theoretical calculations (after Onderdelinden, 1968).

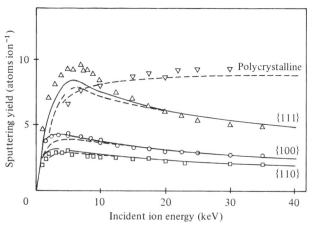

hundred eV most widely used for surface studies, it is substantial at energies of a few keV which are also relevant to surface techniques.

Unfortunately, while our understanding of sputtering in elemental solids is quite good, this is of no real interest in surface science while the sputtering of multicomponent systems, which is of far greater interest, is much less well understood. In particular, if a binary alloy is sputtered, selective sputtering of one species is known to occur and leads to a surface enrichment of the other species. Unfortunately, only a limited number of systematic data are available on this topic so that it is difficult to draw firm conclusions. Examples of studies of this kind are on Cu–Ni alloys (Shimizu, Ono & Nakayama, 1973; Tarng & Wehner, 1971) and on Ag–Pd and Ni–Pd alloys (Mathieu & Landolt, 1975). Using AES these experiments have studied the surface composition of alloys of known bulk composition after various sputtering exposures. For example, fig. 4.42 shows the deduced surface mole fraction versus bulk mole fractions of Pd in the two Pd alloys after bombardment with 2 keV Ar^+ ions. The full lines on the figure correspond to calculations based on a simple model using particular values of the ratio of the sputtering yields of the Pd and the second species. These comparisons indicate that

Fig. 4.42. Surface mole fraction $X_{A,s}$ as a function of bulk mole fraction X_A for 2 keV Ar^+ ion sputtering of two Pd alloys. The circles are experimental data for Pd in Ag–Pd alloys, the triangles for Pd in Ni–Pd alloys. The full lines represent calculations as described in the text (after Mathieu & Landolt, 1975).

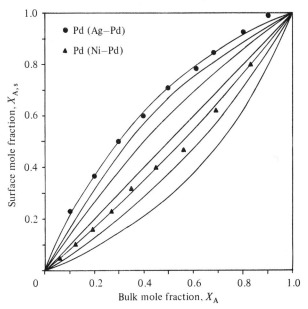

this assumption of a fixed elemental sputtering ratio over the entire alloy range is adequate. Moreover, the values of this ratio obtained, 2.4–2.5 for Pd–Ag alloys and 0.7–0.8 for Pd–Ni alloys, are close to the ratios of the sputtering yields of these species from their own elemental solids (1.89 and 0.63). This behaviour has also been found for Cu–Ni alloys except at very low ion energies (especially below ~ 100 eV) although at the lowest energies close to sputtering threshold (17 eV for pure Cu and 21 eV for pure Ni) one might expect anomalous behaviour. However, while it is tempting to carry over this idea of using elemental sputtering rates to estimate selective sputtering effects, the range of applicability is far from understood. In particular, in view of the sensitivity of element sputtering rates to U_0, it seems likely that the nature of the bonding in mixed species targets could be crucial in dictating selective sputtering effects.

Of course one important feature of selective sputtering from a homogeneous alloy is that, while there will be an initial transient in the surface composition due to the preferred sputtering of one species, a steady state is established (as shown in fig. 4.42) when the surface composition changes by the necessary ratio of sputtering rates to ensure that the composition of the sputtered flux is the same as that of the underlying bulk. One consequence of this is that surface techniques analysing the surface composition passively (such as AES or XPS) will give rather different information from one studying the sputtered particles (SIMS). Thus for a given homogeneous alloy AES and XPS will initially show the true bulk composition but after a transient at the start of sputtering will settle to a steady state showing the new modified surface composition. SIMS, on the other hand, will initially show a false composition due to excess sputtering (and hence signal) of one component, but will settle to a steady state of the true bulk composition (neglecting problems associated with the fact that only the *charged* fraction of sputtered particles is detected).

A rather different type of multispecies target of considerable relevance to surface science is that of adsorbed species on a surface. As in the case of alloys, sputtering of adsorbed species must be investigated using surface techniques rather than the grosser methods used for elemental solids such as measurement of weight loss in the target. Fig. 4.43 shows a set of results from such a measurement, in this case using LEIS to investigate the loss of surface O from various metal surfaces following bombardment by the 500 eV He^+ ions used for the ion scattering itself. In the case of adsorbed species the sputtering yield is not strictly the quantity measured, for, as the surface composition of the adsorbate A decreases, the number of sputtered A atoms per incident ion must obviously decrease. Instead one must define a desorption cross-section σ_d such that if the number of adsorbate atoms is N_A and the incident ion current density is i_0 (in ions cm^{-2} s^{-1}) then

$$-\frac{dN_A}{dt}=i_0\sigma_d N_A(t) \tag{4.31}$$

Note that for an elemental solid where the surface 'composition', i.e. the number of surface atoms per unit area N_0, is constant with time, the sputtering yield is simply $\sigma_d N_0$. The exponential form for N_A deduced from equation (4.31)

$$N_A(t)=N_A(0)\exp(-i_0\sigma_d t) \tag{4.32}$$

is verified by the results of fig. 4.43. Implicit in this remark is the assumption that the LEIS O signal is indeed linearly related to N_A but this can be established by separate calibration experiments using another surface technique (e.g. AES) during adsorption. One further potential artifact of this measurement is that the LEIS signal will only monitor the adsorbate concentration in the top one or two atom layers so that any adsorbate atoms which recoil into the solid will be added to the truly desorbed atoms. However, this effect is probably not large.

In view of our earlier finding that U_0, the surface binding energy, is an important parameter in determining elemental solid sputtering rates, fig.

Fig. 4.43. 500 eV He$^+$ ion dose dependence of 500 eV He$^+$ LEIS scattering signal for O adsorbed on various substrates. The lines are least squares fits for the data points. The bracketed species lead to essentially coincident lines, but their data points are not included. Incidence is at 30° from the surface normal (after Taglauer, Heiland & Beitat, 1979).

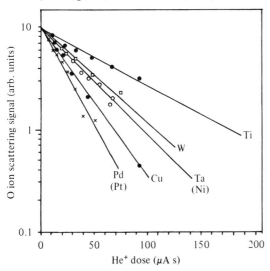

4.44 shows the values of σ_d deduced from the data of fig. 4.43 plotted against the inverse of the binding energy of O atoms to these various surfaces, E_b. While this figure shows considerable scatter about a straight line, and this best line appears not to pass through the origin, it does appear to establish a general E_b^{-1} dependence for the data. This trend appears to dominate over other effects which might be expected to influence the nature of the collision cascade such as the mass of the substrate atom species (the substrates Ti to Pt representing a wide range). Thus, while it is clear that far more work needs to be done in this area to provide reliable standards for surface sputtering work, the underlying trends emerging from the multispecies targets together with the extensive measurements for elemental solids do provide a basis for estimating sputtering effects in a wide variety of systems.

One final important aspect of sputtering in surface science concerns its role in depth profiling. As we have already remarked, depth profiling attempts to combine the surface specificity of surface spectroscopies with the ability of ion sputtering to remove single atom layers to obtain

Fig. 4.44. Experimental desorption cross-sections for O from various substrates using 500 eV He$^+$ ions incident 30° from the surface normal, compared with the inverse surface binding energy (after Taglauer *et al.*, 1979).

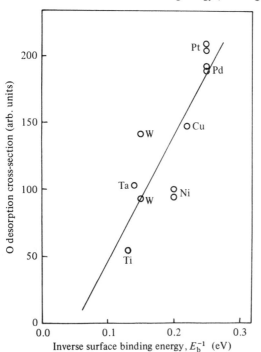

composition analyses of thin films with high spatial resolution perpendicular to the surface. As the method is frequently used to study interfaces below the surface such as a corrosion or protection layer on a substrate or metal films on semiconductor devices, the extent to which the sputtering process broadens these sharp interfaces is of some importance. This depth resolution in depth profiling is influenced by a number of instrumental factors (including the sampling depth of the surface spectroscopy itself) but also contains two major effects resulting intrinsically from the sputtering process itself.

One rather simple effect results from the statistical nature of the sputtering process. As soon as the first atom layer is partly removed, the second may be eroded, and so on. Thus, if we assume that in a mixed species target no preferential sputtering exists and that sputtering takes place only from those parts of the different layers exposed at the same rate, then if $\theta_{i,n}$ is the concentration of species i in the nth layer (before sputtering), the effective concentration of the exposed surface (showing layers $n=0$ to $n=N$) after sputtering for a time t is given by (Benninghoven, 1970)

$$\theta_i(t) = \sum_{n=0}^{N} \theta_{i,n} \frac{1}{n!} \left(\frac{t}{\tau}\right)^n \exp\left(\frac{-t}{\tau}\right) \tag{4.33}$$

where τ is the time taken to remove a single layer at the initial constant rate; i.e. $\tau = (1/i_0\sigma_D)$ in the terminology of equation (4.32). Now if we consider an idealised sample in which $\theta_{i,n}$ contains a step function – e.g. a perfectly flat interface between two different elemental species, we see that depth profiling will lead to a spreading of this step whose width is defined by a Poisson distribution of standard deviation $\sigma = \pm n^{\frac{1}{2}}$ (Hofmann, 1976). For a large number n this will tend to a Gaussian with similar standard deviation. This implies that if we depth profile to a mean depth x and if the single layer spacing is a, we find $x = na$ and the width $(\Delta n/n) = 2/n^{\frac{1}{2}}$ so that

$$\Delta x/x = 2a^{\frac{1}{2}}/x \tag{4.34}$$

Thus, although the absolute depth resolution Δx increases with sputtering depth, the proportional value $\Delta x/x$ improves as $(x)^{-\frac{1}{2}}$. This effect of increasing Δx with increasing depth is shown in the data of fig. 4.45 which shows AES O signals for depth profiling through different thicknesses of an oxide layer on Nb metal. Indeed, we see that the value of Δx increases by about a factor of 2 for an increase in x of a factor of 4 as predicted by equation (4.34). However, it is easy to see from equation (4.33) and the underlying assumptions that the effect it describes is essentially one of a change in surface morphology and that it implies that when the average depth profiled is $x = na$ there will still remain fragments of the original unsputtered surface of fractional coverage $\exp(-n)$. Sputtering is known to produce very rough surfaces in some cases and spectacular examples of surfaces showing unsputtered cones on the surface have been seen while in other cases faceting may occur, presumably due to variations of U_0 with surface orientation. Generally, however, this simple statistical model probably exaggerates the effects of morphology changes, particularly after long sputtering times, by neglecting the correlated nature of sputtering as described in the multiple scattering model we have outlined (Andersen, 1979). Moreover, the interface widths in fig. 4.45 should strictly be corrected for the effects of the inelastic electron mean-free-path in the AES, which is comparable with the width, as well as corrected for sputtering rate variations through the interface (Kirschner & Etzkorn, 1983). On the other hand, the sputtering process we have described, originating as it does from the collision cascade, will produce a mixing of layers even before the surface intersects a region of interest.

This mixing has been considered by Andersen (1979) as due to direct recoil implantation and to 'cascade mixing' involving much smaller displacements of low energy atoms in the collision cascade. In general, recoil implantation is thought to be rather unimportant and the cascade mixing can be estimated using Sigmund's (1969) multiple scattering sputtering theory. Firstly we note that the effect exists while a region of interest exists in the collision cascade; i.e. for continuous sputtering, for

Fig. 4.45. Experimental depth profile, taken using AES, through two different thicknesses of oxide layer on an Nb substrate (after Hofmann, 1976).

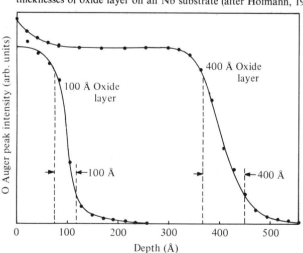

the time from the surface first being within the initial range of the cascade until the surface finally intersects the region. Provided we concern ourselves with depths greater than the extent of the initial cascade this depth resolution is therefore constant and independent of x (although it will, of course, depend on x for smaller depths). By considering the isotropic cascade mixing of low energy recoiling atoms of range R_d, Andersen shows that this constant depth resolution is then

$$\Delta x = 2R_d \left(\frac{E}{4E_{d_{min}}} \frac{1}{S(E)} \right)^{\frac{1}{2}} \tag{4.35}$$

$S(E)$ being the sputtering yield discussed earlier, while $E_{d_{min}}$ is the threshold energy for displacement of atoms in the bulk. Substituting for the low energy sputtering yield formula (equation (4.29)) with T_{max} substituted from equation (4.2) gives

$$\Delta x = 2R_d \left\{ \frac{U_0}{4E_{d_{min}}} \left[\frac{\pi^2(1+A^2)}{\cdot 3A\alpha} \right] \right\}^{\frac{1}{2}} \tag{4.36}$$

where the term in square brackets depends only on the mass ratio of ion and atoms. For the high energy form of $S(E)$ (equation (4.30)), we obtain

$$\Delta x = 2R_d \left(\frac{U_0}{4E_{d_{min}}} \frac{E}{\beta} \right)^{\frac{1}{2}} \tag{4.37}$$

Calculations of these depth resolutions are shown for Si and Cu in fig. 4.46 assuming a value for R_d of 10 Å, the discontinuity in the graphs being due to the difference in the results for the high and low energy approximations. One feature of both formulae is the dependence on the ratio $(U_0/4E_{d,min})$, i.e. the ratio of surface binding energy to internal displacement energy. This is basically because, if the surface binding

Fig. 4.46. Calculated depth resolutions for He$^+$, Ar$^+$ and Xe$^+$ depth profiling in Si and Cu as given by equations (4.36) and (4.37) (Andersen, 1979). Some experimental points, ϕ, are also shown using SIMS studies of a Si/Al$_2$O$_3$ interface with Ar$^+$ ions from Wittmaack (1978).

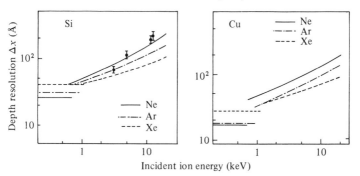

energy is high and the displacement energy low, far more displacements are suffered while the surface is slowly sputtered away. Materials having a high value of this ratio are therefore more difficult to depth profile effectively; unfortunately, as fig. 4.46 shows, Si and indeed other semiconductors, of great interest in depth profiling studies of electronic devices, do fall into this category.

4.5   Secondary Ion Mass Spectrometry (SIMS)

We have already remarked that the particle sputtered during ion bombardment contain information on the composition of the material being bombarded and the masses of the charged component of these sputtered particles are determined in SIMS using conventional mass spectrometers (magnetic or quadrupole instruments). Indeed, the early development of SIMS was motivated not by surface studies, but by the wish to provide a means of applying mass spectrometry to composition analysis and identification of solid samples, and to study thin film compositions by the method of depth profiling. Evidently SIMS is particularly well suited to depth profiling as the profiling beam provides the SIMS signal and, as we have noted, the sputtered particles (ions and neutrals together) reflect the true chemical composition of a bulk solid even when selective sputtering occurs (apart from an initial transient region). Moreover, ion beams (like electron beams in AES) can be readily focussed and deflected on a sample so that chemical composition imaging is possible; commercial instruments using this principle for essentially bulk studies have been available for many years and substantially pre-date modern scanning AES instruments. However, as we noted in our discussion of sputtering (see for example fig. 4.34), the sputtered particles do largely originate from the top one or two atom layers of a surface so that SIMS is a surface specific technique and should provide information on a depth scale comparable with other surface spectroscopies. Of course, a potential disadvantage of SIMS is that it is *intrinsically* destructive in a way which most other methods are not (although they may have an incidental destructive effect). On the other hand, it is possible, using digital techniques and defocussed incident ion beams, to obtain analyses of surfaces in time scales (or, strictly, incident ion flux doses) corresponding to the removal of only a small fraction of a monolayer from the surface. This method has been referred to by some of its exponents as 'static SIMS'. The name, while clearly incorporating some exaggeration, distinguishes the method from high incidence flux methods for bulk analysis involving quite high rates of material removal but also (as a result) providing the potential for extreme compositional sensitivity. For example, typical conditions in the two methods might be incident beam current densities of 1 nA cm^{-2} for static SIMS (e.g. 10 pA into a 1 mm spot) compared with up to 1 mA cm^{-2} for 'dynamic SIMS' (e.g. 10 pA into a 1 μm spot or 10 μA into a 1 mm spot). Approximate material removal rates in the two cases might be $\sim$1 Å per hour compared with 10 μm per hour. Incident ions might typically be Ar$^+$ in the energy range of a few hundred eV to 10 keV or higher, although in some cases chemically active rather than inert species are used.

Two basic questions are of interest in assessing SIMS as a surface technique. Firstly, as the mass analysis of sputtered particles clearly permits element identification, we would like to be able to relate the relative abundance of detected ions to the composition of the initial surface. As we have already seen that the sputtering process, while not fully quantified for adsorbates and multicomponent systems, is moderately well understood, this quantification rests on a good understanding of the relationship between the *ion* yield and the *total* sputtered particle yield. Secondly, we might hope to obtain local structural information from SIMS. In chemical mass spectrometry considerable information on molecular structure can be obtained from the characteristic 'fingerprint' of different fragment abundances in 'cracking patterns'. Similar fingerprinting of sputtered fragments (particularly ionised clusters of atoms) might lead to information on the local structure on the original surface. In addition it has been proposed that local structural information is contained in the angular dependence of the ion emission.

The basic information of a SIMS experiment is the secondary ion mass spectrum of either positive or negative ion fragments. Such a pair of spectra are shown in fig. 4.47 for a LiF surface under the bombardment of 1.3 keV Ar$^+$ ions at low incident flux. A striking feature of these spectra is the wealth of information present due to the many different ion fragments found from a surface having only a small number of elemental species present. On the other hand, for this highly ionic material the spectra show the features one might anticipate; the positive ion spectra is dominated by the electropositive Li and similar impurities, in some cases combined with neutral fragments having the bulk stoichiometry (Li$^+$(LiF), Li$^+$(LiF)$_2$, etc.) while the negative ion spectrum comprises electronegative species and similar combinations (particularly F$^-$, (LiF)F$^-$, (LiF)$_2$F$^-$). Moreover, in these spectra the larger fragments, as may also be expected, have decreasingly smaller yields (note the logarithmic yield scale). However, as we shall see, some of this simplicity of charge and cluster combination is lost in less ionic environments. Moreover, these particular spectra do not show any multiply charged ion fragments; such fragments are commonly observed although in general multiple charge is only found on ions of a single atom; thus an Al surface yields not only Al^+, Al_2^+, Al_3^+, etc., but also Al^{2+} and Al^{3+}. One further point of note from fig. 4.47 is that a rather unique feature of SIMS relative to most other surface analytic techniques, is the ability to distinguish different isotopes of a species; note, for example, the presence of $^6Li^+$ as well as $^7Li^+$ as to be expected from the natural abundance of these isotopes. Potentially this could be of value in the study of surface chemical reactions using 'labelled' reactants and some preliminary studies of this kind have been reported (e.g. Benninghoven, 1975).

As we have already remarked, extracting quantitative compositional information from ion mass spectra such as those shown in fig. 4.47 requires an understanding of the relationship of the relative ion yield to the total (ion plus neutral) sputtering yield, and experimentally this is

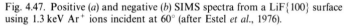

Fig. 4.47. Positive (*a*) and negative (*b*) SIMS spectra from a LiF{100} surface using 1.3 keV Ar^+ ions incident at 60° (after Estel *et al.*, 1976).

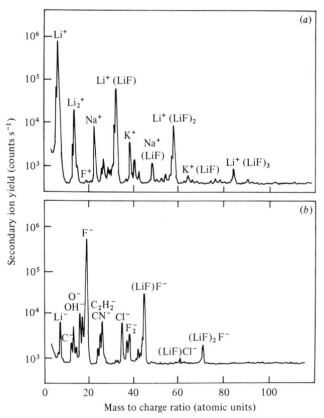

found to be sensitive to the incident ion species and energy, and to the method of measurement of the mass spectrum. For example, fig. 4.48 shows the yields of various sputtered components from a Si sample (doped with 400 ppm of B) as a function of the energy of the bombarding Ar^+ ions. The experiments were performed in a very low partial pressure of O and thus an SiO^+ peak is also shown. Also shown on this figure is the (total) sputtering yield from this target. We see that in general the sputtered ion peaks do not follow the sputtering yield in a simple fashion. In particular, the Si^+ yield shows a substantially stronger energy dependence than the sputtering yield, although the Si_2^+ yield curve is quite similar to that of the sputtering yield. Thus the conditions of the measurement are important in determining the charge and fragmentation state of the sputtered particles. Moreover, the method of measurement can influence the results. The mass spectrometers used to analyse the ion spectrum are sensitive to the kinetic energies of the ions passing through them and it is usual to place a crude electrostatic energy analyser in front of the mass spectrometer to allow some optimisation of the ion detection and measurement system. Evidently this means that the ion masses and abundances shown in a spectrum such as those in fig. 4.47

Fig. 4.48. Energy dependence of the secondary ion intensity of various ions emitted from Ar^+ ion bombarded Si (Wittmaack, 1977).

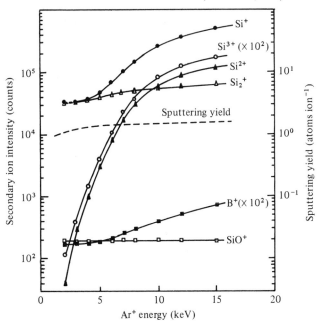

relate to a specific (range of) ion kinetic energies, so that the relative abundances will be affected by any variations in the ejected ion energy spectrum of the different ion species. The data presented in fig. 4.49 show some measurements of ion energy distributions from an Al target bombarded with 10 keV Ar^+ ions. While all three ion fragments represented in the data show the characteristic energy spectrum dominated by low energy ions to be expected in sputtering (except under conditions producing direct recoil sputtering) there are substantial differences in the spectra of the three fragments. Thus the relative yields of the Al^+ and Al_6^+, for example, would appear quite different for a mass spectrometer set to detect ions in the 0–5 eV energy range rather than 5–10 eV (as used, for example, for the data of fig. 4.48). Finally, the incident ion mass can affect significantly the relative yield of different ion fragments, although it appears to have little effect on their kinetic energy distributions. For example, at a constant incident ion energy of 12 keV the relative ion signals of Al_4^+ to Al^+ from an Al target in changing from Xe^+ to He^+ ions fall by a factor of more than 1000.

Perhaps the greatest variations in total and relative ion yields, however, are found for differing chemical environments. This has been investigated variously by using incident electropositive (e.g. Cs^+) or electronegative (e.g. O_2^+) ions, or by studying the effect of these species on the surface, or in the amgient atmosphere during bombardment by noble gas ions. By far the most investigated and exploited species affecting ion yields is O. Fig. 4.50 shows how the yields of various positive ion species from an Si surface vary with ambient O partial

Fig. 4.49. Normalised energy distributions of various ions sputtered from Al using 10 keV Ar^+ ions (Wittmaack, 1975).

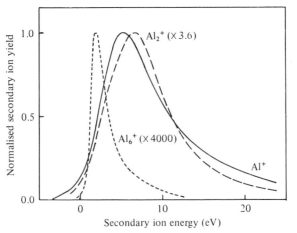

pressure. In an experiment of this kind there is a transient variation in yields as the O partial pressure is changed but the data of fig. 4.50 represents the steady state conditions after these transients. Note that not only does the SiO^+ yield rise steeply with O partial pressure, but the Si^+ (and B^+) follow essentially the same behaviour although the multiply charged Si ions and clusters of Si atoms are relatively unaffected. Evidently this overall picture is not inconsistent with a 'bond-breaking' model for the ion emission which is suggested by the ionic surface results of fig. 4.47. As O is highly electronegative, breaking a Si—O bond might be expected to lead to rather efficient Si^+ production, thus accounting for the increase of this species with O coverage. The transient behaviour suggests that the O having the greatest effect is not adsorbed *on* the surface but is embedded *within* the surface due to the effects of recoil implantation, a picture also consistent with the bond-breaking picture of ion production. The massive enhancement of Si^+ yield with saturation O coverage relative to the clean surface represented by fig. 4.50 (a factor of $\sim 10^4$) shows rather clearly why an O_2 partial pressure or O_2^+ ion bombardment is widely used to give high sensitivity in SIMS studies of bulk composition. Of course, this mode of operation can

Fig. 4.50. Steady state intensity of various secondary ions emitted from 4 keV Ar^+ bombardment of B-doped Si as a function of the O partial pressure in the chamber (Maul, 1974).

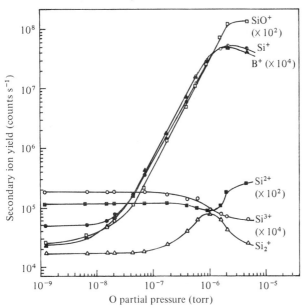

severely disturb the chemical state of the surface under study and is unlikely to be useful in true surface investigations.

The effects and parameter dependences shown in figs. 4.47–4.50 indicate that a proper description of the ion production process is likely to be complex and as yet no satisfactory theory exists which allows the phenomena to be understood quantitatively. One model which indicates how the high kinetic energies of the incident ion and recoiling atoms can lead to electronic excitation and ionisation is the so-called kinetic model of Joyes (1973). If two atoms are forced close together (during an elastic collision), strong overlap of their respective atomic orbitals forces splitting of levels into molecular-like states. A schematic representation of the effect on the atomic 2p- and 2s-levels of two Al atoms brought close together is shown in fig. 4.51. Note that at zero separation the two Al atoms become an Fe atom; in particular the 2p-state (the shallowest fully occupied state of the free atom) splits into four states, one of which crosses the conduction band levels at relatively large internuclear spacing (relative to closest approach distances in nuclear collisions). An impact of two Al atoms of sufficient energy could therefore lead to 2p-level ionisation. Evidence that collisions of this type are important in secondary ion production is provided by experiments of Brochard & Slodzian (1971), who found that the intensity of Al^{2+} ions emitted from Cu–Al alloys was approximately proportional to the square of the Al

Fig. 4.51. Schematic representation of electron promotion in Al–Al collision (Wittmaack, 1977).

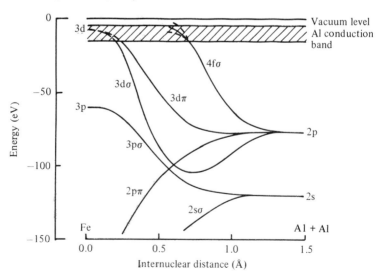

concentration, indicating the ions were produced from symmetric Al–Al recoil collisions rather than primary ion–Al collisions.

An alternative (or additional) view of the ionisation process is that it occurs as the sputtered particles are leaving the surface and passing through the surface–vacuum interface. Mechanisms for the neutralisation of a slow-moving ion close to a surface have already been discussed in earlier sections of this chapter and are, of course, also important in SIMS in that they can *remove* ions from the detected signal due to neutralisation on their outward path from the surface. However, very similar processes can cause ionisation rather than neutralisation. Indeed, we saw that alkali metal atoms could well be resonance ionised at a surface. In addition, Blaise & Slodzian (1970) suggest that creation of autoionising atomic states may be particularly important in creating secondary ions. Fig. 4.52 shows how such an autoionising state in Cu $(3d^94s5s(^4D))$ may be populated (dashed lines). Evidently any theory based on the detailed electronic structure in the surface region may well be capable of accounting for the considerable effect of electropositive or electronegative (especially O) species in the surface region. However, while both of these theoretical approaches are valuable in providing physical insight into possible mechanisms of secondary ion production, they do not lend themselves to quantitative theories which are readily applied to any material and so are not likely to provide a basis for a generally applicable development of quantitative SIMS in the foreseeable future.

Fig. 4.52. Electronic structures of a Cu atom close to a Cu surface. Dashed lines indicate possible transitions to the autoionising state $3d^94s\,5s(^4D)$. Notice that the mean energy E_e of the 4s, 5s atomic levels lies below the Fermi level of the solid, while the $3d^94s\,5s(^4D)$ lies 0.1 eV above the vacuum level. The dash-dot line indicates a de-excitation transition (after Blaise & Slodzian, 1970).

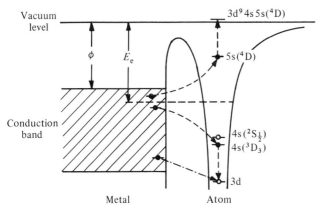

By contrast, an alternative approach which is far less specific in its exact physical basis provides the basic equations used for quantitative SIMS in more intrinsically bulk studies (such as depth profiling). This approach follows the model of Andersen & Hinthorne (1973) of the sputtering region as a plasma in local thermodynamic equilibrium. Thus one may apply the law of mass action to the equation

$$X^0 \rightleftharpoons X^+ + e \tag{4.38}$$

to obtain an equation relating the concentration of ions (n_{X^+}), electrons (n_e) and neutrals (n_{X^0}) of

$$\frac{n_{X^+} n_e}{n_{X^0}} = K_{n^+} \tag{4.39}$$

The constant K_{n^+} may be obtained from statistical mechanics to yield the Saha–Eggert equation

$$K_{n^+} = \left(\frac{2\pi}{h^2} m_e k_B T\right)^{\frac{3}{2}} \frac{2B_{X^+}}{B_{X^0}} \exp\left(-(I_p - \Delta E)/kT\right) \tag{4.40}$$

m_e being the electronic mass, k_B Boltzmann's constant, T the absolute temperature, h Planck's constant and I_p the ionisation energy of the free neutral atom which is reduced by ΔE in the plasma due to Coulomb interactions. The B_X are the internal partition functions for the ions and neutral atoms, the factor of 2 being the partition function for the electrons. This equation provides a method of calculating the ionisation coefficient for the sputtered particles (n_{X^+}/n_{X^0}), using two parameters, the electron density and the plasma temperature. Fig. 4.53 indicates how potentially successful this approach can be. The ion yields of a large number of species in a glass sample, normalised to their known concentrations in the bulk of the sample and corrected by the partition function ratio (B_{X^+}/B_{X^0}) are plotted logarithmically against their individual ionisation potentials. The straight line follows the expression of equation (4.40) for a suitable choice of n_e and T. Of course, we should note that the logarithmic yield axis does allow substantial scatter to appear unimportant; some points in fig. 4.53 have errors of factors of 3 or 4 in their yield relative to the straight line. Nevertheless, the Saha–Eggert equation does provide a basis for quantitative bulk analysis, for if several species concentrations are known these can be used to establish the appropriate values of the parameters n_e and T used for the determination of other concentrations; by the use of iterative procedures to investigate the effect of optimising ionisation energies in the plasma, errors in quantitative studies can be reduced below a factor of 2 and in favourable cases as low as $\pm 20\%$.

The approach is not without its difficulties, however, as the values of the n_e and T are sensitive to the matrix and must be determined using appropriate standards. Moreover, the physical interpretation of the model is not clear. Evidently, the collision cascade does produce rather high kinetic energies in the atoms which can be likened to a plasma and the most probable energy can be related to an effective temperature. However, this effective temperature is not found to equal that evaluated from the use of equation (4.40). Some rationalisations of these differences have been made (Werner, 1978) but it is clear that this method of quantitative analysis must be seen as a pragmatic success despite its rather unclear physical basis.

Despite the successes of the local thermodynamic equilibrium model in the use of SIMS for essentially bulk studies, it can have little value for true surface studies. For example, a major interest in surface adsorption studies is the change in chemical bonding which can occur during adsorption which, of course, can lead to matrix effects influencing the proper choice of quantification parameters. On the other hand, the large changes in yields and spectra seen in the presence of O, in particular, indicate that SIMS may be useful for following oxidation reactions by fingerprinting different adsorption states. Numerous examples of such studies exist (e.g. Benninghoven, 1975; Wittmaack, 1979). To illustrate some of the effects we consider two examples. Fig. 4.54 shows some

Fig. 4.53. Saha–Eggert plot of the logarithm of reduced ion current i_M^+ divided by the known atomic concentration c_M against the ionisation potential I_p, from a glass sample. The straight line corresponds to a 'plasma temperature' of 13 730 K (Morgan & Werner, 1977).

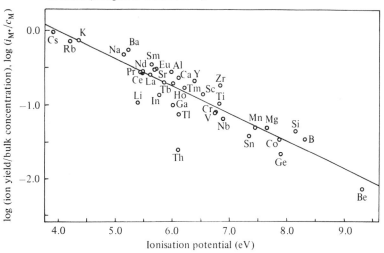

results for a study of O adsorption on polycrystalline Mo, taken from the work of Dawson (1977). Fig. 4.54(*a*) shows the exposure dependence of various positively charged secondary ion peaks using a low incident ion flux ('static' SIMS). While all peaks (both 'oxide' clusters and pure Mo ions) rise rapidly with exposure initially, at higher coverages the Mo_2O^+ and Mo^{2+} peaks fall off indicating a change of adsorption state which could be correlated with other techniques. Fig. 4.54(*b*) shows the results of a further experiment in which the ion flux was increased after achieving the high coverage state to remove material from the surface and the behaviour of the various peaks with sputtering indicates that the adsorption sequence may be reversed. Indeed, plotting the data from both figs. 4.54(*a*) and (*b*) against a common abscissa of MoO^+ yield leads to the remarkably reversible behaviour shown in fig. 4.55. The apparent reversal of adsorption states produced by sputtering relative to the adsorption sequence is surprising but does suggest that the SIMS fingerprint may be a valuable monitor. In surface studies, however, the nature and energy of the incident ion flux appear to be important in determining the degree of surface specificity and hence the type of adsorption states monitored. Some evidence for this is illustrated in the data of fig. 4.56 on the yield of O^+ and O^- ions from a $W\{100\}$ surface which has suffered various exposures of O. In these figures the abscissa is the fraction of the maximum O^+ yield using incident Ne^+ ions of fixed energy (2 keV in (*a*) and 0.5 keV in (*b*)). Results are shown for various energies of incident Ne^+ ions as low as 150 eV. Note that for the lowest

Fig. 4.54. 'Static' SIMS analyses of O uptake (*a*) on Mo and sputter removal (*b*) of an oxidised surface layer using 500 eV Ar^+ ions. Some scaling (Wittmaack, 1979) of the original data (Dawson, 1977) has been applied to the data in (*b*).

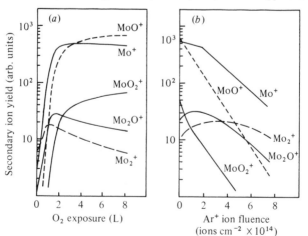

Fig. 4.55. Ion yield intensities on SIMS observed during oxidation (full lines) and sputter removal (dashed lines) of O on Mo (fig. 4.54) plotted as a function of the MoO^+ yield intensity (Wittmaack, 1977).

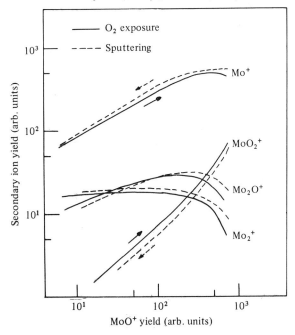

Fig. 4.56. O^+ and O^- yields from Ne^+ bombardment of $W\{100\}$ with various coverages of O. The abscissa are given as the O^+ yield at a fixed Ne^+ ion energy of 2 keV (*a*) and 0.5 keV (*b*) relative to the maximum such yield (after Yu, 1978).

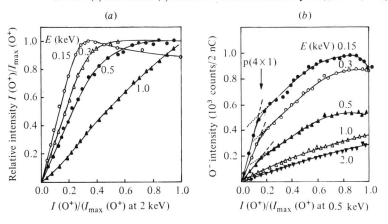

energy ions, the O^+ yield saturates at lower coverages than at 2 keV, indicating that the low energy incident ions allow sampling of only the most surface specific species while higher ion energies probe O atoms absorbed more deeply as 'inner oxide'. In many ways the O^- yield appears to be even more sensitive to incident ion energy and increases overall as the ion energy decreases. Moreover, while for high (e.g. 2 keV) incident ions the yield shows no structure at low coverage, for low incident ion energies the low coverage behaviour appears to show two approximately linear sections with a break in gradient corresponding to the coverage at which a well-ordered (4×1) LEED structure forms. This ordering transition is therefore monitored by the O^- yield providing the incident ion energy is sufficiently low.

A somewhat different application of the SIMS fingerprinting approach to surface studies is illustrated in fig. 4.57 from the work of Barber, Vickerman & Wolstenholme (1977) on adsorption of CO on a polycrystalline Fe surface. The SIMS spectrum in fig. 4.57(a) was

Fig. 4.57. SIMS data from polycrystalline Fe exposed to $\sim 1\,L$ of CO at 195 K. (a) shows the positive ion spectrum, (b) the relative yields of $FeCO^+$ and FeC^+ as the sample is subsequently heated (Barber *et al.*, 1977).

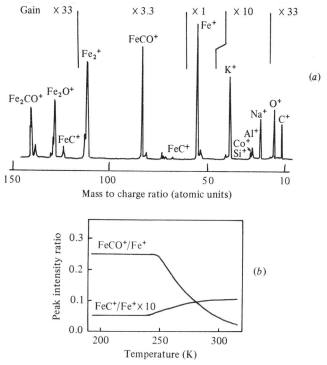

obtained after exposure to about 1 L of CO at a sample temperature of 195 K while fig. 4.57(b) shows how the ratio of the $FeCO^+$ and FeC^+ to Fe^+ peaks vary as the sample was heated. The relative increase of FeC^+ and decrease of $FeCO^+$ yield found above about 250 K strongly suggests that a molecular dissociation of the CO occurs above this temperature. While this kind of interpretation seems rather clear, the spectra themselves, such as that in fig. 4.57(a), appear to contain much more information and it is tempting, for example, to try to extract structural information from the fragment yields. For example, it has been suggested that the $FeCO^+$ ions originate from CO molecules bonded to a single Fe surface atom while Fe_2CO^+ originate from bridge bonded sites. Indeed, there is circumstantial evidence to support this kind of assignment in some cases. However, the incident probe is a very disturbing one, both electronically and in terms of atomic displacements and kinetic energies, and in the case of weakly bound surface species correspondences of this kind are unlikely to be valid generally.

One quite different method proposed for obtaining structural information from SIMS which has received attention lately is to study the angular dependence of the secondary ion yields. The existence of angular effects in sputtered particles from bulk solids has been known for many years and can be understood in terms of low index directions in the bulk leading to 'channelling' and 'blocking' in a fashion essentially similar to the effects found for different incident beam directions (see section 4.3.4). However, in the case of secondary ions of surface species it has been suggested, on the basis of model sputtering calculations, that the preferred directions of ion yield are dependent on adsorption site. The existence of this effect has been demonstrated experimentally by Holland, Garrison & Winograd (1979) and the results are shown in fig. 4.58(a). This shows the measured Cu^+ and O^- yields from a $Cu\{100\}$ $c(2 \times 2)$–O structure as a function of crystal azimuth (relative to a $\langle 100 \rangle$ direction) at a fixed polar angle of 45° to the surface normal, the incident ions being 1500 eV Ar^+. In fig. 4.58(b) four-fold averaged versions of these data are compared with model calculations for a particular geometry of adsorption (with the O atom 1.2 Å above the top Cu layer in four-fold coordinated sites). Although the model calculations have not been performed for an exhaustive range of trial structures (and the actual structure used in fig. 4.58(b) is not supported by studies using other methods) the azimuthal effects are found to be sensitive to structure so these data certainly indicate that the method has a structural capability. It should be noted, however, that the calculations do not consider the charge state of sputtered particles so that substantial crystallographic variations in charge state would invalidate the comparisons. If such an effect does exist, our present understanding of ion yields in SIMS is not likely to be adequate to model it, although there is little in existing theories to suggest significant anisotropies should occur. Clearly, this aspect of SIMS requires further exploration before a balanced assessment of the value of angle-resolved studies in structure determination can be made.

In summary, therefore, SIMS as a method of bulk composition analysis is a well-established technique which, with proper calibration and the use of a simple semiempirical theory, is capable of providing quite good quantitative analysis with considerable sensitivity to some species. As a technique for surface studies, however, our understanding

Fig. 4.58. (a) Cu^+ and O^- yields from 1.5 keV Ar^+ ions as a function of crystal surface azimuth (relative to $\langle 100 \rangle$) from a $Cu\{100\}c(2 \times 2)$–O surface structure (b) shows the same data four-fold averaged and with a background subtraction, compared with a theoretical prediction (dashed lines) based on a specific O–Cu coordination site in a four-fold hollow 1.2 Å above the top Cu atom layer (Holland *et al.*, 1979).

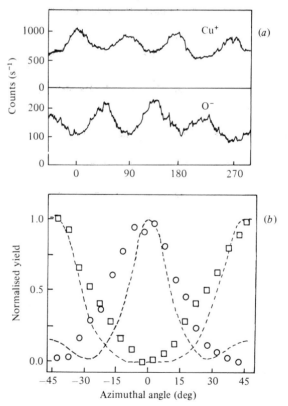

of the processes leading to ion emission are inadequate to provide the same kind of broad analytic capabilities. On the other hand, SIMS spectral 'fingerprints' do provide a means of characterising different adsorption states and, particularly using low incident ion energies, can prove extremely surface specific.

In closing this chapter on incident ion techniques two further effects or techniques deserve mention. Both have some interest in their own right but also relate to an understanding of ion emission in SIMS in that they involve the de-excitation of excited electronic states which are created during the bombardment and sputtering process. The first of these is the detection of Auger electron emission following ionisation of surface atoms by the incident energetic ions. In our brief discussion of the kinetic model of ion emission we postulated a mechanism for the creation of 2p holes in Al, and the existence of such holes could be monitored by the Auger transitions associated with their refilling for surface atoms, thus providing information on the excitation processes induced by incident ions. As a primary technique for AES, on the other hand, primary ions are inefficient ionisers relative to incident electrons and generally do far more damage.

The second decay process which can be followed results from the decay of much less energetic excited atom states giving rise to optical emission in the vicinity of the visible spectrum. This emission typically emerges from *outside* the surface as it results from the de-excitation of states of lifetimes compatible with flight times of microscopic rather than atomic distances for sputtered particles. Again the detection of this emission allows the state of excitation in the sputtering process to be studied, but also provides a spectroscopy for the determination of surface composition. Like SIMS, but unlike most electron spectroscopies, the technique has the potential to detect surface H. So far the method has not been widely used but a number of groups have studied the effect (e.g. White, Thomas, van der Weg & Tolk, 1977). Similar light emission from excited species leaving the surface can be seen in stimulated desorption studies.

Further reading

A collection of articles from a conference entitled *Inelastic ion–surface collisions* (Tolk, Tully, Heiland & White, 1977) contains quite a number of valuable reviews and summaries of relevant background material to this chapter. In particular there are articles on basic neutralisation effects and their role in INS, LEIS and HEIS by Hagstrum, Heiland & Taglauer, Buck, Rusch & Erickson and Tully &

Tolk, the last two articles being specifically concerned with the effects of oscillatory ion yields (a more recent review of this general area is Woodruff (1982)). In addition, the volume contains a review of some aspects of sputtering theory by Sigmund, an article on secondary ion production by Wittmaack and one on light emission from ion bombardment by White *et al.* More general reviews of LEIS are given by Taglauer & Heiland (1976) and of HEIS by Saris & van der Veen (1977) and Saris (1982). Useful background material to sputtering can be found in many SIMS reviews while an interesting general review of ion bombardment (mainly at higher energies and concerned with 'bulk' effects) may be found in the article by Dearnaley (1969). Many reviews of SIMS exist and some varying views on the subject are illustrated in the articles by Benninghoven (1975), Werner (1978) and Wittmaack (1979).

5 Desorption spectroscopies

5.1 Introduction

Many surface techniques involve some damage, or destruction of the surface being investigated but, with the exception of the SIMS technique described in chapter 4, this is an incidental side-effect rather than a primary feature of the technique. In the case of SIMS, the destruction of the surface is by the rather brutal method of sputtering, and the analysis of the sputtered, charged fragments is carried out primarily with the object of determining the surface composition. In this chapter, we discuss two very different types of desorption spectroscopy, in which adsorbed species specifically, are desorbed from the surface in an attempt to learn about the nature of the adsorbate–substrate bonding. Information on surface composition (or more often surface *coverage* of an adsorbed species) may also be obtained, but this is usually incidental.

The two general methods of desorption are by thermal and by electronic stimulation. Any species adsorbed on a surface must be bound to the surface with some specific amount of energy and will desorb at a rate determined by a Boltzmann factor. Heating the surface will increase this desorption rate, and the desorbing species may be detected in the gas phase by conventional mass spectrometers. Evidently, a study of the temperature dependence of the desorption rate can lead to information on the binding energy states of the adsorbate (or, more strictly, on the *desorption* energies). Of course, the binding energy states investigated will relate to those occupied at the temperature at which the desorption rate is substantial; these *need* not be the same as the states occupied at lower temperatures. One particular feature of thermal desorption spectroscopies is that they are totally destructive; at the end of a heating cycle one has usually desorbed all those surface species which can be desorbed in this way, and thus the technique is usually only of interest when heating returns the surface to a 'clean' state. Although, as we will show, detailed quantitative analysis of thermal desorption spectra is not trivial, and contains many potential pitfalls, such spectra are often readily obtained in conventional surface science instruments (equipped with a mass spectrometer for vacuum 'trouble-shooting'), and can provide a valuable 'fingerprint' of adsorption states in much the same way that qualitative LEED (chapter 2) is widely used for characterising the general structural state of a surface.

The information most readily obtained from electronically stimulated desorption methods is quite different. Incident electrons or photons lead to excitation of the adsorbed species to a new electronic state (ionic or antibonding) in which they find themselves in a repulsive potential and are desorbed. In principle, the threshold energies for these processes, and the energy distributions of the desorbed species, provide the same binding state information available in thermal desorption. In practice, this is not so, because the detailed description of the process is not fully resolved, and the nature of the final state is intimately involved and not of such general interest. Moreover, there appear to be very large variations (several orders of magnitude) in the desorption cross-sections for different adsorption states, sometimes on the same surface. For example, it appears that species adsorbed at defect sites are often desorbed with greatly enhanced efficiency. This means that electronically stimulated desorption may show signals dominated by minority species on the surface, but of course the yield may then be a valuable means of monitoring the occupation of such states. These features mean that electronically stimulated desorption techniques have far less general appeal than thermal desorption techniques. Apart from the use of Electron and Photon Stimulated Desorption (ESD and PSD) as a means of characterising certain adsorption states, there is some evidence that the techniques may be capable of providing detailed *structural* information on adsorbate–substrate coordination, through studies of the angular dependence of desorbed ionic species or by a particular modification of SEXAFS (see chapter 3).

In the following sections more detailed discussion of these two general methods are presented.

5.2   Thermal desorption techniques
5.2.1   *Introduction*
When a metal sample is heated rapidly in a vacuum, gas is desorbed from the surface. Experimentally, it is observed that the rate of gas evolution changes markedly with temperature and, in addition, there may be several temperatures for which the evolution rate goes through a relative maxima. Since the initial work of Urbach (1920), rate measurements at continuously changing temperatures have been widely applied, not only in the study of surfaces, but in other unrelated areas such as thermoluminescence, chemical transformation in solids, and annealing of defects. As the temperature of the surface increases, the rate of evolution of gas increases as well, resulting in a rise of the instantaneous gas density. From this increase it should be possible, in principle, to derive information on the nature and number of adsorbed species as well

as on the kinetics of their evolution.

The experiment may be done in two distinct ways:

(1) The temperature rise may be carried out rather rapidly, typically in less than half a second, so that the desorption rate is very much greater than the rate at which gas is pumped out of the system. This procedure is commonly called flash desorption and the analysis is the same as if the desorption were carried out in a closed system with no pumping.

(2) The temperature rise may be carried out rather slowly, over perhaps 15 seconds to a few minutes and here the gas evolved rapidly at a particular temperature is removed by pumping as the temperature continues to rise. The desorption of a particular binding state now results in a peak in the pressure–temperature curve rather than a plateau. It is convenient to think of the effect of the continual pumping as differentiating the flash desorption curve and this latter type of curve is often referred to as a thermal desorption or a Temperature Programmed Desorption (TPD) spectrum. The difference between the two approaches is brought out clearly in fig. 5.1 which shows CO desorbing from W under the two regimes.

Although flash and thermal desorption are simple in concept their execution is fraught with difficulties which can invalidate quantitative and even qualitative conclusions, but before these problems can be understood a general analysis is required to derive the rate law of desorption from evolution curves and indicate the effects to be expected when different binding states are present at a surface.

5.2.2 *Qualitative analysis of pressure–time curves*

At any instant the gas density N in a flash desorption cell of volume V is dictated by the competition, on the one hand, between the rate of supply of gas (either through flow from the gas source or cell walls F_A, or from the experimental filament F_F) and, on the other hand, by the rate of depletion by adsorption on the filament NS_F and on the walls of the vessel, by escape to the traps, or through pumping by the ionisation gauge. The total of these last three is represented by NS_E. Mass conservation immediately yields the restriction

$$V\frac{dN}{dt} = F_F - (NS_F) + F_A - NS_E \qquad (5.1)$$

Under the usual experimental conditions the rate of supply F_A, as well as the pumping speed out of the system S_E, is maintained constant and independent of time. All other quantities on the right-hand side of equation (5.1) vary with time after the temperature has been displaced. The way in which the gas density N changes on heating the sample depends on the rate law obeyed by the desorption, the heating curve for the sample, the pumping speed out of the cell, and the rate law for adsorption.

Provided that the rate of adsorption NS_F as well as the flow from the reservoir F_A are negligible during the time interval over which the filament temperature is raised, then the density N is given, as a function of time, by

$$\Delta N = \exp\left(-\frac{S_E t}{V}\right) \int_0^t \frac{F_F}{V} \exp\left(\frac{S_E t}{V}\right) dt \tag{5.2}$$

Fig. 5.1. Comparison of (*a*) thermal desorption and (*b*) flash desorption curves showing the α, β_1 and β_2 states of CO on W (Goymour & King, 1973; Ehrlich, 1961b).

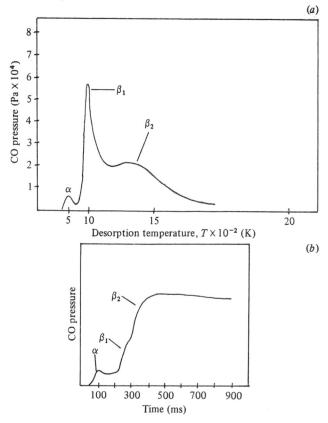

If we now apply the flash filament condition, namely that the pumping speed is so small throughout the time interval during which F_F has a significant value that we satisfy the condition $1 \gg S_E t/V$, we can simplify equation (5.2) to

$$\Delta N = \int_0^t \frac{F_F}{V} \, dt \qquad (5.3)$$

The rate of evolution from a surface of area A may be represented by an Arrhenius equation

$$F_F/A = -dn/dt = n^x v_x \exp(-E_d/RT) \qquad (5.4)$$

where n is the surface concentration, v_x is the frequency factor and x is the order of the reaction. Generally, we might expect x to be 1 or 2 in thermal desorption. Thus, for a molecular adsorbate, such as CO, the rate of desorption will depend linearly on the number of surface species available for desorption, i.e. we expect a first order desorption rate ($x = 1$). If a diatomic molecule is dissociatively adsorbed but desorbs reassociatively (e.g. O_2 or even CO again), then a requirement for desorption is that there are two suitable atoms adjacent to one another, the probability for which depends on the square of the surface concentration of the atoms. This is an example of a second order reaction ($x = 2$). For a heating curve which has the form $1/T = a + bt$ we obtain

$$\frac{\Delta N}{\Delta N_\infty} = 1 - \exp(-X) \quad \text{for } x = 1 \qquad (5.5)$$

or

$$\frac{\Delta N}{\Delta N_\infty} = \frac{n_0 X}{1 + n_0 X} \qquad \text{for } x = 2 \qquad (5.6)$$

X in all cases is of the form $C(\exp(Bt) - 1)$ where C and B are constants. ΔN_∞ is the density change after infinite time.

Evolution curves for first and second order desorption with the same activation energy and heating curves are shown in fig. 5.2. They differ qualitatively in shape and in their dependence on the initial concentration of adsorbed gas n_0. The second order curve is shifted towards lower times, that is, lower temperatures, as the initial concentration is increased, while the first order desorption curve is of course independent of initial concentration. When the activation energy for desorption is itself a function of concentration, this simple dependence is removed, and first and second order desorption processes can only be distinguished by the fact that the first order reaction terminates at the same time, regardless of initial conditions.

During flash desorption, as already pointed out, the temperature of the sample is raised rapidly compared with the rate processes determining the concentration of adsorbed material. At any instant the system is far removed from a steady state and the rate of evolution determines the surface concentration. To evaluate the experimental data the pressure–time curve must first be corrected for any net gain of gas from the reservoir or, more particularly, loss by pumping. This is done by graphical integration of $\int_0^t (F_A - NS_E) \, dt$ and it is to this corrected curve of the instantaneous surface concentration n, as a function of time and pressure, that quantitative analysis may be applied.

The instantaneous slope of the curve of surface concentration as a function of time is just

$$-\mathrm{d}n/\mathrm{d}t = n^x v_x \exp\left(-E_\mathrm{d}/RT\right) \tag{5.7}$$

In the situation where both E_d and v_x are independent of temperature

Fig. 5.2. Evolution curves for first and second order desorption with the same activation energy, and identical heating curves (Ehrlich, 1961c).

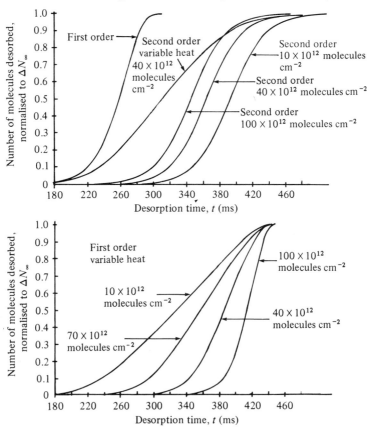

and concentration, the slopes of a single curve at different times (temperatures) yield the rate law of desorption; the correct concentration dependence may be determined by plotting $\ln(n^{-x}\,dn/dt)$ versus $1/RT$. A linear relationship with slope $-E_d$ is obtained for the correct value of x, either 1 or 2. Substitution of E_d and n^x in equation (5.7) then gives the pre-exponential term.

None of the above is true if the evolution curve is not represented by equation (5.7), and this can be checked by noting that the change in surface concentration during a small time increment $\Delta t = t_2 - t_1$ is given by

$$\int_{t_1}^{t_2} \frac{dn}{n^x} \approx k_x \,\Delta t + \frac{1}{2}\frac{dk_x}{dT}\frac{dT}{dt}\,\Delta t^2 + \cdots \tag{5.8}$$

In the limit, $\Delta t \to 0$ yields

$$\frac{1}{\Delta t}\int \frac{dn}{n^x} = k_x = -\frac{1}{n}\times\frac{dn}{dt} \tag{5.9}$$

and is independent of the rate of heating dT/dt. This independence must be ascertained experimentally by determining dn/dt at different heating rates. This method is valid for any arbitrary heating curve and can be applied to systems in which the heat of adsorption is itself a function of the surface concentration. When this latter situation prevails the semilog plot of dn/n^x against $1/T$ will not yield a straight line for the appropriate value of x. Instead a whole family of desorption traces must be obtained at different initial concentrations and possibly at different heating rates. Each curve of this family then yields a value of the slope dn/dt for a fixed surface population n, but at different temperature. A plot of $\ln(dn/dt)_{n=\text{const}}$ against $1/T$ gives, as usual, the activation energy for desorption at a fixed surface concentration, without, however, any indication of the order of reaction. This can be deduced from any given desorption trace using previously obtained activation energy values, by fitting the slopes at different temperatures to try the appropriate concentration terms n^x.

The analysis of experimental data obtained under the alternative experimental regime, that is, where the pumping rate is very large compared with the desorption rate, can be carried out by noting that the pressure in the system (pressure rise) is simply proportional to the desorption rate, and a peak in the thermal desorption spectrum at some temperature T_{max} indicates a maximum desorption rate at that temperature.

Starting with the Arrhenius equation (5.7) we obtain the condition for the maximum desorption rate by differentiating and setting $d^2n/dt^2 = 0$

at $T = T_{max}$. For a first order reaction we obtain the expression

$$E_d/RT_{max}^2 = v_1(dt/dT) \exp(-E_d/RT_{max}) \qquad (5.10)$$

The heating function dt/dT (the inverse of the heating rate) is an experimentally accessible parameter and, if we assume a value for v_1, we can solve equation (5.10) for E_d. It should be noted that the factor n, the surface density of adsorbed molecules or atoms, is not involved in equation (5.10). This means that the peak temperature is not dependent on the surface coverage. However, a shift in peak temperature with coverage can be used to demonstrate a second order process, since differentiating the second order rate equation gives an expression containing n directly

$$E_d/RT_{max}^2 = 2n_{max}v_2(dt/dT) \exp(-E_d/RT_{max}) \qquad (5.11)$$

Generally a second order desorption curve is fairly symmetric about T_{max} so that n_{max}, the surface density at T_{max}, is just $n_0/2$ where n_0 is the initial density, thus we can write

$$\Delta E_d/RT_{max}^2 = n_0 v_2(dt/dT) \exp(-E_d/RT_{max}) \qquad (5.12)$$

The difference between first and second order behaviour is clearly demonstrated by the spectra shown in fig. 5.3, the first peak labelled β_1 is clearly first order since it does not shift with concentration whilst peak β_2

Fig. 5.3. Thermal desorption curves for H on W{100} showing the β_1 and β_2 states. Each curve corresponds to a different initial coverage (Madey & Yates, 1970).

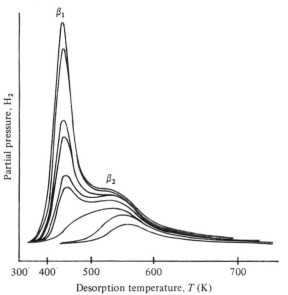

moves to lower temperatures with increasing coverage. This analysis is only appropriate for simple spectra where the rate constant is coverage independent. For cases where E_d is a function of n, T_{max} will vary with n_0.

The activation energy for desorption and the pre-exponential factor can be determined from experiments in which the heating rate is varied. For a first order process, or a second order process with constant initial coverage, differentiation of equations (5.11) or (5.12) gives

$$\frac{\text{d} \ln (T_{max}^2 \, \text{d}t/\text{d}T)}{\text{d} (1/T_{max})} = \frac{E_d}{R} \tag{5.13}$$

Thus the activation energy for desorption can be determined from the slope of a plot of $\ln (T_{max}^2 \, \text{d}t/\text{d}T)$ versus $1/T_{max}$. The pre-exponential factor v can then be found by substituting E_d into the rate equation (5.7).

For a second order process E_d can also be determined by measuring the shift in T_{max} with n_0 at constant heating rate. A plot of $\ln (n_0 T_{max}^2)$ versus $1/T_{max}$ will have a slope equal to Ed/R. In principle, it is possible to obtain E_d directly from equation (5.10) by assuming the ideal value for v of 10^{13} s^{-1}. This is a potentially dangerous way to proceed since the pre-exponential contains an entropy term $\exp (\Delta S/R)$ which can change v by several orders of magnitude.

The shapes of desorption peaks have also been analysed, notably by Redhead (1962). For an arbitrary desorption peak, fig. 5.4, with full width at half maximum ω and low and high temperature half-widths τ

Fig. 5.4. An arbitrary desorption peak with full width at half maximum ω, and low and high temperature half-widths τ and δ.

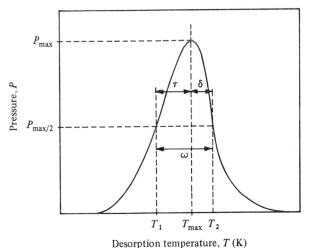

Desorption temperature, T (K)

and δ we can use the approximation

$$n_{max} = \int_T^\infty c\left(\frac{dt}{dT}\right)P\,dT \approx c\,\frac{dt}{dT}\,P_{max}\delta \qquad (5.14)$$

where the high temperature half of the peak is approximated by a triangle of the same height and half-width, where c is a constant and P is the pressure at time t. However, P_{max} is related to the maximum desorption rate so that we can write

$$-(dn/dT)_{max} = cP_{max} = (dT/dt)E_d n_{max}/xRT_{max}^2 \qquad (5.15)$$

$$E_d = xRT_{max}^2/\delta \qquad (5.16)$$

Thus we have another way to determine the activation energy, provided the desorption peak has the theoretical shape.

Redhead (1962) has shown that for $10^{13} > T_{max}\,dt/dT > 10^8$, the relationship between T_{max} and E_d is linear to a good approximation, so

Fig. 5.5. Activation energy of desorption E_d as a function of T_p for a first order reaction and a linear temperature sweep, assuming $v = 10^{13}$ s^{-1} (Redhead, 1962).

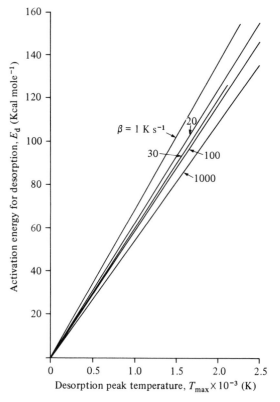

that the half-width should also be proportional to E_d. Despite the caution advised above, the simplicity of this linear relationship found by Redhead (see fig. 5.5), coupled with an assumed constant value for v_1 of 10^{13} s^{-1}, forms the basis of many analyses of TPD spectra by relating the desorption energy directly to the temperature of the desorption peak in the spectrum. The relationship found by Redhead is

$$E_d/RT_{max} = \ln (v_1 T_{max}/\beta) \times 3.64 \qquad (5.17)$$

with $\beta = dT/dt$. There is little doubt that this procedure can lead to significant errors and may well be applied when the desorption is not truly first order. The logarithmic dependence on v_1 does, however, help to reduce the sensitivity of the results of this analysis to error and useful *estimates* of desorption energies often result.

The shape of the desorption peak depends on the order of the desorption process and can, in principle, be used to determine the reaction order, thus the areas of the low and high temperature halves of the desorption peak are almost equal for second order desorption and the peak is therefore symmetrical $\tau = \delta$. The first order peak, however, is unsymmetrical with $\delta < \tau$; these shapes are shown in fig. 5.6. It must be realised that the peak shape methods can only be applied to well-resolved peaks that are undistorted by experimental artifacts such as uneven temperature distributions across the sample or inadequate pumping speed.

In general, surface reactions or even just surface heterogeneity can lead to the desorption of several different products or different binding states of the same adsorbate. The sorts of complexity which can be

Fig. 5.6. The theoretical shapes of first and second order desorption peaks (Redhead, 1962).

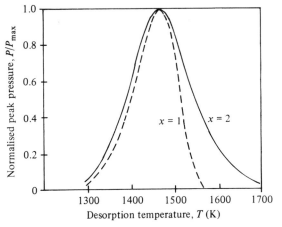

observed are shown in fig. 5.7 for H_2 and CO on single crystal planes of W, Mo and Ni. This leads to the question of peak resolution for peaks with discrete desorption energies. A simple criterion for good resolution is that the peaks must be separated by twice the sum of their half-widths, that is

$$T_{max} - T'_{max} = 2(\delta + \tau') \qquad (5.18)$$

For a first order reaction with desorption energies of 2.6 eV, peaks separated in energy by 0.35 eV will be well resolved. Various experimental factors control the resolution, the most important being the

Fig. 5.7. Typical thermal desorption spectra for H_2 and CO from single crystal planes of W, Mo and Ni. Data are for saturation coverage at 78 or 300 K (Schmidt, 1974).

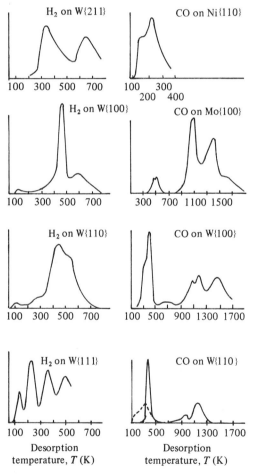

heating rate. A reduced heating rate increases the resolution, but it also reduces the peak height so that there must be a compromise with the sensitivity of the detection system. The resolution will be decreased, on the other hand, by inadequate pumping speed or non-uniform specimen temperature which both act to broaden the peaks.

For activation energies of desorption, which are uniformly distributed between limits E_{d_1} and E_{d_2}, we can consider that desorption is comprised of a superposition of peaks for discrete energies in this range. In the case of first order kinetics we have

$$-\,\mathrm{d}n/\mathrm{d}t = -\sum_i \mathrm{d}n_i/\mathrm{d}t = \int_{E_{d_2}}^{E_{d_1}} \nu_1 \exp\left(-E_d/RT\right)n_i\,\mathrm{d}E_{d_i} \qquad (5.19)$$

where n_i is the number of adsorbed species on sites for which $E_d = E_{d_i}$ at time t and is given by

$$n_i = \frac{n_0}{E_{d_2} - E_{d_1}} \exp\left[-\nu\int_0^t \exp\left(-E_{d_i}/RT\right)\mathrm{d}t\right] \qquad (5.20)$$

Now T_{max} is roughly linear with E_d and the result is a fairly constant desorption rate and therefore a broad flat topped peak between T_{max_1} and T_{max_2}.

5.2.3 Experimental arrangements for flash desorption and TPD

Although flash desorption was the first desorption technique to be deployed experimentally, notably by Ehrlich (1961a), it is TPD which has most frequently been employed experimentally. The experimental implementation of the flash filament and TPD techniques appears at first sight to be simple and straightforward; nevertheless, there are a number of experimental pitfalls which can be encountered. It would be convenient to assume that what comes off the surface is what existed at the initial temperature of interest. Unfortunately this assumption is not always valid, for a number of reasons, which are detailed below.

A large class of experimental difficulties are constituted by wall effects. Vacuum chamber walls may selectively pump gases with rates which may vary during a desorption experiment. Unless a pressure gauge is placed in the desorption chamber, or on a dead end communicating with the chamber via a high conductance tube, there may be serious errors introduced by adsorption on the walls between the chamber and the gauge. In addition, the radiation from the heated sample can cause material already adsorbed on the chamber or cell walls to desorb, making its own spurious contribution to the measured pressure. Another wall effect which is much more subtle is the displacement of one gas adsorbed on the system walls by another. Lichtman (1965), for example, has shown that CO desorbed from a filament adsorbed on the cell walls displacing H previously adsorbed there, again causing a spurious peak. This latter effect highlights the importance of identifying the desorbed gas so that one does not assume that what is desorbed is that which was originally adsorbed. A mass spectrometer is generally necessary and it must be so placed that it examines only gas molecules emerging from the sample surface directly, rather than molecules which have made collisions with the cell walls, en route. For example, a 'snout' made of some relatively inert material, such as glass, which gives line-of-sight for the detector of the central part of the sample only, can greatly reduce wall and support effects.

There are other experimental pitfalls which tend to broaden the desorption trace half-width and thus render line shape analysis hazardous, or useless. The most common of these is temperature inhomogeneity across the sample during the temperature cycle. This is normally due to the conduction cooling effect of the heavy support rods required if resistive heating is used. These support rods can also act as a gas source and introduce spurious peaks; an example of this is to be seen in the three peak trace for H_2 on $W\{211\}$ (Adams, Germer & May, 1970), which has been attributed by Rye & Barford (1971) to this problem. An

Fig. 5.8. Early type of flash filament cell. A, G, H, Bayard Alpert gauges; D, E, magnetically operated valves; F, sample filament (W); L, glass covered Mo leads; B, metal valve (Ehrlich, 1961a).

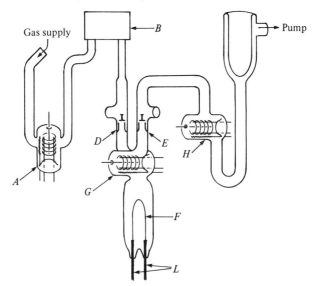

additional difficulty arises if the adsorbate exists in both molecular and atomic states on the surface and interconversion occurs during the flash. Unless the adlayer is fully equilibrated, desorption traces will not necessarily reflect the adsorbate distribution prior to the flash. Lastly, care must be taken to ensure that the pressure rise during the heating cycle is not so high that significant readsorption can occur. With the above experimental restrictions in mind the design of a flash filament or TPD system is reasonably straightforward. A reasonably typical flash desorption system is shown schematically in fig. 5.8. In this system the pressure measuring gauge G is built so that it lies directly in the path of gas atoms or molecules desorbing from the filament sample F. By this means a very high conductance path is established between the two. The cell itself is designed in such a way that it and the associated ion gauge can be immersed in a temperature bath. Fig. 5.9 shows a more elaborate system used for TPD studies, which incorporates a quadrupole mass spectrometer, together with the usual ion gauge. Generally, it is

Fig. 5.9. Typical thermal desorption system: *IG*, ionisation gauges; *FEM*, field emission microscope; *QMS*, quadruple mass spectrometer; *DV*, 'Dekker' ball and socket ground glass valves; *R*, gas supply bulb (Goymour & King, 1973).

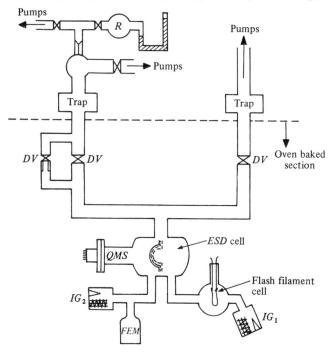

necessary to monitor or determine the composition of the desorbed gas, even when the adsorption–desorption of a single component is under investigation. In the latter case the measurement provides a means of monitoring the system purity. In addition, experiments exploiting isotopic mixing can be carried out. Such experiments can prove invaluable in determining the nature of adsorbed species and particularly of potential reaction intermediates. As a rather trivial example, if O_2 gas is adsorbed and O_2 gas desorbed, it is not clear whether the O is adsorbed in a molecular form or whether the molecule dissociates on the surface and is then reassociatively desorbed. If, however, a mixture of $^{16}O_2$ and $^{18}O_2$ are adsorbed, only the same species would be desorbed from a molecularly adsorbed species, while dissociative adsorption would give rise to isotopic mixing and a substantial fraction of desorbed $^{16}O^{18}O$. Similar methods can be used to identify far more complex surface species arising from reactions of several molecular species or dissociation of larger molecules. Of course, such results, obtained from desorption spectra, can only indicate the state of a surface intermediate immediately prior to the desorption. In the case of the O_2 example, for instance, the O may be adsorbed molecularly at low temperature and might only dissociate in the heating cycle used to investigate the problem.

Among the important features of a TPD experiment already outlined above, a key feature is the means by which the sample is heated and its temperature measured. Clearly, the heating method adopted must depend, in the first instance, on the nature of the sample. Filaments or ribbons may be heated resistively and the sample resistance is also a measure of its temperature, which is convenient. However, heat conduction along the sample support rods can result in a non-uniform temperature along the length of the sample leading to errors in the recorded desorption spectra where the peak shapes are to be analysed, as well as yielding faulty temperature data. A four-lead method using very fine wires to measure the voltage across the central more-uniform temperature section is sometimes employed (Peng & Dawson, 1971).

A better method of heating, in which the sample can be mounted almost adiabatically, is the use of electron bombardment from a filament mounted close to and behind the sample. Here the sample temperature distribution can be made almost uniform. Sample temperature is now conveniently measured by means of a thermocouple. Circuits have been developed in which the thermocouple output, or the potential across the central section of a resistively heated sample, are used to control the heating power (Redhead, 1962).

5.2.4 *Flash desorption and TPD spectra*

Most flash desorption studies have been carried out on polycrystalline filaments although there do exist some studies carried out on single crystal samples. Inevitably, the majority of this work has been carried out on W or Mo samples. An early example of flash desorption from W surfaces with N or CO as the adsorbate are shown in fig. 5.10(*a*) and (*b*). The flash traces show very clearly the existence of multiple binding states which reveal themselves as plateaux or points of inflection on the flash curves. The character of the different binding states, i.e. atomic or molecular, is readily established by examining the desorption curves as a function, say, of initial concentration, as described previously. These data are due to Ehrlich (1961b). Flash desorption curves obtained using single crystal samples and electron bombardment heating are shown in fig. 5.11(*a*), (*b*) and (*c*) (Delchar & Ehrlich, 1965), and show again some of the features seen for the W filament, namely, for N, the α, β and γ binding states again identified by their desorption kinetics according to their atomic or molecular constitution.

The data obtained by thermal desorption, i.e. where the pressure trace is effectively differentiated to yield pressure peaks, is generally much more informative than the data obtained from flash desorption. The multiple peaks often seen from polycrystalline adsorbents and ascribed to the action of different crystal planes, take their place alongside similarly complex desorption spectra obtained from single crystal samples, where single peaks might perhaps have been expected. Some examples of this type of behaviour are shown in fig. 5.7.

Multiple pressure peaks produced from a single crystal plane during the thermal desorption schedule do not, however, necessarily imply several distinctly different adsorption sites. They often arise from direct, or even indirect, lateral interactions between chemisorbed species. This

Fig. 5.10. Flash desorption spectra for (*a*) N_2 on W and (*b*) CO on W, showing different binding states (Ehrlich, 1961a, b).

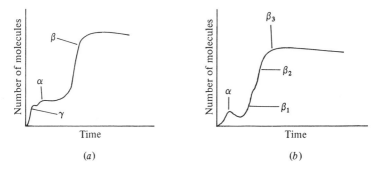

should not be surprising since LEED data has produced many instances in which ordered structures have been observed in the adlayer at submonolayer coverages. Double-spaced structures for O, CO and N on a variety of transition metal crystal planes imply repulsive interactions between nearest neighbours (while an 'islanding' tendency in some of these systems indicates longer range attraction). Similarly, the desorption traces for H adsorbed on a W{100} surface with increasing amounts of co-adsorbed CO added subsequently, reveal interactions in which the β_2 and β_1 states observed for the H_2–W system alone are gradually converted into a new set of states v_1, v_2 and v_3 (fig. 5.12) with a lower desorption activation energy; this lowering of the apparent desorption activation energy is attributed to adsorbate–adsorbate interactions.

These examples, coupled with the earlier presentation of the background theory and a discussion of some experimental difficulties, serve to highlight the great potential complexity of the thermal desorption technique. Proper quantitative analyses of the reaction order and desorption energy associated with desorption peaks is complex and

Fig. 5.11. Flash desorption traces from single crystal surfaces: (a) γ-N on W{110} at 130 K, (b) β-N on W{100}, (c) α-N on W{111} (Delchar & Ehrlich, 1965).

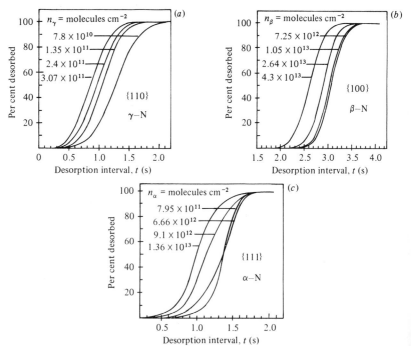

presents pitfalls, while even the identification of the number of distinct adsorption states (as well as the distinction between first and second order peaks) is susceptible to adsorbate–adsorbate interactions. Nevertheless, the techniques do present a wealth of useful data, and with caution extremely valuable qualitative and semiquantitative conclusions can be drawn from essentially simple experiments, regarding the nature of the adsorbed species, their interaction and their approximate desorption energies. While the possibility that the heating cycle itself may influence the nature of the adsorption has already been mentioned, one particular case in which this effect is put to use deserves a brief mention. This is the idea of temperature programmed *reaction* spectroscopy. In the foregoing analysis it has been assumed that the desorption is limited by the activation barrier for desorption from the surface which has been implicitly linked to the binding energy with which the species is held on the surface. In more complex systems this need not be so. Thus, a

Fig. 5.12. Desorption traces for H_2 absorbed on $W\{100\}$ at ~ 100 K with co-adsorbed CO showing binding state shift. CO exposure (a) 0, (b) 3.5×10^{-2} torr s, (c) 69×10^{-2} torr s, (d) 56×10^{-7} torr s, (e) 66×10^{-6} torr s (Yates & Madey, 1971).

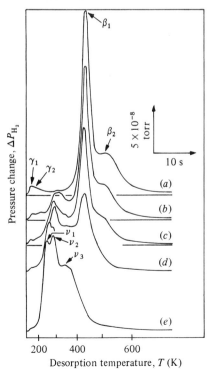

molecule weakly adsorbed on a surface may overcome a dissociation barrier during the heating cycle and at least one of the dissociation products may be created at a temperature well above that at which its desorption peak would have occurred. The desorption of this species is thus limited by the dissociation step rather than the desorption one. More generally, a surface *reaction* may be the rate limiting step. Such an effect can prove most valuable in its own right, providing information on the surface reaction kinetics. An example of such a set of data is presented in fig. 5.13 and relates to the desorbing reaction products following adsorption of formic acid on a Cu{110} surface from the work of Madix (1979). The adsorbed species was actually HCOOD (i.e. with the —OH group deuterated) although some undeuterated HCOOH was present. The data provide clear evidence for the formation of a surface formate species (HCOO). Thus the essential reactions are

$$HCOOD_{(a)} \rightarrow HCOO_{(a)} + D_{(a)}, \quad 2D_{(a)} - D_{2(g)}, \quad 273\ K$$

or

$$HCOO_{(a)} \rightarrow H_{(a)} + CO_{2(g)}, \quad 2H_{(a)} - H_{2(g)}, \quad\quad 473\ K$$

Note that the coincidence of the H_2 and CO_2 desorption peaks at 473 K indicates that the desorption of these species is limited by the dissociation of the formate species itself. The fact that the high temperature H desorption is essentially entirely of H_2, while the low temperature peak is dominated by D_2 also indicates that the formate is formed by the

Fig. 5.13. Temperature programmed reaction spectra for the products evolved from a Cu{110} surface exposed to HCOOD (with a small amount of HCOOH) at 200 K (after Madix, 1979).

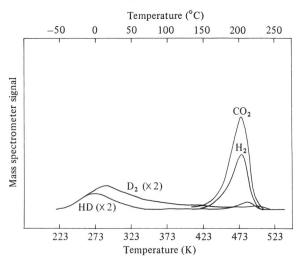

dissociation of the —OH species of the acid. Strictly, the evidence of fig. 5.13 would be consistent with formic acid dissociation occurring at a temperature lower than 273 K, with the D_2 evolution being desorption limited. Other experiments, which can distinguish between formic acid and formate (UPS and vibrational spectroscopies) show that in fact the rate limiting step controlling the D_2 emission is the formic acid dissociation.

5.3   **Electronically stimulated desorption**
5.3.1   *Basic mechanisms*
Desorption of adsorbed species via electronic excitation can occur through stimulation by either incident electrons or photons, but the use of incident electrons is far more common and historically significantly older, so we shall concentrate on this type of excitation, but add some details on the special aspects of photon stimulation.

Many workers concerned with the operation of UHV gauges had noticed the spurious pressure effects caused by the desorption of material from the electrode structures; this is a practical example of electron stimulated desorption of ions. The term Electron Stimulated Desorption (ESD), is used in a general way to denote those physical and chemical changes caused in the surface region of a solid by electron bombardment with low energy electrons (typically less than 500 eV). Although thermal effects can occur, it is the effects arising from electronic transitions within the adsorbed layer with which we will concern ourselves.

Quite a variety of electron stimulated processes have been seen to occur on surfaces including conversion from one binding state to another (including dissociation) and desorption of ground state and excited neutral as well as positively charged and negatively charged atomic and molecular species. Because of the relative ease of detection of charged particles of opposite signs to the incident electrons (and the emitted secondary electrons), by far the most work has been performed on positive ion emission. All these effects are observed to occur at low electron power densities (10–10^{-1} W m^{-2}) so that surface heating effects are both too low to be measurable, or to cause thermal desorption.

Early workers (Rork & Consoliver, 1968) observed that the rate of liberation of ions in ESD is linearly related to the electron-bombardment current, so that the collision process between an electron and an adsorbate species is an isolated event. Also, ejected ions are observed to have most probable kinetic energies as high as 8 eV (Redhead, 1967). Possible electron ejection mechanisms include direct momentum transfer between the bombarding electron and the adsorbate species, or an electronic excitation and dissociation of the adsorbate as in gas phase ionisation–dissociation processes. From the laws of conservation of energy and momentum, it can be seen that the direct transfer of momentum between a low energy, < 500 eV, electron and adsorbate is too small to account for most of the observed processes. The maximum kinetic energy E transferred to a free particle of mass M upon collision with an electron of mass m_e having initial kinetic energy E_S is

$$\Delta E = 4E_S[m_e M/(m_e + M)^2] \approx 4E_S m_e/M \qquad (5.21)$$

Bonding of the particle M to the surface will change its effective mass slightly, but it is clear that even for light atoms and molecules, the energy transfer is small for the electron energies normally employed in ESD. To take an example, E for 100 eV electrons bombarding chemisorbed H_2 molecules is 0.11 eV; thus the strongly bound H atom with a binding energy of 2.3 eV will clearly remain unaffected, and it is only the much more weakly bound physisorbed H_2 molecule which might be affected. As chemisorption energies typically fall in the range 1–8 eV, this conclusion is clearly quite general.

Several workers have proposed, independently, mechanisms for ESD based on the models for electronic excitation and dissociation of gaseous molecules (Menzel & Gomer, 1964; Redhead, 1964) and the general model has come to be known as the Menzel–Gomer–Redhead or MGR model. All of these mechanisms have as their basis the Franck–Condon principle, which states that, during an electronic transition in a molecule, the nuclear separation and relative velocity are essentially unchanged; that is, the electronic transition takes place quickly compared with the time required for appreciable nuclear motion.

By analogy with a diatomic molecule we can represent the potential energy of interaction between a metal substrate surface M, and adsorbate atom A, by a potential curve of the type shown in fig. 5.14. This curve represents the lowest bound state of the metal–adsorbate system; E_A is the binding energy of the atom A. The upper curve represents the state $M + A^+ + e^-$; at large distances from the surface, the curves are simply separated in energy by the ionisation potential of the free atom $E_i(A)$. Following bombardment with electrons, transitions may occur from the bonding $M + A$ curve to the repulsive part of the upper curve. The range of possible internuclear separations in the ground state is represented by the width of the shaded area, while the probability of a particular internuclear separation is the square of the wavefunction for the ground state M–A oscillator. The Franck–Condon principle states that the range of internuclear separations allowed for the M–A system is unchanged immediately after the transition, and the final states on the upper curve are distributed over the Franck–Condon region. For example, the A^+ ions formed at the repulsive part of the $(M + A^+ + e^-)$ curve may desorb with a range of kinetic energies, as indicated in the inset to the figure. The actual yield of positively charged ions, and indeed the energy distribution, will, however, be modified by reneutralisation. In particular, charge exchange can occur by either Auger or resonance neutralisation as described in detail in the previous chapter. Reneutralisation of the ion results in a transition back to the bonding $M + A$ curve. Depending on the kinetic energy gained by the ion before neutralisation, the atom A may be trapped in a vibrationally excited electronic ground state, or may desorb as a neutral.

While fig. 5.15 shows an arrangement of potential energy curves in which the ionic state is repulsive over the range of separations found for the initial ground state system, this need not be so. Moreover, transitions to other possible states of M–A systems are possible. Fig. 5.15 illustrates a possible set of these in the case in which a direct transition to the $M + A^+ + e^-$ system would not lead to desorption (fig. 5.15(a)). Illustrated are possible additional potential curves for an antibonding neutral state and an excited neutral state of the absorbed species. For the special case of rare gas atoms, Hagstrum (1954) has presented evidence that the

Fig. 5.14. Schematic potential curves for interaction between a surface M and an atom A, and between M and the ion A^+. A possible electronic transition resulting in ESD at A^+ is indicated by the shaded region.

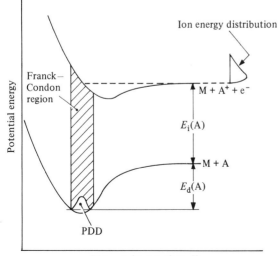

Distance from surface, X

potential energy curves are displaced along the distance axis such that the equilibrium separation distance between substrate and adsorbed atom lies in the order $d_{ion} < d_{ground\ state\ neutral} < d_{metastable}$ as shown in fig. 5.15(b). In this case desorption of metastable neutral A*, positive ions (through ionisation of A*) and ground state neutrals (via neutralisation or direct transition to the antibonding state) are all possible.

In fact, a simple theory based on excitation to a repulsive ionic state, plus the possible effects of reneutralisation, can account for *most* of the basic observations of ESD. These are that (1) generally many more neutrals than ions are observed; (2) the cross-sections of neutral desorption are usually orders of magnitude smaller than comparable gas phase processes, cross-sections of ionic desorption are correspondingly smaller; (3) different modes of bonding of electronegative adsorbates exhibit different ESD cross-sections; (4) ESD is not observed for metallic adsorbates on metal substrates.

Referring to fig. 5.14, excitation from the ground state to the repulsive part of the ionic state is followed either by desorption as an ion, or by reneutralisation. If neutralisation occurs before some critical distance X_c, recapture results. On the other hand, if reneutralisation occurs after the ion has passed X_c, the neutral thus formed will have sufficient energy to escape.

Fig. 5.15. Additional schematic potential curves for interaction between a surface M and an adsorbed atom A. Possible excited states including an ionic state $(M + A^+ + e^-)$, a metastable atomic state $(M + A^*)$ and an antibonding state $(M + A)^a$ are shown. The shaded zones show possible electronic transitions from the ground state to the various excited states.

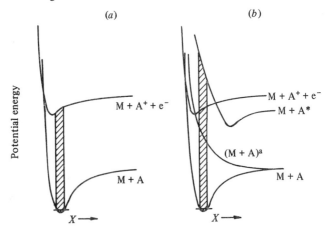

Distance from surface, X

The probability of desorption of an ion formed by electronic excitation of an adsorbed neutral species at a distance X_0 from a surface is given by

$$P_1(X_0) = \exp\left(-\int_{X_0}^{\infty} \frac{R(X)}{v}\,dX\right) \tag{5.22}$$

where $R(X)$ is the rate of neutralisation of the ion at a distance X (see chapter 4); v is the velocity of the ion along the excited state potential curve. The velocity of the ion at X is

$$v = \{2[V(X_0) - V(X)]/m\}^{\frac{1}{2}} \tag{5.23}$$

where $V(X)$ is the excited state potential function and m is the mass of the ion. Combining these two equations we get

$$P_1(X_0) = \exp\left\{-m^{\frac{1}{2}}\int_{X_0}^{\infty} \frac{R(X)}{\{2[V(X_0) - V(X)]\}^{\frac{1}{2}}}\right\} \tag{5.24}$$

Alternatively, the total probability of desorption, regardless of mode, is given by

$$P_T(X_0) = \exp\left\{-m^{\frac{1}{2}}\int_{X_0}^{X_c} -\frac{R(X)\,dX}{[V(X_0) - V(X)]^{\frac{1}{2}}}\right\} \tag{5.25}$$

where X_c is the critical capture distance for the ion excited at X_0.

It is clear from the last two expressions that $P_T(X_0) \gg P_1(X_0)$ and that more neutrals than ions are expected in ESD, as supported by direct experimental evidence. Further, the dependence of ion yield as predicted by this approach has been confirmed by isotope experiments (Madey, Yates, King & Uhlaner, 1970).

It is possible to make some simplifying, but not too crude, assumptions for the foregoing theory. These are that for the region of interest repulsive terms dominate the curve $M + A^+ + e^-$ and that we may write

$$V(X) = B\exp(-bX) \tag{5.26}$$

where B and b are constants. The neutralisation rate $R(X)$ may be expressed (see chapter 4) in the form

$$R(X) = A\exp(-aX) \tag{5.27}$$

where A and a are constants. Equation (5.23) can now be integrated following the insertion of the above quantities to yield

$$P_1(E) = \exp\left[-\frac{A}{b}\left(\frac{m}{2B}\right)^{\frac{1}{2}}\left(\frac{E}{B}\right)^{(a/b-\frac{1}{2})} F\left(\frac{a}{b}, \infty\right)\right] \tag{5.28}$$

where $P_1(E)$ is the probability of escape for an ion formed at X_0, having kinetic energy E at large distances from the surface, and where

$$F(a/b, \infty) = \frac{\pi^{\frac{1}{2}}\Gamma(a/b)}{\Gamma(a/b+\frac{1}{2})} \tag{5.29}$$

$\Gamma(a/b)$ is Gamma function a/b, π has its usual value.

The cross-section for excitation to the ionic state should be similar or comparable with that of similar processes in atoms or molecules if the electrons involved in the transition are reasonably localised. That is to say, the cross-section for excitation should be $\sim 10^{-20}$–10^{-21} m^2 at 100 eV. The ion desorption cross-sections observed in ESD experiments are much smaller than this, the difference being attributable to the rapid neutralisation effect. The magnitude of the cross-section for excitation of the adsorbed species to the ionic state has not been measured directly, although values of this cross-section can be inferred from indirect evidence. For example, data on the isotope effect are consistent with a value of 10^{-20} m^2 for the desorption of O$^+$ from W. It is clear that configurational changes in an adsorbed layer which might alter bond

Fig. 5.16. Energy level diagram illustrating the energy levels involved in the core hole Auger decay mechanism for O desorbing from TiO$_2$ (Knotek & Feibelman, 1978).

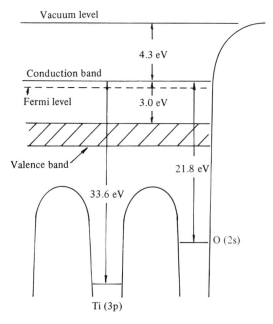

lengths, or the parameters in $V(X)$ and $R(X)$ would alter the escape probability for an ion very dramatically. Thus multiple binding states observed in the chemisorption of atoms and molecules are likely to exhibit different ESD cross-sections (Menzel & Gomer, 1964). Further, if the adsorbed layer exhibits a multiplicity of binding states then the effective displacements of the curves of the type of fig. 5.16 may result in a multiplicity of peaks in the ion–energy distributions. Conversion between binding states is clearly possible since on de-excitation the particle may return to a state differing from its initial state, and this sort of effect has indeed been observed.

Turning now to the observations of low ESD cross-sections for desorption of metallic adsorbates, we note that these species form a rather highly delocalised electronic bond on metal surfaces. Thus the valence level of alkaline and alkaline-earth atoms broaden appreciably on interaction with the surface, as discussed in chapter 4, leading to excellent charge equilibrium between the adsorbate and substrate. In such cases, therefore, reneutralisation and recapture should be very efficient. The very low cross-sections ($\ll 10^{-25}$ m^2) expected for ESD of metallic adsorbates on metal surfaces have been verified experimentally. Thus for Cs on W the cross-section is reported to be $<6 \times 10^{-26}$ m^2, whilst for Th on W the value reported is as low as 10^{-30} m^2.

In the case of thermal desorption we saw that the desorption energy entered the theory in a rather direct way and could be estimated from the data in a relatively straightforward way, even if an accurate evaluation is more troublesome. The potential energy curves of figs. 5.14 and 5.15 show that the adsorption energy is involved in determining the energy required to initiate the process, although some knowledge of the excited repulsive states may also be needed. In the simple case we have outlined above, however, with direct excitation to a repulsive ion state, the threshold energy for the excitation is essentially the sum of the ionisation energy $E_i(A)$ of the adsorbate species and the binding energy on the surface $E_d(A)$. Of course, if the desorbed ions have some minimum kinetic energy this must also be added, but it is convenient to define a threshold with the ion energy zero when the ion is at infinity. The threshold energy will also depend on the destiny of both the bombarding electron and the electron originally on the adsorbed atom, the ionising electron. We can distinguish three cases as follows:

(a) The ion, the bombarding electron and the ionisation electron are all released with zero kinetic energy at $X = \infty$ (Redhead, 1964). In this case, the threshold energy is given by

$$V_T^I = E_i(A) + E_d(A) \tag{5.30}$$

Threshold measurements made for ESD of O on Mo and CO on W and Mo yield values which are consistent with the known binding energies E_d and ionisation potentials E_i according to equation (5.30).

(b) An alternative viewpoint is that at the threshold the ionisation electron is promoted to the Fermi level of the substrate while the bombarding electron and the ion are released with zero kinetic energy at $X = \infty$ (Menzel & Gomer, 1964). This predicts a threshold lower by an amount proportional to the substrate work function ϕ so that

$$V_T^{II} = E_i(A) + E_d(A) - e\phi \tag{5.31}$$

(c) Finally, the bombarding electron and the ionisation electron may go to the Fermi level of the substrate with the ion released having zero kinetic energy at $X = \infty$ (Nishijima & Propst, 1970). The threshold in this case is given by

$$V_T^{III} = E_i(A) + E_d(A) - 2e\phi \tag{5.32}$$

and is termed the absolute threshold. If the probability for transitions at the absolute threshold is zero or extremely small, the experimentally observed threshold will be greater than this absolute value. Sensitive measurements of the threshold values, typically for example for CO on W, show values close to V_T^I whilst measurements for O on W yield values lying between those predicted by V_T^{II} and V_T^{III}. Experimental measurements of ion kinetic energies in ESD indicate that in fact there are essentially no zero kinetic energy ions released upon electron impact; indeed, the minimum kinetic energy may be as high as 3–4 eV.

Threshold measurements can clearly be valuable in clarifying excitation mechanisms operating in ESD. Note that within the restricted model we have used above, the threshold for neutral species desorption would be identical to that for ion desorption, because the neutrals are simply reneutralised desorbing ions. This is not always the case, and some examples show desorption thresholds for neutrals significantly lower than for ions, indicating that an additional mechanism, such as excitation to an antibonding state (fig. 5.15(b)), must be operative. Note that one feature of the general model of thresholds we have presented is that the lowest energy ions should show a somewhat lower threshold than high energy ions. This has been demonstrated (Nishijima & Propst, 1970) for O^+ ions desorbed from O layers on W. Using bombardment energies below 30 eV (threshold occurs at around 20 eV), the peaks in the ion energy distribution are shifted to lower ion energies. Most of the systems examined experimentally show thresholds which lie in the range 5–20 eV.

In several cases, the ion current for ionic desorption has been observed to increase with increasing temperature, that is to say, the cross-section increases with *T*. This increase in cross-section with temperature arises partly from the increased equilibrium distance of the adsorbed species from the surface in excited vibrational levels. The probability of ionic desorption may either increase or decrease with distance from the surface, depending on the values of the parameters *a* and *b* of equations (5.26) and (5.27). However, the total ion current can decrease with increasing temperature as well, if the neutralisation rate of the ions increases. With increased temperature this could occur due to the increased population of energy levels above the Fermi level. An effect of this type has been noted by Madey & Yates (1969).

Another effect of increases in temperature on the desorption process reveals itself as an energy broadening of the ion energy distribution. Thus at higher temperatures the ion energy distribution is smaller than maximum amplitude, but wider in extent, than the distribution for the relatively cool surface. The fractional change in the half-width of the ion energy distribution can be shown to be roughly proportional to the fraction of oscillators in the first excited vibrational state. The temperature dependence of this latter quantity, when plotted as a function of reciprocal temperature, gives as the slope of the plot, the separation between the ground state and the first excited vibrational state of the surface oscillator.

A relatively recent development in ESD, which was highlighted by studies of ionic components, has shown that a somewhat different mechanism may be of considerable importance in many adsorption systems. The ideas are best appreciated by considering the results from some of the ionic systems which have been studied. Thus in the case of TiO_2 and WO_3 substantial O^+ desorption is seen, but at least for high cross-sections the threshold energies are of the order of 30 eV or higher, compared with typical thresholds for the usual MGR model of no more than 20 eV. One obvious problem in this situation is how O^+ is desorbed when the O in the compound is in a (multiply) negatively charged state; apparently some two or three electrons are removed from the O, while we have so far tended to regard the MGR model as essentially a one-electron excitation.

A model which encompasses these experimental features has been proposed by Knotek & Feibelman (1978) and Feibelman & Knotek (1978) and is based on core hole Auger decay, since Auger processes readily involve two or even three electrons. The essence of the model may be seen by reference to fig. 5.16 which depicts the energy levels which are appropriate to the desorption of O from TiO_2. The processes envisaged are as follows. Firstly, a hole is created in the Ti(3p) level some 34 eV below the conduction band. Since in TiO_2 there are no valence electrons on the Ti atoms, the dominant channel for 3p hole decay is an interatomic Auger process. If the two electrons necessary for such a decay come from a surface O then this O will be electrically neutral and easy to desorb. Since one of the remaining valence electrons can be removed by a double Auger process or may be shared between the O and another Ti atom (in equilibrium the O atom is neither O^- or completely O^{2-}), there is a strong probability that the O will desorb as O^+. This mechanism is able to account for the high threshold energies found for oxide materials.

Note that this mechanism does not require any modification to the idea that the Franck–Condon principle is involved with excitation to a repulsive ion state as in fig. 5.14; however, it provides a multielectron mechanism for achieving the transition, which has significant implications for the threshold energies which now correlate with core level ionisation thresholds of either the desorbing species itself or of the metal atom to which it is bonded. Knotek and Feibelman have argued that this latter threshold should only be conspicuous when the bonding forms a maximum valency ionic compound so that intra-atomic Auger decay becomes impossible (or at least ineffective) while the interatomic process must then take over. One obvious and interesting implication is that the detection of core level thresholds can indicate to which species a desorbing ion was bonded in a multicomponent system.

The Auger decay of an ionised core level has subsequently been shown to be an important mechanism in many adsorption systems. Fig. 5.17 shows the results of some measurements on O^+ desorbing from an O exposed $W\{100\}$ surface using both ESD and PSD. It is clear that the electronic excitation processes we have described (and particularly ionisation) should be achievable by incident photons (of sufficient energy) as well as by incident electrons, and many experiments, including the early work of Knotek, Jones & Rehn (1979) and the data presented in fig. 5.17 (Woodruff *et al.*, 1981) have illustrated the equivalence of the two processes in threshold energies and ion angular dependences. PSD has some advantage over ESD in determining threshold energies in that photoionisation cross-sections generally rise very sharply at threshold and then fall off with increasing energy, while electron ionisation cross-sections rise far more slowly. This basic effect has been discussed in chapter 3. Moreover, while electron ionisation cross-sections do fall off (after peaking at about three times the threshold energy), the effect of the increasing number of low energy secondary electrons for a species on a solid surface means that the peak in cross-section is often not seen in the yield; at least for low threshold energies, the secondary electrons can also induce substantial ionisation. Thus PSD yield curves show thresholds far more clearly than ESD yields, although an electron energy derivative ESD yield can show the same fine detail, as is seen in fig. 5.17. The figure shows that the W 5p and 4f core level ionisation thresholds are clearly seen in the O$^+$ PSD and derivative ESD yields, these levels being seen in an optical transmission spectrum of a W foil. Notice that some O$^+$ desorption occurs at lower excitation energies, but the cross-section in this region is low.

One aspect of ion desorption which we have not considered is the angular dependence of the emission. So far, we have simply proposed that an adsorbed species can be excited by some mechanism to a

Fig. 5.17. PSD yield curve of O$^+$ emission from O adsorbed on W{100} compared with the energy derivative ESD from the same surface and with the optical absorption spectrum of a W foil. All curves are normalised to constant incident photon or electron flux, while the ESD is offset in energy by 5 eV as an approximate work function correction to give the same excitation energy relative to the Fermi level (after Woodruff *et al.*, 1981).

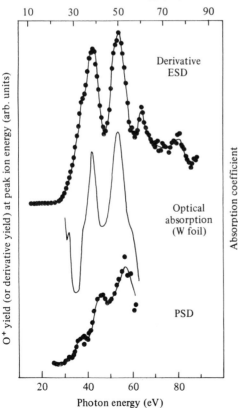

repulsive potential which we have simply represented in one dimension (perpendicular to the surface). In fact the repulsive potential is clearly three dimensional in character and will reflect the symmetry of the adsorption site. This is particularly easy to appreciate in the case of positive ion desorption of an electronegative species from an ionic crystal surface in which we can visualise the desorption as a 'Coulombic explosion' with repulsive forces centred on each electropositive neighbour. The argument, however, is no less valid for more covalent bonding; if the potential is repulsive it must be directed away from the previously bonded neighbours. Consider, for example, the case of CO bonded on a surface by the C end. If O^+ desorption is seen, we expect the desorption direction to indicate the CO bond axis orientation. CO^+ desorption, on the other hand, should reflect the symmetry of the CO-substrate bond. The fact that experiments do show strong angular dependences in ESD and PSD means that there is an important potential for surface structure determination in the two techniques. Moreover, ESD Ion Angular Distributions (ESDIAD) from stepped surfaces have tended to highlight the sensitivity (or properly, specificity) of ESD to adsorption located at defect sites on surfaces. Some examples of these effects are presented in the following section.

5.3.2   *Instrumentation and measurements*

We may divide the experimental methods used for ESD investigations into three categories. The first category comprises those techniques which rely on the detection of a measurable change in the physical or chemical properties of a surface which has been bombarded by slow electrons. Almost any other technique described in this book can be involved, although perhaps most results are centred on techniques which may be used, in the system studied, to establish coverage changes and thus to monitor total desorption cross-sections. The second category of methods relies upon the direct detection of ions, neutrals or metastables as they are released by electron bombardment of the surface. Here are included mass spectrometric analysis of the ions and neutrals released, measurements of ion currents produced by ESD, measurement of the ion kinetic energy distributions and finally combinations of both mass and energy analysis together. The last category concerns itself primarily with measurements of the angular distribution of ions following ESD. Many of the studies in the first category have utilised techniques which use incident electron beams and thus may not have had ESD measurements as their primary objective. Nevertheless, they provide a valuable source of information. Thus LEED studies have revealed striking changes in the observed diffraction pattern as a result of the electron bombardment. Similarly, AES shows that, in some cases, the yield from a particular surface species decreases with time and can be directly related to the changes in surface coverage with electron beam exposure. In addition, desorption studies have shown not only coverage changes with electron bombardment, but also the conversion of the state of a surface species such as molecular dissociation. Provided that the monitoring technique can be used to determine surface coverage, in the bombarded area, in a reasonably reliable fashion, the cross-sections for electron induced processes can be calculated.

If the surface density (coverage), of adsorbed species is, say, n m^{-2}, then the first order rate equation for the rate of change of coverage during bombardment by electron flux n_e (electrons m^{-2} s^{-1}) is

$$-dn/dt = n_e Q n \tag{5.33}$$

where Q is the total cross-section for the electron induced process and has the units, m^2, n is the total coverage of adsorbed species or merely the coverage of species existing in a specific binding state. A first order equation is valid since the experimental evidence (Moore, 1961) shows that the collision between an electron and an adsorbate species is an isolated event. On integration of the above equation one obtains the coverage as a function of time $n(t)$ as

$$n(t)/n_0 = \exp\{-(JQ/e)t\} \tag{5.34}$$

where J is the current density in A m^{-2}, e is the electronic charge in coulombs, and n_0 is the initial surface coverage. In general, it is a question of translating the surface property measured into its form equivalent to $n(t)/n_0$. The cross-section for ESD may then be determined from a semilogarithmic plot of $n(t)/n_0$ against time under conditions where readsorption is negligible. For a true ESD process the value of Q should be independent of the electron current density. All of the indirect methods are limited in sensitivity by the fact that they depend on measuring a surface property associated with small coverage change. This type of measurement frequently involves the small difference between large numbers and is thus easily rendered inaccurate. As a consequence, one can specify a practical lower limit on the size of the cross-section which can be measured by any of the indirect methods; this is $Q \sim 10^{-26}$ m^2. In addition, it is important to note that equation (5.34) involves the incident electron *current density* which requires a knowledge not only of the total current but also the area illuminated. In techniques using directed beams (such as LEED and AES) this conversion can lead to very significant errors in the resulting value of the absolute cross-section.

From the foregoing account it is clear that the direct methods involving detection of ions and neutrals have advantages over the indirect methods. One of the earliest approaches, using a mass spectrometer to detect ionic species liberated upon electron bombardment, was that of Moore (1961). Ions released from his Mo target, following electron bombardment, were accelerated into the analyser region of a magnetic deflection mass spectrometer. More elaborate systems using mass spectrometers have been devised. For example, in one form of mass analyser experiment, the surface under study is scanned by a focussed electron beam, whilst the desorbed ions are mass analysed. The desorbed ion signal of any particular mass to charge ratio is used to modulate the intensity of an oscilloscope driven synchronously with the electron beam, thus creating an image of the distribution of gases adsorbed on the surface (Rork & Consoliver, 1968).

Quadrupole mass analysers have been employed as detectors by a number of workers (Sandstrom, Leck & Donaldson, 1968). An example of the sort of arrangement adopted is shown in fig. 5.18. In this simplified diagram of the electrode system used, a flux of electrons from the filament strikes the target; positive ions emitted by the target travel in the direction opposite to the electrons, are accelerated past the filament and through the grids G_2 and G_3 into the source region of the quadrupole mass analyser. The potentials are arranged so that an ion ejected from the target with near zero kinetic energy can just surmount the potential barrier at G_3; this arrangement of potentials, first suggested by Redhead (1964), completely discriminates against the detection of ions formed in the gas phase by the electron beam.

Practically all of the systems using mass analysers are extremely sensitive and give unambiguous mass identification of ionic and sometimes neutral desorbed species. However, neither the ion transmission probabilities and sensitivities of the analyser–detector systems nor the angular distribution of the ejected ions are known, so that absolute measurements of ionic desorption cross-sections cannot be determined using these instruments. This problem can only be overcome in an apparatus designed to collect all of the ions and neutrals liberated from the sample surface, rather than the small fraction characteristic of the mass spectrometer analysers described above.

While the use of conventional (magnetic or quadrupole) mass spectrometers allows measurements to be made on both ionic and neutral desorbed species, the detection of positively charged ions is generally much easier than neutrals and their energy distributions can be measured by any of the electron energy analysers described in chapter 3. A set of concentric hemispherical grids centred on the sample and used as a variable energy high pass filter is particularly simple and, because the energy range of ions to be detected is small (normally less than 10 eV) and the widths of peaks in the energy distributions are relatively large ($\gtrsim 2$ eV), this simple detector has much to offer (fig. 5.19). Indeed, conventional LEED optics (see chapter 2) can be used, although the subtended angle needs to be increased for true total ion desorption cross-section measurements.

In general, of course, the use of an energy analysing spectrometer to

Fig. 5.18. Apparatus used to detect and mass analyse ESD products. (*a*) Schematic diagram of electrode system showing potentials for electron bombardment and ion extraction; (*b*) simplified diagram of the same electrode system showing positive ion path from target ribbon to multiplier. Note, to improve clarity, grid G_1 is omitted from the diagram (Sandstrom, Leck & Donaldson, 1968).

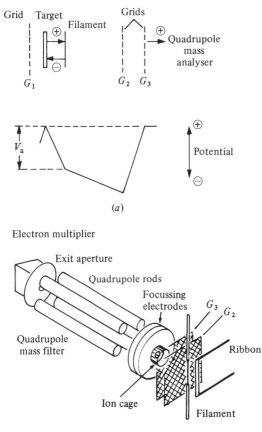

(*a*)

(*b*)

determine the energy distribution of the desorbed ions fails to determine the mass of the desorbed species (except in cases where the energy distribution presents a sufficiently unique value as in the W–CO case described below). For dispersive analysers, however, the mass can also be determined using a *time-of-flight* technique. The time-of-flight of a charged ion from desorption to collection depends on the kinetic energy and the mass of the ion through the trivial kinetic energy equation. In ESD at least it is a simple matter to pulse the incident electron beam

Fig. 5.19. (*a*) Retarding potential apparatus for study of ESD. *X* is a W single crystal sample, *H* is a sample heater (radiative and electron bombardment heating), *F* is a thoria-coated electron bombardment filament, *TC* are thermocouple leads, G_1 and G_2 are both 98% transparent grids, *C* is the ion collector, *GC* conductive coating on the glass envelope. (*b*) Potential distribution in the tube showing application of the retarding voltage V_r for measurements of ion energy. V_e is the electron potential (Madey *et al.*, 1970).

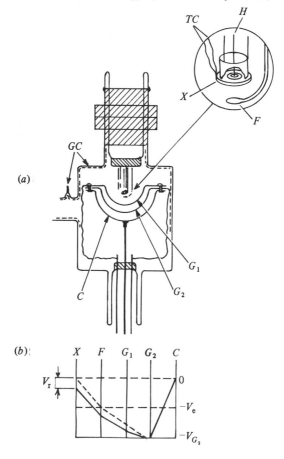

(either by periodically deflecting it or by applying a suitable voltage step to the emission stage of the electron gun), so that time gating of detection allows the flight time to be determined. Channel electron multipliers and channel plate multipliers can detect single emitted ions with excellent time resolution ($\ll 100$ ns). Thus a very simple detector (which is angle-resolving) is simply a channel electron multiplier placed at a suitable distance from the sample and electrically screened to prevent disturbance of the emitted ions distribution, and indeed such a device has been used with some success (Niehus & Krahl-Urban, 1981). However, in this form only the *velocity* of the ion is found and thus the energy and mass information are not separable without some additional information. If the flight time of the ion through a dispersive electrostatic energy analyser is measured, however, the energy and mass can be found independently and thus any analyser of this type (see chapter 3) can be used to measure both simultaneously (Traum & Woodruff, 1980). The method relies on the fact that the *trajectory* of an ion through such an analyser depends on its energy (or strictly on its energy to charge ratio) and *not* on its mass, while the flight time for this known energy is then determined by the mass. As an example of timing requirements, using a double-pass CMA with the desorbing ions accelerated to pass through the analyser at a fixed energy of 78 eV, the velocity of H^+ along the axis of the analyser is 9.03×10^4 m s^{-1} giving a flight time, for a 0.31 m axial length of 3.4 μs (Traum & Woodruff, 1980). The flight time increases as the square root of the mass, so, for example, O^+ ions have a total flight time of 13.7 μs. Timing experiments in the μs range are not very demanding and by varying the pre-acceleration voltage (for fixed pass energy) at constant flight time gating, the energy distribution of a particular species can be measured, while a mass spectrum at fixed energy can be obtained from a time-of-flight spectrum. An example is given in fig. 5.20.

The time-of-flight method, in a rather specialised form, has also been used to detect the mass (but not the energy) of desorbed ions in PSD. Almost all PSD experiments to date have been performed using synchrotron radiation, primarily because one of the strengths of PSD lies in the potentially rich threshold energy information available using tunable radiation in the energy range above about 20 eV (see fig. 5.17). One feature of synchrotron radiation is that it is intrinsically pulsed and, in the case of a single 'bunch' of electrons in the storage ring, the periodicity of the light pulses corresponds to the orbiting time of this electron bunch. This period is invariably short ($< 1 \mu$s) but extremely well defined. In order to utilise this time structure to determine the masses of desorbing ions it is clear, from the discussion above, that the ions need to be accelerated substantially in order that flight times should be comparable to the radiation cycle time. This has been done by accelerating the ions to ∼2–3 keV. In doing so, the initial ion energy (≲ 10 eV) becomes negligible so that the flight time is determined by the mass and the imposed acceleration, and not by the initial desorption energy. Some early results using this method are shown in fig. 5.21 using

Fig. 5.20. ESD data from a contaminated W{100} surface, the emitted positively charged ions being selected by a CMA with the output measured at fixed time intervals after a stimulating incident electron pulse. (*a*) shows the time-of-flight spectrum recorded at pass energies corresponding to different initial ion energies (dash-dot lines represent data collected at an ion energy of 1.75 eV; full lines correspond to an energy of 8.0 eV). (*b*) shows the ion energy spectra recorded at fixed time delays corresponding to different emitted species (after Traum & Woodruff, 1980).

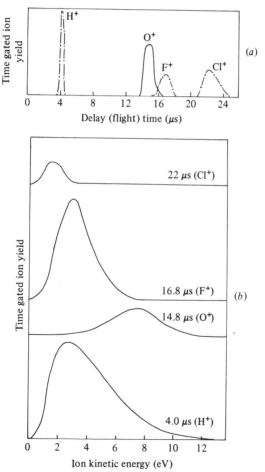

the synchrotron radiation source at Stanford in the USA which has a cycle time of 780 ns in these experiments. Note that the heaviest ions still have a flight time longer than this and so must be correlated in the periodic time-of-flight spectrum with an initiating pulse earlier than the triggering one. The schematic diagram in fig. 5.21(a) illustrates this. More recent results have substantially improved the time resolution of these methods.

Fig. 5.21. Time gated (mass resolved) PSD data for desorption from a TiO_2 surface. (a) shows schematically the time structure of the experiment with the regular photon pulses and the ion collection at fixed time delays after this stimulation. Note how the heavier species (e.g. OH^+) correlate with an earlier photon pulse and so appear in the spectra close to much lighter species. (b) shows actual data taken at photon energies of 22 eV (upper trace) and 30 eV (lower trace). Two different acceleration potentials (2.6 kV and 3.0 kV respectively) account for the shift in the peaks in time for these two spectra (after Knotek, Jones & Rehn, 1979).

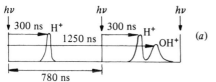

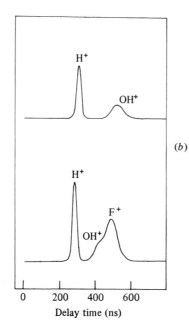

Measurements of the angular distribution of desorbing ions ideally requires this information in addition to the mass and energy distribution although the simplest experimental arrangement concentrates on the angular information alone, and is shown in fig. 5.22. The desorbing ions are emitted into a field free region between the sample and G_1 and are then accelerated onto the first of a pair of microchannel plate multipliers. These give rise to pulses of electrons for each ion reaching them while maintaining the positional information. The electrons striking the phosphor thus give rise to an image which is a projection of the ion angular emission pattern. The acceleration of the ions between grids G_1 and G_2 produces some distortion of this projection but allows the collection of a larger range of emission angles than would otherwise be possible. The only real alternative to this imaging instrument is to use a movable angle-resolving detector and is generally far more complicated, although at least one study has been performed using a simple movable channel electron multiplier as a time-of-flight detector. One virtue claimed for this method is that the time gating allows one to exclude from the measurements the effects of photon detection (due to inverse photoemission at the sample or to ion desorption at the grids screening the electron multiplier) due to scattered electrons from the sample. Both of these processes occur at extremely low time intervals after the ESD stimulation. In the display apparatus these spurious signals can lead to a substantial background signal when the real ion emission signal is low.

Fig. 5.22. Schematic diagram of UHV apparatus for measurement of ion angular distributions in ESD. The crystal is bombarded by a focussed electron beam; ions liberated from the crystal are accelerated by G_1 and G_2 to the microchannel plates (*MCP*). The secondary electron output from *MCP2* is displayed visually on the phosphor screen (Madey, Czyzewski & Yates, 1975).

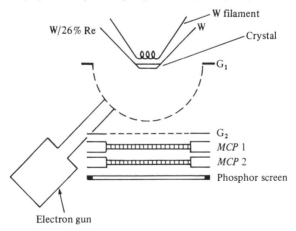

5.3.3   *Some applications and results*

The general utility and limitations of ESD and PSD can best be appreciated by reviewing a few results of experiments using these techniques. One particular system which has already been mentioned in the context of threshold energy measurements (fig. 5.18) is the O–W system which has received an enormous amount of attention using ESD over an extended period of time. The results are complex, as there appears to be quite a large number of surface phases which are formed at different coverages and temperatures. A number of general features emerge. The first is that the yield of the desorbed species, O^+, is very different at low coverages and at high coverages. For example, for adsorption on the $\{100\}$ surface at ~ 77 K, yields of only about 5×10^{-9} ions electron^{-1} are obtained for coverages below $\sim 8 \times 10^{14}$ atoms cm^{-2}, while above this coverage the yield increases to $\sim 10^{-6}$ ions electron^{-1}. The magnitude of the O^+ yield is, therefore, clearly not a good measure of surface coverage, although it may be a good measure of the occupation of one particular state. The desorption of O^+ ions from W$\{100\}$ was also the first system studied by ESDIAD which showed that the emission occurs in quite well-defined 'lobes', with one such emission peak being along the surface normal, while a set of four (symmetrically equivalent) off-normal lobes are seen at 50° to the surface normal. In the case of the W$\{100\}(3 \times 1)$–O structure, these off-normal lobes split, and fig. 5.23 shows an attempt to simulate this ESDIAD pattern, together with the model structures used to generate the simulation. The calculation involved (Preuss, 1980) is a rather sophisticated one which involves constructing a repulsive potential to match not only the directions of the emission lobes but also their widths and the ion energy distributions. Nevertheless, the main features of the results, which can be appreciated from a comparison of the ESDIAD and the associated structural models, is that off-normal emission has to be associated with low symmetry adsorption sites, and, indeed, the off-normal lobe splitting is associated with an additional symmetry lowering achieved by shifting the top layer W atoms within the surface plane (as shown by the arrows in figs. 5.23(*c*) and (*d*)). Emission from any high symmetry site (i.e. atop, bridge, or four-fold hollows on W$\{100\}$) leads to emission along the surface normal. One further feature which the calculations highlight is the difficulty of including reneutralisation in a meaningful way. In particular, a model of the kind we have discussed so far, which is based on a neutralisation rate which increases steeply as the ion–surface separation decreases, leads to the conclusion that a far higher neutralisation probability is incurred in ion emission from a site low in the surface (such as a four-fold hollow) than for a site relatively high above the

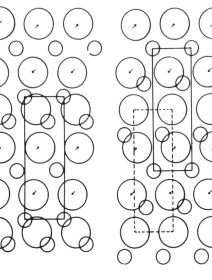

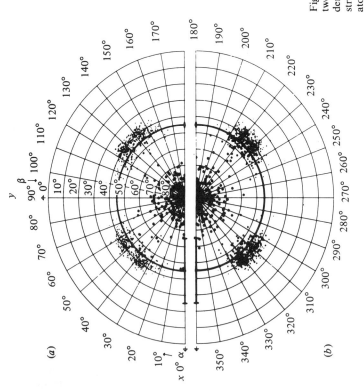

Fig. 5.23. Computer simulations of ESDIAD patterns for O⁺ desorption from two different models of the $W\{100\}(3 \times 1)$–O structure. (a) shows the pattern derived from the structure in (c); (b) the pattern from the structure in (d). In the structural models the small circles represent O atoms and the large circles W atoms in a plan view of the surface. The arrows show the direction of displacements of the top layer W atoms from their unreconstructed positions. These displacements are necessary to generate the splitting of the off-normal emission lobes in the simulation (after Preuss, 1980).

surface (such as an atop site). Whether this is actually true is unknown due to lack of detailed results from known adsorption structures and lack of combined desorbed ion and neutral species studies. What does seem likely is that a proper theoretical description of the effect needs a more sophisticated and local picture of the neutralisation. Indeed, local neutralisation effects could even contribute significantly to the ESDIAD patterns themselves, although a recent simple calculation suggests this may not be too important (Woodruff, 1983).

Another result from the O–W system, but in this case for surface orientations around the {110} face, is illustrated by the montage of ESDIAD patterns shown in fig. 5.24 taken from the work of Madey (1980). O^+ desorption patterns were studied from a sample which was cut to display five different flat faces; in the centre the face is {110} type, while around it are faces cut at 6° and 10° from the low index face in the two principal symmetrically distinct azimuths. For the conditions used to generate fig. 5.24 (4 L O exposure at 300 K) the {110} face shows essentially no O^+ emission, but the vicinal surfaces all show strong emission both along the direction normal to the {110} terraces and in a down-step tilted direction. These data strongly suggest that ESD is occurring, with high cross-section, only from sites of low symmetry in the vicinity of the step edges. They therefore not only reinforce the notion that ESDIAD can provide rather unique structural information, but also the idea that ion desorption cross-sections can be strongly dependent on the state or site of the adsorption. Two possible mechanisms for this specificity can be found in our discussion. One of these is that the Knotek–Feibelman interatomic Auger excitation mechanism, which appears to be important in O^+ desorption from the high coverage $W\{100\}$–O state (see fig. 5.17) may only be possible when a local 'oxide' is formed with the necessary 'high valency'. An alternative view is that the differences in measured ion desorption are associated with differences not in initial excitation, but in site-dependent reneutralisation. Whether either of these explanations is correct is not known at the moment.

As we have already inferred, desorption from molecularly adsorbed species can occur either by breaking an intra-molecular bond or by breaking the molecule–surface bond. Thus in CO adsorption, which has also been studied extensively on W surfaces, both CO^+ and O^+ species are observed to desorb. In this case there is a marked difference in ion kinetic energy, with CO^+ having a peak energy of only about 1.3 eV while O^+ ions have a peak energy of 7.9 eV. There is evidence in some cases that these two desorbing species originate primarily from different adsorption states of the CO on the surface. CO also leads to down-step desorption ion lobes from vicinal $W\{110\}$ surfaces.

One particularly interesting result from ESDIAD studies of adsorbed molecular species arises from work on the adsorption of NH_3 on Ni{111} surfaces, and the effect of pre-adsorbed O (Netzer & Madey, 1982). In particular, H^+ desorption is seen from the surface NH_3, and on the otherwise clean surface, these desorbing H^+ ions form a 'hollow ring' ESDIAD pattern which is attributed to the NH_3 being bonded to the surface with the three N—H bonds tilted away from the surface normal in a three-fold symmetric fashion, but having no azimuthal ordering.

Fig. 5.24. Experimental O^+ desorption ESDIAD patterns produced from individual vicinal surfaces of a multifaceted W crystal around the (110) surface after an O exposure of 4 L at room temperature. Individual patterns are superimposed on a plan view of the crystal to indicate the face to which they relate (after Madey, 1980).

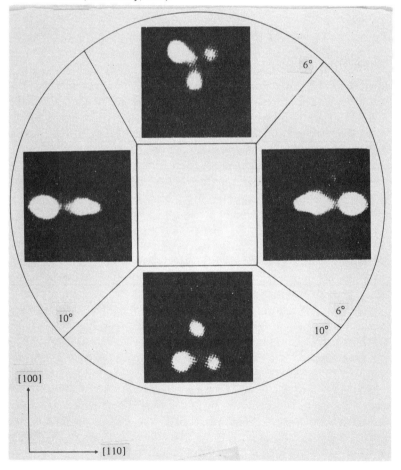

However, in the presence of pre-adsorbed O, this ring pattern is replaced by a three-spot pattern. Apparently the NH_3–O surface interaction leads to azimuthal ordering of the N—H bonds despite the absence of long range order. Similar effects have also been seen for the H_2O–O surface interactions.

These recent results highlight the fact that it is the ESDIAD technique in particular which may prove to be the most useful aspect of electronically stimulated desorption effects, through its structural information. There is, however, one further way in which this form of desorption, particularly in photon stimulation, may lead to surface structural information. In chapter 3 we describe briefly the technique of SEXAFS, in which the modulation of a core level photoionisation cross-section with photon energy above photoionisation threshold can be used to obtain what is, in essence, the radial distribution function around the ionised species. We now see in the case of PSD initiated through the Knotek–Feibelman Auger process, that the photodesorbed ion yield is a monitor of the relevant core level photoionisation cross-section. Thus, the variations of the ion yield with photon energy will also show SEXAFS. Consider, now, the specific case of emission dominated by the interatomic Auger mechanism. The desorbing ions will then provide a monitor of the SEXAFS for the atoms to which these ions were bonded. One experiment to exploit this idea has been performed involving the desorption of O^+ ions from O adsorbed on a $Mo\{100\}$ surface. In this case the SEXAFS should reflect the structural environment of those top layer Mo atoms bonded to the adsorbed O, but not of the substrate Mo atoms. The data from this study (Jaeger *et al.*, 1980) do indicate that this is the case although the fact that the electron backscattering (which causes SEXAFS) is weak from O atoms and strong from Mo atoms meant that the structural information was dominated by the nearest Mo neighbour atoms in the underlying substrate. So far this potential to use PSD to study SEXAFS has not been realised further. One obvious limitation is that it does require the dominance of the Knotek–Feibelman mechanism and also requires that the final state of the photoelectron is not important in the photodesorption process. Measurements on molecular adsorbates, however, suggest that this latter assumption may not be generally valid and that much work remains to be done to understand these processes thoroughly.

In summary, therefore, we see that ESD and PSD measurements do provide considerable information on the electronic and structural aspects of adsorbates on surfaces, and can provide a 'fingerprint' of specific adsorption states. However, the fact that most of the desorption may originate from minority states, and that desorption cross-sections vary enormously from species to species and surface to surface means that they do not provide techniques of general utility. A really detailed quantitative description of the techniques is clearly not available, but nevertheless, ESDIAD, in particular, appears to be able to offer valuable, semiquantitative structural information.

Further reading

A useful review of ESD and PSD can be found in *Desorption Induced by Electronic Transitions* DIET I (Tolk *et al.*, 1983). A review of thermal desorption from metal surfaces may be found in King (1975).

6 High field techniques

6.1 Field emission

The field emission of electrons from a cold metallic cathode in the presence of a large surface electrical field was first reported by Wood (1899). Classical theory fails completely to describe field emission and it is to quantum mechanics that one must turn. Quantum mechanics were first applied to the field emission of electrons from a metal by Fowler & Nordheim (1928).

A simplified view of their result may be obtained by considering a potential energy diagram for electrons in a metal and the adjoining vacuum, in the presence and absence of an external electric field, fig. 6.1. The energy of the highest filled level in the metal, measured from the potential minimum in the metal, is called the Fermi energy E_F, and is equal to the chemical potential of electrons in the metal. The energy difference between the Fermi level and the potential energy of electrons in the vacuum is the thermionic work function ϕ. (A more complete discussion of this is to be found in section 7.2.) The number of quantum states near the top of the Fermi sea is much larger than near the bottom, so that most electrons can be considered to be accommodated in energy levels near E_F, and tunnelling can be assumed to take place largely from the Fermi level.

In the absence of an external field, electrons in the metal are confronted by a semiinfinite potential barrier so that escape is possible only over the barrier, as for example in thermionic emission. The presence of a field F V Å^{-1} at and near the surface modifies the barrier as shown in fig. 6.1. Electrons in the vicinity of the surface now face a finite potential barrier, so that tunnelling can occur for sufficiently low and thin barriers.

The probability P of barrier penetration is given by the Wentzel–Kramer–Brillouin (WKB) method as

$$P = \text{const} \times \exp\left[-2^{\frac{3}{2}}m^{\frac{1}{2}}/\hbar \int_0^l (V-E)^{\frac{1}{2}} \, \mathrm{d}x \right] \qquad (6.1)$$

where m is the mass of the particle; $\hbar$ Planck's constant divided by 2π; E and V the kinetic and potential energy respectively; and l the width of the barrier. Fig. 6.1 shows that the barrier term $(V-E)^{\frac{1}{2}}$ is approximately triangular in shape and hence the area represented by this term is approximately given by

$$A \approx \tfrac{1}{2}\phi^{\frac{1}{2}} \times \phi/F \approx \tfrac{1}{2}\phi^{\frac{3}{2}}/F \tag{6.2}$$

For electrons at the top of the Fermi sea the transmission probability P becomes

$$P = \text{const} \times \exp\left[(2^{\frac{1}{2}}m^{\frac{1}{2}}/\hbar)\phi^{\frac{3}{2}}/F\right] \tag{6.3}$$

Multiplication of P by the number of electrons arriving at unit surface in unit time gives the field emission current density J. The rigorous derivation by Fowler & Nordheim (1928) yields an equation exhibiting a similar form and dependence, namely

$$J = 6.2 \times 10^6 (E_F/\phi)^{\frac{1}{2}}(E_F + \phi)^{-1}F^2 \exp\left(-6.83 \times 10^7\ \phi^{\frac{3}{2}}/F\right) \tag{6.4}$$

The simple potential model of fig. 6.1 can be improved by taking into account the image forces experienced by an electron leaving the metal, for both a clean and adsorbate covered surface, fig. 6.2. It can be seen that the image potential decreases the effective barrier area by giving a reduction in barrier height (decrease in work function) of $e^{\frac{3}{2}}F^{\frac{1}{2}}$. Nordheim (1928) has corrected the exponential part of equation (6.4) to take account of the classical image force to give

$$J = 1.54 \times 10^{-6}F^2/\phi t^2(y) \exp\left[-6.83 \times 10^7\ \phi^{\frac{3}{2}}f(y)/F\right] \tag{6.5}$$

Fig. 6.1. Schematic potential energy diagram for electrons in a metal with and without an applied field. The metal is assumed clean and the image potential is neglected. ϕ, work function; μ, depth of the Fermi sea; E and V are kinetic and potential energies respectively as a function of distance from the metal surface.

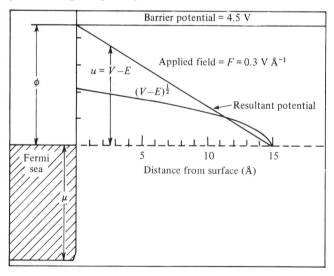

where $f(y)$ and $t(y)$ are slowly varying elliptic functions of the dimension-less parameter

$$y = e^{\frac{3}{2}}F^{\frac{1}{2}}/\phi \qquad (6.6)$$

both $f(y)$ and $t(y)$ are available in tabular form.

Equation (6.5) has been tested experimentally over a wide range of J, notably by Dyke & Trolan (1953), and verified quantitatively. Equation (6.5) can also be written as

$$I/V^2 = a \exp\left(-b\phi^{\frac{3}{2}}/cV\right) \qquad (6.7)$$

where a, b and c are constants, I the emission current and V the applied potential, related to F by

$$F = cV \qquad (6.8)$$

It is apparent that a graph of $\ln(I/V^2)$ versus $1/V$ should be linear and have a slope proportional to $\phi^{\frac{3}{2}}$; this graphical test has been employed in a wide variety of field emission experiments and forms the basis of work function measurements by the field emission method (see section 7.6). Although equation (6.7) is satisfied by experimental results, strictly speaking, it applies only to a clean emitter surface. In the presence of adsorbed material the triangular or pseudotriangular barrier should be modified to account for the presence of the adsorbate at the surface by including a potential well in the barrier.

An exactly solvable one-dimensional pseudopotential model of this form has been used by Alferieff & Duke (1967) to calculate the field

Fig. 6.2. Schematic potential energy diagram based on the image potential for electrons in a metal with and without an applied field. Barriers are shown for clean metal and metal with a dipole layer of adsorbed gas atoms (N). ϕ, work function; μ, depth of Fermi sea; V_A here represents the contribution of the dipole layer to the overall potential.

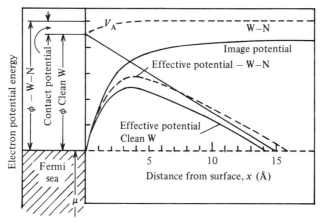

emission probability and current, from a free-electron metal, through both metallic and neutral absorbates. The modified potential used to describe emission from a metal in the presence of an adsorbate is shown in fig. 6.3. The theory developed by Alferieff & Duke is interesting in that it predicts some unexpected effects, namely resonances in the emission probability for metallic adsorbates together with an additional peak or shoulder in the energy distribution of emitted electrons and enhancement of the emission current and reductions in the slope of the Fowler–Nordheim plot at fields >0.5 V Å^{-1}.

Neutral adsorbates, on the other hand, can be divided into those without bound states, and those with bound states below the metallic conduction band. The former lead to reductions in both emission probability and current, together with simple scaling of the Fowler–Nordheim energy distributions. The latter lead to enhancement of the current for loose binding; tight binding leads to a reduction of the current and strong-field reductions of the slope of the Fowler–Nordheim plots. This latter prediction has been used to interpret the experimental data for N adsorbed on W{100} or {411} surfaces where there is an apparent, simultaneous reduction of the work function *and* emission current (Delchar & Ehrlich, 1965; van Oostrum, 1966). Thus, to summarise, the adsorbate can act as an energy source (sink) for the tunnelling electrons. The energy distribution of field emitted electrons is inherently more sensitive to the potential near the surface than the

Fig. 6.3. Schematic diagram of the one-dimensional, one-electron pseudo-potential used to describe field emission from a metal in the presence of an adsorbed atom. d, adatom spacing from the surface; V_R, potential difference between the bottom of the Fermi sea and the adatom ground state; $\psi(E, X)$, electron wavefunction at a distance X from the surface.

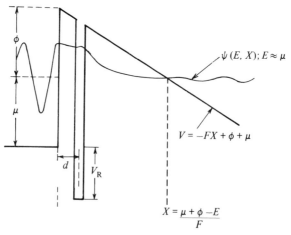

current density, which is an integral over all energies. The model demonstrates that resonance transmission of tunnelling electrons will occur in field emission when the incident electrons from within the metal have the same energy as a virtual level (atomic band) in the adsorbed atom.

A useful aspect of the theory outlined above is that it shows that measurements of the total energy distribution of field emitted electrons provide information on the 'virtual levels' of adsorbate atoms. An early example of this type of measurement is provided by the work of Plummer & Young (1970), fig. 6.4.

6.2. The field emission microscope

The field emission microscope itself was invented by Müller (1936). This instrument approached, for the first time, the ideal of being able to view a surface on a scale that approached the realm of atomic dimensions and yet simultaneously allowed one to follow rapid changes at the surface. In addition, it gave for the first time a direct indication of the cleanness of a surface.

Fig. 6.4. Comparison of experimental and total energy distributions in field emission from a W{100} surface showing the effect of (probably) CO contamination. Curve 1, clean surface. Curve 2, after 10^{-2} L adsorption at 77 K. Curve 3, same after warming to 300 K showing elimination of hump (after Plummer & Young, 1970).

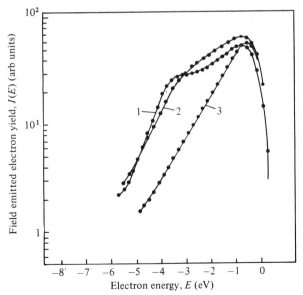

In its simplest form it embodies a wire etched to a very sharp point, placed at the centre of a spherical bulb coated with a conducting fluorescent screen (fig. 6.5). Once the system has been evacuated to pressures of the order of 10^{-7}–10^{-11} torr, the emitter is heated by passage of electricity through the supporting loop until the metal is outgassed and a smoothly rounded tip produced. The radius of this tip varies from 10^{-7} to 10^{-6} m depending on the melting point and ease of outgassing of the metal used. If a potential of the order of 10^4 V is applied between tip and screen, cold emission occurs, since the field at the tip is

$$F = \frac{V}{rk} \tag{6.9}$$

where k is a constant ~ 5, and r is the tip radius, so that $F \sim 10^9$–10^{10} V m^{-1}. Electrons leave the tip with very low initial kinetic energy and will therefore follow paths parallel to the lines of force, at least initially. Since these enter the metal tip perpendicularly, electron paths like those in fig. 6.6 result. The image on the fluorescent screen is thus an electron emission map of the tip, magnified by an amount $D/\delta = x/r$, or more exactly cx/r where c is a compression factor ~ 0.6. Linear magnifications of the order of 10^5–10^6 are possible. The resolution is limited to ~ 20 Å by the tangential velocity of the electrons in the free-electron gas. Their kinetic energy of motion parallel to the interface amounts to ~ 0.1 eV and remains unchanged in the act of tunnelling. It is thus not possible to detect individual adsorbed atoms, but only larger

Fig. 6.5. Schematic drawing of one form of the field emission microscope. *E*, glass envelope; *S*, phosphorescent screen; *B*, tin oxide backing; *A*, anode connector; *T*, emitter tip.

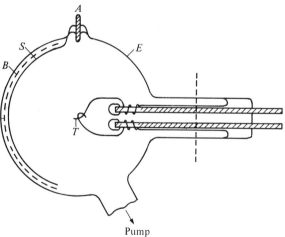

Pump

aggregates. The resolution is, however, sufficient to specify emission changes occurring on regions of known crystal orientation.

Although the emitter is generally made from a polycrystalline wire, the wire drawing process tends to orientate the individual microcrystals, this, coupled with the fact that the emitter tip is very small, normally results in the emitting surface being formed in an individual single crystal with a preferred orientation along the wire axis. This is easily seen in the case of W emitters where the wire axis is normally perpendicular to the {110} plane. The result is that the projected work function map of the emitter is centred around the {110} plane with four {211} planes set symmetrically about it. The close packed planes of the b.c.c. lattice, for example {110} and {211}, have higher work functions than their surrounding planes and appear as dark spots in a more brightly emitting background (fig. 6.7). With the aid of a standard orthographic projection the orientation of the emitter and the identity of the planes can be deduced from the symmetry of the pattern. Thus a four-fold axis occurs only in the ⟨100⟩ and a three-fold only in the ⟨111⟩ direction in cubic crystals. Examination of crystal models shows which planes are close packed in a given structure and may, therefore, be expected to have the highest work function. Once two or three principal directions have been identified by inspection, their angular separation on the image can be compared with the theoretical values and the compression factor c determined.

6.2.1 *Factors governing operation*
The operating conditions of the field emission microscope impose fairly stringent requirements on actual emitters. Visual or

Fig. 6.6. Schematic diagram showing the optics of the field emission microscope. r, radius of curvature of the tip, x, tip to screen distance. A region of linear dimension δ will be magnified to appear as D on the screen.

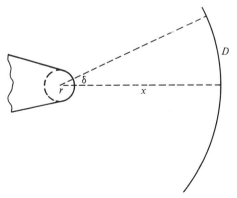

photographic observation of emission patterns requires current densities of $\geqslant 10^{-5}$ A m^{-2} at the screen. For the usual tip to screen distances, currents of 10^{-8}–10^{-7} A are needed. In order to obtain useful magnification and to keep the applied voltages below, say, 20 kV, tips of 10^{-7}–10^{-6} m radius must be used, so that minimum current densities of 10^6–10^7 A m^{-2} on the tip are required. The narrow section of an emitter shank is usually 10^{-3}–10^{-4} m long. This length is restricted by the need to reach high tip temperatures for cleaning purposes, which is impossible if the shank is too long, since radiation cooling limits the tip temperature. In addition, if $i = 10^{-7}$ A then the emitter material must have a resistivity in the range 0.03–0.3 Ω m if there is not to be an excessive voltage drop along the emitter, and this excludes poor semiconductors or indeed insulators.

The maximum current density obtainable from a good conductor in high vacuum is limited in practice by resistive heating of the tip and

Fig. 6.7. Field emission image from a clean W field emitter showing the symmetry typical of such patterns (Delchar, unpublished data).

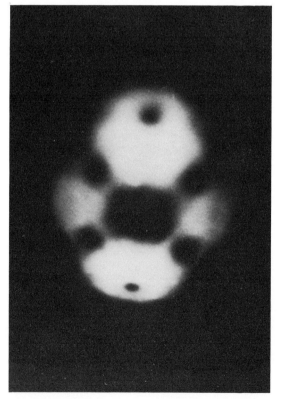

shank. If this is excessive, vaporisation occurs, the evaporated atoms are overtaken and ionised by emitted electrons and the ions are attracted back to the tip where their presence increases the field, neutralises the space charge and leads to even more electron emission. The result is vacuum arcing and destruction of the tip. With a refractory metal like W, currents of $\sim 10^{-4}$ A corresponding to a current density of $\sim 10^9$ A m^{-2} can be obtained. A similar result occurs if the ambient vacuum is allowed to increase much above 10^{-7} torr.

In addition to the current requirements outlined above, the fields required for minimum visible emission are of the order of $(3\text{–}6) \times 10^9$ V m^{-1}. Calculation shows that the stress over the circular section of a hemispherical field emitter is such that it must withstand stresses of the order of $10^8\text{–}10^{10}$ N m^{-2}. Field emission is thus restricted to relatively strong materials, unless tips are made from highly perfect crystals such as whiskers (Melmed & Gomer, 1959). In practice, tips can be fabricated from wires if $T_m > 1300$ K (where T_m is the melting temperature of the tip) or, alternatively, in some cases by epitaxial growth of the desired emitter on a W substrate (Melmed, 1965).

6.2.2 *Practical microscope configurations*

Field emission microscopes, in practice, take a more complex form than the schematic of fig. 6.5 would suggest. Each microscope design depends critically on the type of measurement which it is intended shall be carried out. Some examples of this diversity are shown in figs.

Fig. 6.8. Cross-section through a field emission tube designed primarily for surface diffusion studies. The nozzle N allows gas to be projected at the side of the emitter whilst the cold Dewar surfaces B and C trap molecules which have not interacted with the emitter. S, viewing screen (after Ehrlich & Hudda, 1961).

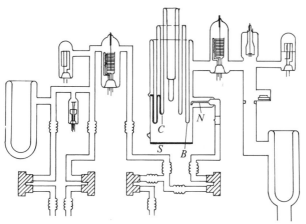

6.8, 6.9 and 6.10 which show, respectively, microscopes designed for surface diffusion measurements; emission measurements from epitaxially grown deposits; total energy distribution measurements of emitted electrons. The significant differences between the various microscope configurations are not necessarily obvious. Thus, in fig. 6.8, the design is intended to ensure that gas for surface diffusion studies can be deposited, via the inlet tube, onto one side of the cold emitter. The second Dewar is cooled with liquid H_2 or He and gas molecules not condensed on the field emitter assembly are trapped on colliding with the glass walls. The diffusion of the gas deposited on the emitter tip can then be studied by raising the emitter temperature. Fig. 6.9 differs from the usual field emission microscope in the provision of an auxiliary metal source, Cu in this instance, which allows vapour deposition and epitaxial growth of Cu crystals on the W field emitter. In this way field emission images can be obtained from rather weak metals. In fig. 6.10 is demonstrated a further variation from the standard pattern where a retarding potential energy analyser is fitted below a probe hole in the phosphor screen. The total energy distribution of the field emitted electrons can thus be measured and used to gain information on the electronic energy levels of adsorbed atoms.

Fig. 6.9. Schematic of field emission microscope for forming emitters by epitaxial growth (after Melmed, 1965).

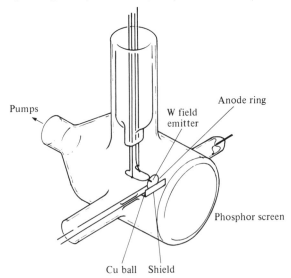

6.2.3   *Experimental results from Field Emission Microscopy (FEM)*
One of the principal uses of the field emission microscope has been in the measurement of work functions and work function changes, particularly by the probe hole technique. This sort of measurement is described more fully in the next chapter. It is, however, appropriate to consider the sort of data which can be obtained concerning other aspects of the adsorption process, namely surface diffusion and thermal desorption. No attempt will be made to provide an exhaustive overview; the example chosen provides a more unusual combination of materials, to demonstrate the scope of the technique.

Although most field emission experiments have involved gases adsorbed on W emitters, materials other than gases can be studied, thus the adsorption of a semiconductor, Si, onto a Mo emitter surface has been described by Venkatachalam & Sinha (1974). The results obtained by these workers for Si on Mo exhibit several interesting features, for example the average work function of a Si covered Mo field emitter decreases, with a simultaneous reduction in the total field emission current. This suggests resonance tunnelling of the field emitted electrons

Fig. 6.10. Schematic representation of a field emission tube fitted with a retarding potential energy analyser and probe hole screen for field emission total energy distribution measurements (after Plummer & Young, 1970).

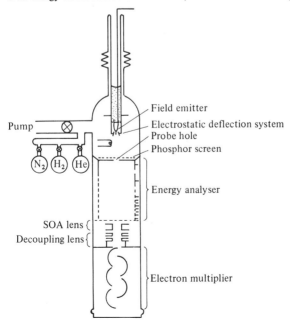

in the manner suggested by Alferieff & Duke (1967). With low coverage, boundary free surface diffusion occurs at 565 K on the {111} zones. Above 585 K diffusion occurs with a sharp boundary and an activation energy of 2.1 eV. The diffusion process can be visualised directly on the emitter tip and is shown in fig. 6.11. It can be seen that diffusion proceeds in the ⟨211⟩ → ⟨100⟩ direction. The activation energy for surface diffusion of Si in the ⟨211⟩ → ⟨100⟩ direction was measured by determining the temperature dependence of the spreading rates. Adsorption of Si on Mo is found to be anisotropic and if the Si-laden emitter is annealed at 1000 K a Si-rich surface phase is produced with new crystal

Fig. 6.11. Surface diffusion of Si on Mo. Si evaporated from the left. (a) Clean Mo tip. (b)–(d) Progressive migration of Si at 610 K. (e)–(f) Further migration at 640 K. Note the dark Si deposits near the {211} planes on the left-hand side of the patterns (b) and (c) (after Venkatachalam & Sinha, 1974).

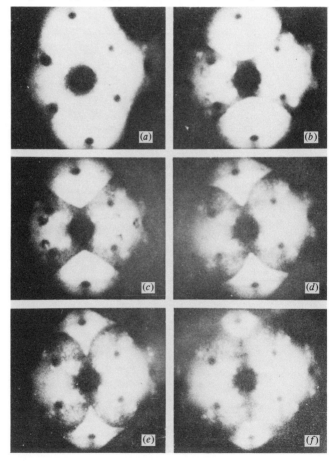

planes, see fig. 6.12; contrast with the clean Mo emitter image of fig. 6.11(*a*). The activation energy of thermal desorption can be determined by measuring the desorption rates as a function of temperature. In this case the measurements were performed by noting the time t for a chosen plane to disappear at a given temperature. If thermal desorption were uniform over the whole emitter it would be sufficient to measure the emission current as a function of temperature T, but this condition will not generally be true. The plot of log t versus $1/T$ for the thermal desorption of Si from Mo is shown in fig. 6.13 for the $\{411\}$ and $\{111\}$ planes, which were shown to have desorption energies of 5.3 eV and 2.7 eV respectively.

6.3   Field ionisation

It was predicted theoretically by Oppenheimer (1928) that a H atom in a high electric field had a finite probability of being ionised by tunnelling. In other words, the electron of a H atom has a finite lifetime in

Fig. 6.12. Field emission pattern from a Si enriched surface phase. Contrast · with fig. 6.11(*a*) (after Venkatachalam & Sinha, 1974).

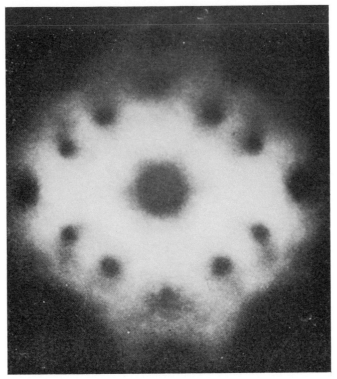

an intense electric field. This prediction remained untested for many years until the brilliant pioneering work of Müller, beginning in 1936, showed first how the necessary high fields could be attained and culminated in the demonstration of field ionisation (Müller, 1951). The stage was then set for an investigation of field ionisation at surfaces.

The current theoretical view of the ionisation of atoms and molecules at surfaces, in high electric fields, is basically that due to Gomer (1954). An electron in a free atom finds itself in a potential well, fig. 6.14(a), where an energy E_i, the ionisation energy, must be supplied to excite the electron and ionise the atom. In an electric field, the potential barrier is reduced, fig. 6.14(b) and (c), and under certain conditions tunnelling of the electron through the barrier, without excitation, can occur. The tunnelling probability is only high when the electric field is sufficient to reduce the barrier to a width comparable with the de Broglie wavelength of an electron inside the atom. The penetration probability for a one-dimensional barrier can be found by the WKB approximation to be

$$D[E, V(x)] = \exp\left\{-(8m/h^2)^{\frac{1}{2}}\int_{x_1}^{x_2}[V(x) - E]^{\frac{1}{2}}\,dx\right\} \quad (6.10)$$

Fig. 6.13. Plot of log t versus $1/T$ for thermal desorption of Si from Mo. Curve 1, thermal desorption from the {411} plane, $E_d = 5.3$ eV; curve 2, thermal desorption from {111}, $E_d = 2.7$ eV (after Venkatachalam & Sinha, 1974).

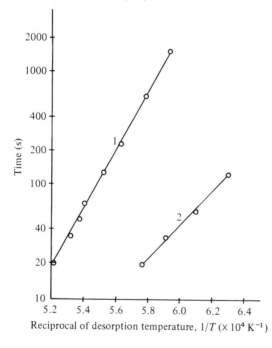

Reciprocal of desorption temperature, $1/T$ ($\times 10^4$ K^{-1})

Fig. 6.14. (*a*) Schematic diagram of the potential well for a free atom. (*b*) Schematic diagram of the potential well for an atom in free space in a high electric field. (*c*) Potential well for the same atom adjacent to a metal surface; ϕ, work function of metal; E_i, ionisation potential of atom; x_c, distance from image plane at which electron level in atom lines up with Fermi level.

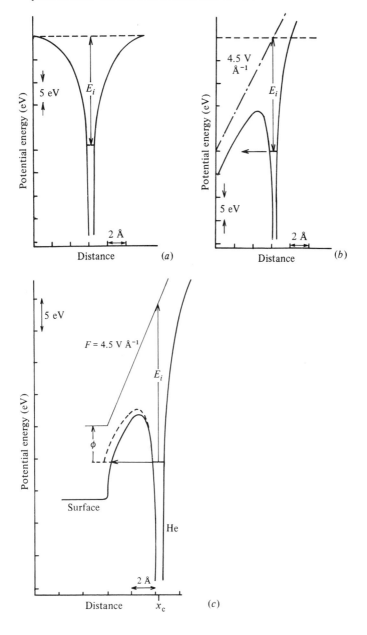

where $V(x)$ and E are the electron's potential and kinetic energies, m its mass, $\hbar = h/2\pi$ where h is Planck's constant, and x_1 and x_2 denote the edges of the barrier at the energy level E. When a gas atom is near the surface of a conductor the penetration probability can be written in an analytical form only if one makes some simplifying assumptions about the effect of the close proximity of the surface upon the potential barrier for the electron. A reasonable approximation is

$$V(x) = -e^2/(x_i - x) + eFx - e^2/4x + e^2/(x_i + x) \qquad (6.11)$$

in which the first term represents the Coulomb potential due to a positive ion of charge e located at a distance x_i from the plane surface of the conductor. The potential energy of the electron due to the applied field is given by the second term, and the third and fourth terms describe the potential energies due to the electron's positive image charge, and the negative image charge of the ion behind the conductor surface respectively. This image potential concept is not valid very close to the metal surface.

The penetration barrier narrows as the atom approaches the metal surface. This reduction of the barrier is caused by the image force or the appropriate exchange and correlation interactions which may contribute to the potential. Nevertheless, the closest distance x_c at which the atom can still be field ionised is determined by the condition that the ground state of the atom must be just above the Fermi level of the metal. At a distance smaller than x_c (fig. 6.14(c)) the metal cannot of course accept an electron since the ground state of the atom will then fall below the Fermi level, and all states below the Fermi level are occupied, at least at $T = 0$ K. Field ionisation by tunnelling of the electron into the metal therefore cannot occur at values of x less than x_c and simultaneously the distance of greatest penetration probability is now obtained by equating the energy of the atomic and ionic states, thus

$$eFx_c = E_i - \phi - e^2/4x_c + \tfrac{1}{2}F^2(\alpha_a - \alpha_i) \qquad (6.12)$$

where ϕ is the work function of the metal and α_a and α_i are the polarisabilities of the gas atom and the ensuing ion respectively. The third term is just the image potential of the electron and the fourth term represents the difference in the polarisation energy before and after ionisation. These latter terms are generally very small compared with E_i and ϕ so that for most purposes equation (6.12) reduces to

$$x_c = (E_i - \phi)/eF \qquad (6.13)$$

Some idea of the magnitude of x_c can be gained by taking the case of He. When E_i and ϕ are expressed in eV and F in V Å^{-1} then x_c is in units of Å. For He with $E_i = 24.5$ eV being ionised, say, at a W surface with $\phi =$

4.5 eV and in a field of 5 V Å^{-1}, the critical distance $x_c = 4$ Å.

If the potential profile, viz. equation (6.12) is known, then the barrier penetration probability D can be obtained by the superposition of the Coulomb field and the applied field, then one obtains

$$D(x_c) = \exp\left[-6.83 \times 10^7 \, E_i^{\frac{3}{2}}(E_i - 7.6 \times 10^{-4} \, F^{\frac{1}{2}}/E_i)^{\frac{1}{2}}/F\right] \quad (6.14)$$

The ionisation probability of an atom is obtained by multiplying the barrier penetration probability by the frequency with which the electron inside the atom strikes the barrier. This frequency can be calculated for H using the Bohr model, whilst estimates can be made for other atoms using the Bohr approach and the effective nuclear charge. Typical values would then be 4.1×10^{16} s^{-1} for H; 2.4×10^{16} s^{-1} for He and 1.5×10^{16} s^{-1} for Ar.

The preceding treatment has covered the question of the ionisation probability of a single gas atom; however, this is not a quantity which is directly measurable. A more direct test of the validity of the field ionisation theory lies in measurements of the field ion energy distribution. Field ionisation of gas atoms occurs at a distance greater than x_c from the metal surface, where x_c is the critical distance as defined by equation (6.13). An ion formed at the surface of the ioniser has an energy equivalent to the full accelerating voltage, when the ion reaches the screen. An ion formed at a distance x from the surface, on the other hand, has an energy equal to the full acceleration voltage, reduced by an amount $\int eF(x)\,\mathrm{d}x$. A measurement of the energy distribution of field ions will then locate the origin of the ionisation and simultaneously provide information on the field ionisation mechanism and field ion image formation.

Experimental results have established the existence of a critical distance x_c and shown that the distribution half-width at 'best image field' is about 0.8 eV for He atoms on W. This corresponds to a very narrow ionisation zone of 0.18 Å depth, or about one-tenth of the diameter of the He atom. Similar results are obtained for Ne, H and Ar.

In the early experiments of Müller he noticed that the measured field ion current was larger by a factor of 10–20 than could be expected from the complete ionisation of the supply of gas molecules bombarding the tip, even when due allowance is made for the multiplying effect of secondary electrons. This was an unexpected, but nevertheless useful, effect and it is due to the enhancement of the gas supply by gas molecules polarised in the inhomogeneous electric field, and attracted to the tip. The calculation of the enhancement of the gas supply by dipole attraction is not simple and various attempts have been made to solve the problem. Calculations of the supply function to a sphere of radius r_t, for which the field F at a distance r from the centre is

$$F(r) = (r_t/r)^2 F_0 \qquad (6.15)$$

where $F_0 = F(r)$ show that the supply function Z_s is given approximately by

$$Z_s \approx \frac{4\pi r_t^2 P}{(2\pi m k_B T)^{\frac{1}{2}}} \left(\frac{\pi \alpha F_0^2}{2k_B T} \right)^{\frac{1}{2}} \qquad (6.16)$$

where m is the mass of the approaching molecule, P is the pressure and k_B is Boltzmann's constant.

In the discussion thus far nothing has been said about the practical form of field ionisation experiments, that is, how the required electric fields of around 4×10^{10} V m^{-1} are attained experimentally without resort to impossibly large power supplies. This problem was overcome quite simply by recognising that the field at a surface of radius of curvature r_t is very much higher than that at a plane surface for a given applied potential, so that

$$F_t = V/kr_t \qquad (6.17)$$

where V is the applied potential, and k is a constant; exactly analogous to the field emission situation. It is then merely a question of constructing surfaces with very small radii of curvature, typically between, say, 10 and 50 nm in the case of a field ion emitter. Incident gas molecules approaching such a tip are then ionised close to the surface and the positive ions formed are swept away towards an imaging phosphorescent screen with a conducting underlayer. The image on the screen then reflects those parts of the field ioniser which have locally higher radii of curvature, such as steps or edges of atom layers, or even isolated metal atoms.

The discussion thus far has considered field ionisation occurring at an ionising tip operating at room temperature. The random tangential velocity of the imaging ions of room temperature energy, say 0.025 eV, cannot, in principle, upset the operation or disturb the resolution of the field ion microscope. However, Müller concluded that this picture is an oversimplification of the situation in the vicinity of a field ioniser, and that the gas molecules arrive at the tip in a near normal direction with a dipole attraction velocity $(\alpha/M)^{\frac{1}{2}} F_0$, which is very much larger than kT for the gas. Müller realised that it is mostly rebounding molecules that are ionised since these would, after diffuse reflection, stay longer in the ionisation zone, and the ions formed would retain a larger than thermal tangential velocity component up to the full value of the dipole attraction velocity. If, on the other hand, the rebounding molecules could be slowed down by accommodation to a tip operating at cryogenic temperatures, then improved resolution could be entertained. In this way Müller demonstrated the first clear resolution of the atomic lattice (Müller, 1955). The molecules, after having lost part of their energy through collision with the ionising tip, may be trapped in the inhomogeneous field near the tip. Since, for example, He atoms are not adsorbed on W at 21 K, they will diffuse over the tip surface in a hopping motion, slow down to near tip temperature equivalent velocity, and be ionised when they pass through the ionisation zone above one of the more protruding surface atoms. At very low fields the ionisation probability is so small that the removal of ions does not appreciably alter the equilibrium density of the gas near the tip. One can show from statistical mechanical considerations that the gas density at the tip n_t exceeds that far away from the tip n_g by a factor

$$n_t/n_g = (T_g/T_t)^{\frac{1}{2}} \exp (\alpha F_0^2/2kT_g) \tag{6.18}$$

where T_g and T_t are the temperatures of the gas and the tip respectively. Thus this expression is only valid if the gas molecules have accommodated to the tip temperature. The ion current contributed by a volume element $2\pi r^2 \, dr$ is given by

$$di = 2\pi r^2 e n_t v D(r) \, dr \tag{6.19}$$

where vD is the ionisation probability. The total current is obtained by integration over r beyond the critical distance x_c and yields

$$i = 2\pi n_g e \left(\frac{T_g}{T_t}\right)^{\frac{1}{2}} \int_{r_t+x_c}^{\infty} r^2 v D(r) \exp \left(\frac{\alpha F^2 r}{2kT_g}\right) dr \tag{6.20}$$

Energy distribution measurements have shown that the ionisation zone is very narrow so that $D(r)$ can be represented by a step function

$$D(r) = D(x_c) \quad \text{for } r_t + x_c \leqslant r \leqslant r_t + x_c + \Delta x \tag{6.21}$$

$$= 0 \qquad \text{for } r > r_t + x_c + \Delta x \tag{6.22}$$

With $F \approx F_0$, the field at the surface of the tip, equation (6.20) becomes

$$i \approx 2\pi r_t^2 n_g e \left(\frac{T_g}{T_t}\right)^{\frac{1}{2}} v D(x_c) \exp \left(\frac{\alpha F_0^2}{2kT}\right) \Delta x \tag{6.23}$$

It can be seen that the current rises steeply with increasing field due to the product of the exponential function D and the final exponential. Equation (6.23) is valid only in the low field region. Unfortunately, this region and the high field region, where free space ionisation predominates, are of less interest in the image formation process.

As long as the kinetic energy of thermalised or rebounding atoms is less than $\frac{1}{2}\alpha_a F^2$, they must stay close to the surface, executing short hops into the inhomogeneous field surrounding the emitter. Escape into space can take place in two ways only. (a) In hopping to heights larger than x_c

when the atom may become ionised, this will occur rapidly close to x_c in low fields. The ion thus formed is then accelerated away towards the imaging screen and is recorded there. (b) The hopping atoms may wander toward the shank of the emitter, into a region in which the potential gradients perpendicular to the surface are small. There they may evaporate.

Hopping atoms penetrate into a region of higher potential energy and are opposed by a force $\alpha_a F \, dF/dr$. The atom is turned back at the height at which the gain in potential energy just equals the initial kinetic energy. Since the field drops off as $F_r = F_0(r_0/r)^x$, where $1 < x < 2$, the distance traversed is proportional to $kTr_0/\alpha_0 F_0{}^2$, where F_0 is the field at the surface, and k is a constant.

We can summarise the processes occurring at the needle shaped ionising surface by beginning with the tip immersed in the image gas, but without any field applied. When the field is turned on, the rate of supply of gas to the tip increases above the normal rate due to molecular bombardment, as the effect of the field polarising the image gas molecules comes into play. The capture cross-section of the tip is increased in the field, the distance covered by hopping into the field diminishes and the depth of the trapping well, equal to $\frac{1}{2}\alpha_a F^2$, increases. This increase in the surface concentration continues until the field becomes high enough to bring about ionisation. As pointed out before, this will occur at those places on the surface at which the field is highest. These areas have three advantages: the supply from the gas is highest there, as atoms are accelerated into the region of greatest field inhomogeneity; the loss by diffusion is least; the barrier to tunnelling is more transparent.

On raising the field still further, both the hop height and the edge of the ionisation zone are brought closer to the surface, the former is more strongly affected however (varying as $1/F^2$ compared with $1/F$) and the fraction of the hopping atoms that are ionised at these spots therefore diminishes. Other areas with a larger radius of curvature may, at this point, become effective for ionisation and draw upon the supply of trapped He atoms. This will, of course, affect the course of ionisation elsewhere, by lowering the atom concentration at the surface.

At very high fields, the incoming image gas may undergo ionisation before colliding with the surface, that is, at some distance from x_c. The ionisation process loses its dependence upon the immediate atomic structure, and the ion image becomes blurred.

A test of the ionisation scheme, briefly outlined above, and the theory which preceded it, is afforded by measurements of the ionisation current. In principle, ion current measurements provide quantitative data to compare with the theoretical predictions. However, because of the complexity of the situation, full agreement between theory and experiment has not been reached at the present time. One of the basic difficulties of describing actually measured I–V characteristics in terms of simple models, is the inhomogeneity of the field over the emitting region. Not only does the field strength drop gradually towards the shank, but also the tip can itself deviates considerably from a geometry that would provide a homogeneous field. That is to say, particular crystal planes protrude sufficiently for the local field to be some 10–20% higher than in other regions. These areas already emit in the field-proportional, supply-limited mode, while the lower field regions, including the shank, still operate in the strongly field-dependent mode. At low fields the ion current should increase very steeply with field and decrease very rapidly with raising of the temperature. When the ionisation probability is large enough to ionise all the incoming molecules in the imaging region, the ion current must be limited by the gas supply function. The ion current can then be expected to depend on the first power of the field and the inverse first power of the temperature as in equation (6.16). This dependence is not well supported by the experimental data. In general, in the imaging region the ion current is determined by the dynamic equilibrium between the gas supplies from the various sources, namely dipole attraction from the shank of the emitter and the region in front of the tip, the diffusion rate and the ionisation rate. Contribution to the ion current by the gas supply from the shank is supported by the dependence of the slope in the plot ln I versus ln V upon the tip cone angle, and the time delay of the ion current following a temperature change.

The effect of the gas supply by dipole attraction can be distinguished from the effect of the gas–surface interaction on the ion current, by keeping the gas temperature constant whilst varying the tip temperature. The effect of tip temperature turns out to be much less dramatic than might be expected, implying that in the imaging field region the hopping gas molecules are not fully accommodated to the tip temperature. At intermediate fields, the situation is more complicated, since the incoming molecules have velocities in excess of the thermal values because of polarisation and the dipole forces. If the atoms or molecules striking the tip are thermally accommodated there, at least in part, their rebound velocity will be less than the incoming velocity. As a result they will spend more time in the high field region on the rebound where ionisation is more likely to occur.

If the polarisation energy exceeds kT_t, a fully accommodated particle is unable to leave the tip in one trajectory but will perform a series of hops as already described. Although the capture condition is already met at room temperature for even slightly polarisable gases, adequate accommodation is not likely to occur in a single collision for molecules hitting the tip with thermal plus polarisation energy. However, when the tip is cooled, sufficient accommodation to prevent escape becomes more probable. If a molecule fails to escape on the first try, full accommodation to the tip temperature almost certainly occurs in the course of the subsequent hops. If the hop trajectories take molecules into the ionisation zone beyond x_c, the total time spent there may be sufficient to insure complete ionisation of all these molecules, and the current will again depend only on the supply function. Equally, if the temperature is so low that the average hopping height is less than x_c, most incoming molecules will diffuse out of the high field region without becoming ionized.

In the case of polar molecules such as H_2O, the polarisation energy may be so high that condensation occurs on the tip. The resulting liquid or solid film does not usually extend beyond x_c, since the ionisation probability becomes extremely high there, because of the long times spent by the film molecules in the ionisation zone.

At sufficiently high fields all particles approaching the tip become ionised before reaching it. Indeed, ionisation can occur as far away from the tip as 100 Å, so that the current is determined only by the supply function. For a more detailed discussion of the field ion current, the reader is referred to the work of Müller & Tsong (1969).

6.4   Field evaporation and desorption

One of the most important aspects of field ionisation is its role in the field ion microscope, which has enabled us to view well-resolved images of the metal atoms of the ionising tip. This imaging by radial projection, depends critically upon the feasibility of producing a nearly hemispherical and atomically smooth tip surface and this is achieved by field evaporation which occurs when a sufficiently high electric field is applied to the roughly shaped tip. Any protrusions are removed in the form of positive ions in an entirely self-regulating way. The local field enhancement at sharp edges and protrusions causes these to evaporate preferentially until an ideally smooth surface is obtained. This surface is also atomically clean. At room temperature the field necessary for field evaporation is above 5 V Å^{-1}.

Since the probability of ionisation is a sensitive function of field, the greatest current will come from regions where this is highest. The field distribution near the tip is determined by its atomic structure. To a first approximation, the equipotentials can be represented by sections of spheres, centred on the surface atoms. At x_c where ionisation occurs, the overlap of the equipotential spheres causes a certain amount of smearing out and loss of structure compared with the tip. Field anisotropies will therefore be most marked for atoms with relatively few nearest neighbours and will be least for atoms in smooth, densely packed planes. The former will show up on the pattern as points of high brightness, while the latter may not show up at all at the threshold field. Of course, as the voltage is raised even some of the latter regions will begin to emit. At even higher fields ionisation will become appreciable everywhere and the structure of the pattern will disappear. Optimum resolution will therefore be obtained at fields slightly in excess of the threshold for most exposed or protruding regions of the tip surface. This occurs at the 'best image field' F_i.

In the preceding sections the ionisation of a He atom at the surface of a metal in the presence of a high electric field has been outlined. The field desorption of more strongly bound electronegative atoms, for which $E_i - \phi$ is also high, occurs in much the same way. The only real difference is introduced by the greater binding energy. At fields so low that x_c is larger than the range of the forces binding the atoms to the surface, the adatom must first be thermally desorbed. Only then can ionisation occur as it does for He. However, ionisation can be accomplished without thermal activation, by increasing the field so that the ionic curve is depressed almost to the bottom of the well for the atomic level. An isoenergetic conversion from the atomic to the ionic state can now take place at

$$Fex = E_i - \phi - e^2/4x + \tfrac{1}{2}F^2(\alpha_a - \alpha_i) - V(x) \tag{6.24}$$

where $V(x)$ is the atomic potential at the point x. This tends to increase the fields required for ionisation and counteracts the generally lower values of the ionisation potential. The barrier facing the atom can be adjusted arbitrarily, by applying different electric fields.

A different situation prevails when the ionisation potential of the atom is close to the work function of the surface. For an ion at a distance x_e from the surface, the attractive part of the potential energy curve in the absence of a field can be closely approximated by the image potential. When a high field is now superimposed, the resulting curve shows a maximum below the zero field vacuum level, the well-known Schottky saddle, fig. 6.15. If the field is increased still further, the saddle can be lowered sufficiently to coincide with the ground state level of the ion. Evaporation will then occur, even at $T=0$ K, at a field given by

$$Fe^3 = (E_d + E_i - \phi + Fex_e)^2 \tag{6.25}$$

where E_d is the desorption energy of the ion.

When the ground state is atomic, the effect of the field must be taken into consideration and represented as a polarisation term. Evaporation at $T=0$ K will occur when the field satisfies the condition

$$Fe^3 = [E_d + E_i - \phi + \tfrac{1}{2}F^2(\alpha_a - \alpha_i)]^2 \tag{6.26}$$

Since evaporation can be induced at high fields, without heating, a sample surface can be prepared free from thermal disorder. The fields required for this can most conveniently be estimated by assuming that evaporation occurs by depression of the Schottky saddle of fig. 6.15 and then the required fields follow from equation (6.25). The magnitude of the fields required for some common metals is listed in table 6.1.

6.5   The field ion microscope

Although historically the study of field ionisation began with the use of H as the imaging gas, it has been with He that most recent work has been done. Despite the fact that the stable He^+ ion images of the refractory metals are of extremely low intensity, it is just sufficient for direct photographic recording. However, the common transition metals and the noble metals are seen to field evaporate at the best image conditions for He at a rate of around one atom layer per second. The introduction of the use of Ne and H has enabled successful imaging of materials such as Fe and Ni, but although the images are stable they are so weak that even visual observation is unsatisfactory. This problem has been overcome by the use of efficient and convenient image intensifiers which have at the same time made all of the inert gases available for Field Ion Microscopy (FIM) and thus allowed high quality imaging from

Fig. 6.15. Field desorption of an ion. (*a*) Zero field situation showing the position of ionic ground state at a surface relative to the vacuum level. (*b*) High field situation; x_S, position of Schottky saddle; x_e, equilibrium position of ion; F, applied field.

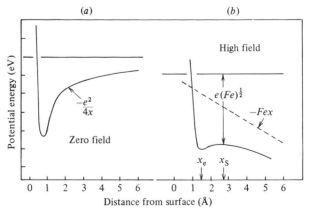

Table 6.1. *Field evaporation at T=0 K*

Metal	Field V Å^{-1}
W	10.2
Ta	9.4
Ir	8.1
Nb	7.8
Mo	6.6
Pt	6.3
Au	4.7
Fe	4.6
Si	4.4
Ni	3.8
Ge	3.4
Cu	3.2
U	2.8

rather weak metals such as Au. Fig. 6.16 shows a schematic diagram of the sort of experimental arrangement used: it also emphasises its extreme simplicity.

For a given set of imaging conditions it is possible to define a characteristic value F_i of the field at the specimen surface at which the most informative and satisfactory field ion images are obtained. This characteristic field, known as the best image field, is primarily a function

Fig. 6.16. Schematic diagram of field ion microscope incorporating an image intensifier.

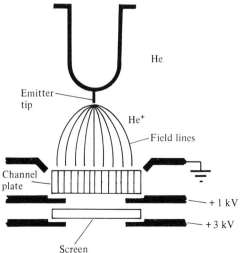

of the imaging gas species and particularly its ionisation energy, and it provides a useful basis for comparing the various possible imaging gases. A second important parameter is the characteristic field F_e, for field evaporation of the surface of the specimen material, which is primarily a function of the tip material, and rather weakly dependent on temperature and other experimental conditions, see equation (6.26). For a surface to yield stable and useful field ion images, it is clearly necessary that F_i should be less than F_e.

A further general requirement is that the imaged surface should have a well-ordered surface structure, at least as a starting point in the case of adsorption or vapour deposition experiments, if the field ion images are to be readily interpretable. Such a surface must generally be produced by the process of field evaporation and should remain stable under the imaging conditions, particularly with respect to changes due to inadvertant adsorption or desorption, the latter commonly by field desorption of an adatom with a substrate adatom at a field which is less than F_e of the clean substrate.

For the latter reason the inert gases are preferred for field ion imaging since they interact only relatively weakly with the specimen surface. Of the inert gases He was the one usually selected when the image detection system was a phosphor screen. This was an obvious choice because of the much lower efficiency, and more rapid deterioration, of phosphors for ions of the heavier gases. However, He has the highest ionisation field F_i at 4.4 V Å^{-1}, its use, therefore, is restricted to the few refractory metals with larger values of F_e, and to the most strongly bound adsorbates. Fig. 6.17 shows an example of the image obtained for He on W. The values of F_i for the inert gases lie in the order He > Ne > Ar > Kr > Xe. The use of the more active gases, such as H, N or CH_4, is usually limited by their strong chemical interactions with metal surfaces under the influence of the high fields required for ion imaging, although H was the imaging gas originally used by Müller. Nevertheless, these interactions have the effect of lowering the field F_e, required for evaporation of the specimen material since a further requirement for satisfactory imaging is that the specimen should not deform or fracture under the large mechanical stress imposed by the electric field; the ability to prepare a surface by field evaporation in H at a reduced level of mechanical stress is a useful asset.

The use of Ne and Ar particularly as imaging gases in the field ion microscope has enabled very high quality images of the very weak materials such as Cu, Au and Al to be obtained. Ar images of Au specimens generally are of poorer resolution than those obtained with Ne. A crude guide to the application of different imaging gases to metals is provided by the melting points of the metals, for example, metals with melting points above 2000 °C may be imaged with He, Ne may be used for metals with melting points between 2000 °C and 1000 °C, and Ar for metals with melting points lying in the range 1000–600 °C. At the lower end of their respective ranges, He and Ne are only useful at 20 K and Ar at 55 K. An example of the use of Ne as an imaging gas is shown in fig. 6.18.

Although the field ion microscope, the practical embodiment of field ionisation, is one of the most powerful microscopic instruments currently known, it is also inherently one of the most simple and unsophisticated experimental arrangements. Indeed, its simplicity is such that, as pointed out by Müller, it could readily have been invented in the late 1800s. Indeed, the German physicist Goldstein managed, in 1876, to image the relief details of a coin in a gaseous discharge tube. Unfortunately, he did not think of taking his experiment one step further

Fig. 6.17. Field ion micrograph for W at 78 K using He as the imaging gas.

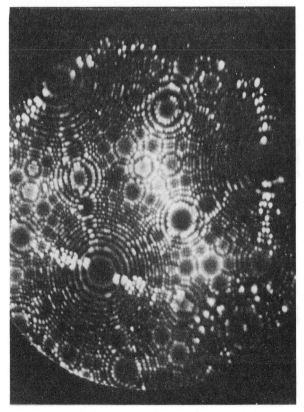

with the use of a convex rather than a flat surface so that the invention of the field ion microscope had to wait a further seventy years.

The first field ion microscope of 1951 was simply a field emission microscope operated in reverse, that is, with the tip positive and the screen negative, and with filling of $\sim 10^{-3}$ torr of H_2. Fortunately, H_2 requires quite a modest field for imaging, some $2\,\text{V}\,\text{Å}^{-1}$, and also excites the usual phosphor screens reasonably well. An elegant example of an all glass system for use with He is based on fig. 6.8. It is fitted with double jackets to facilitate liquid H_2 or liquid He cooling. This particular tube was designed originally, among other tasks, for the study of individual W atoms, vapour deposited, on the $\{110\}$ plane of W (Ehrlich & Kirk, 1968). For this purpose, direct access to the specimen tip is via ports in the concentric cooling jackets, not shown.

The cooling jackets serve to provide thermal insulation of the tip cold

Fig. 6.18. Field ion micrograph for Fe at 78 K using Ne as the imaging gas.

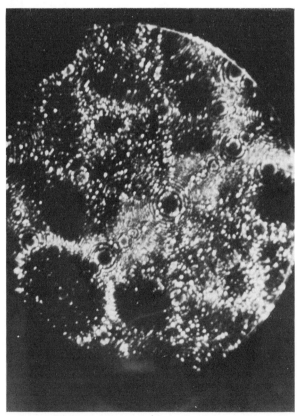

finger and at the same time increase the supply of gas to the ionising tip. The inner walls of the jackets and of the surrounding vessel, including the screen, are coated with a transparent conducting coat of tin oxide to which contact is made by sealed-in wires. The tip specimens are mounted on a heating loop and spot welded to the W leads of the cold finger.

For adsorption studies and related problems it is necessary to attain UHV conditions with the field ion microscope but if the imaging gas is He this restriction can be relaxed. When He is used as the imaging gas the specimen surface is operated with such a high field that it is protected from possible contamination. Molecules of residual gases such as CO, N_2, O_2, etc., all have such low ionisation potentials that they are ionised in free space above the specimen tip and then repelled. Once the specimen surface is cleaned by field desorption and field evaporation, no impurity molecule can reach the imaged region of the tip from the space above it, as long as the field at the tip is kept high enough. The protection of the tip by the high field is only effective with He as the imaging gas, all other imaging gases require the maintenance of UHV conditions.

It is perfectly possible, and in many instances more convenient, to construct the field ionisation system from metal rather than glass. This provides ruggedness together with easy screen and specimen replacement. Further advantages are the improved safety when using inflammable refrigerants such as liquid H, and improved screening of the high voltage leads.

The preparation of the specimen tip is crucial to the whole operation of the field ionisation process and the field ion microscope. The most convenient and effective method of tip specimen preparation is by the shaping of the end of a thin wire, usually 5×10^{-5} m in diameter, to a tapered needle by anodic electropolishing. The various techniques and recipes appropriate to different materials are very fully described by Müller & Tsong (1969).

6.6   The atom probe field ion microscope

There are many variants of the basic field ionisation system which have been constructed, amongst which probably the most striking is the atom probe field ion microscope. Although it is possible to image adatoms such as N, etc., on the surface of a W tip, it has not been possible to state unequivocally the chemical nature of the atom producing the image. That is to say, the N might well have caused a displacement of a surface W atom in such a manner that the W atom protruded from its original site and became a point of high local field. Equally, even one species of atom making up the surface of a pure metal specimen may appear in a wide range of spot sizes, thereby making it impossible to recognise, unambiguously, foreign atoms by their appearance. This problem has been overcome by combining the field ion microscope with a mass spectrometer of single particle sensitivity to form the atom probe field ion microscope. The atom probe field ion microscope identifies the nature of one single atom as seen on the specimen surface and selected at the whim of the observer.

Basically, the atom probe field ion microscope (fig. 6.19) is a field ion microscope modified so that the atom spot chosen for analysis can be placed on a probe hole in the screen. This hole is the entrance to the time-of-flight mass spectrometer section of the instrument. Analysis is carried out by field desorbing the test atom with a high voltage pulse. The resulting ion travels through the probe hole and the drift tube of the mass spectrometer until it reaches an electron multiplier detector. The desorption pulse triggers the horizontal sweep of an oscilloscope. When the ion reaches the detector its output is fed into the oscilloscope. In this way the time-of-flight, t, of the ion to the detector at a distance d is recorded and, since it had acquired its final velocity within a few tip radii away from the emitter, its mass to charge ratio M/n can be calculated quite simply to be

$$M/n = 2eV_c t^2/d^2 \qquad (6.27)$$

where n is an integer between 1 and 4 and V_c is the applied evaporation voltage.

Fig. 6.19. Schematic diagram of the atom probe field ion microscope (after Müller, Krishnaswamy & McLane, 1970).

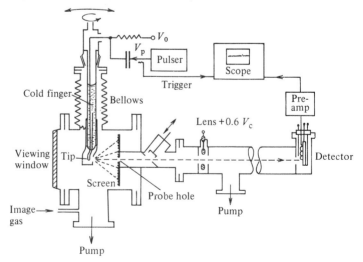

Thus it is possible to identify mass spectrometrically a single atom, making the device the most sensitive microanalytical tool in existence. A further benefit is the extra light thrown on the imaging process and field evaporation. It has been shown by means of the atom probe field ion microscope that molecular ion compounds of the tip metal with the imaging gas are formed and evaporated. Ions of H, He, Ne and Ar, the latter two with double charges and combined with H, are seen. A further interesting aspect is the multiple charge of field desorbed metals. Thus for Mo up to four-fold charging of the field desorbed atom has been found. More interestingly still, the apparently absolutely clean refractory metal tip is in fact covered by an invisible and probably quite mobile adsorption film of the image gas (Müller, Krishnaswamy & McLane, 1970). The accommodation of the incoming image gas of kinetic energy $\frac{1}{2}\alpha_a F^2$ to the low tip temperature is apparently much more efficient than had been thought before, as the collision frequently occurs with adsorbed gas atoms of equal mass, rather than with the heavier metal atoms. The mechanism of local image gas supply to the field ionisation zone above a protruding surface atom must now be viewed rather differently, as the ionised gas atom may be one knocked off from an adsorbed state rather than being the original hopping atom.

Further reading

FEM is described in considerable detail in *Field Emission and Field Ion Microscopy* (Gomer, 1961). FIM is exhaustively considered by the master of the technique, Erwin Müller, in *Field Ion Microscopy* (Müller & Tsong, 1969).

7   Work function techniques

7.1   Introduction

The measurement of work functions or, more precisely, work function changes (surface potentials), has been widely used in the study of adsorption processes on metal surfaces. The technique has been used both on its own and in conjunction with other techniques such as LEED, infrared spectroscopy or flash desorption to elucidate the mechanism of surface reactions.

The measurement of the work function change or surface potential is useful in that it provides a relatively simple method of monitoring the state of the surface. Any adsorption on the surface will, in general, produce a change in the work function of the surface as will any further change in the state of the adsorbate and/or adsorbent. The method is very sensitive, since adsorption of a monolayer on a surface produces surface potentials which are usually in the range 0.1–1.5 V and, since surface potentials can be measured to within ± 0.001 V, very small amounts of adsorption can be measured in a way which causes little or no disturbance to the surface.

A number of techniques are available for measuring surface potentials. In principle, any method which will measure work functions or Contact Potential Difference (CPD) may be used for measuring surface potentials, although some techniques may interfere with the adsorption process to a limited extent. The most commonly used methods are diode, capacitor methods (vibrating and static), field emission and photoelectric methods.

Before considering the measurement of surface potentials the precise meaning of the terms work function and CPD must be determined for single and polycrystalline surfaces in the presence and absence of electric fields.

7.2   Single crystal surfaces

The work function ϕ of a surface may be defined as the difference between the electrochemical potential $\bar{\mu}$ of electrons inside the metal and the electrostatic potential Φ_o of electrons just outside the surface.

$$e\phi = -e\Phi_o - \bar{\mu} \tag{7.1}$$

$$\phi = -\Phi_o - \bar{\mu}/e \tag{7.2}$$

where $\bar{\mu}$ is defined as

$$\bar{\mu} = \left(\frac{\partial G}{\partial n_e}\right)_{T_1 P} \tag{7.3}$$

Difficulties arise when one considers the meaning of the statement 'just outside the surface'. The electrostatic potential energy of an electron just outside a metal surface is very dependent on the distance from the surface. In the absence of applied electric fields and at distances from the surface which are not very small, a reasonable approximation to the potential experienced by an electron is given by the image potential,

$$V(r) = -e/16\pi \mathscr{E}_0 r \tag{7.4}$$

where r is the distance from the surface and $\mathscr{E}_0$ is the permittivity of free space. Since this potential, which corresponds to Φ_0 in equations (7.1) and (7.2), is a function of r it is necessary to define the point at which Φ_0 is defined.

For a single crystal sample, in the absence of applied electric fields, the potential $\Phi_0 \to 1$ as $r \to \infty$ and hence the point just outside the surface at which Φ_0 is defined should be infinity. In practice the potential does not change significantly for distances greater than 10^{-7} m from the surface.

In the presence of an applied, accelerating field the work function is less than at zero field (see fig. 7.1), this reduction is known as the Schottky effect. The reduction in the work function $\delta\phi$ and the value of r_0 may be calculated if the form of the variation of the electrostatic potential at the metal surface is taken to be of the image potential type. The potential at distance r from the surface is then

$$V(r) = -Fr - e/16\pi\mathscr{E}_0 r \tag{7.5}$$

Fig. 7.1. The effect of an accelerating electric field on the work function of a metal. The full line shows the image potential, the dashed line shows the potential due to an applied electric field, the chain line represents the total potential.

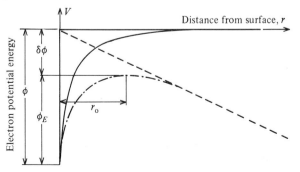

and

$$\delta\phi = -V(r_o) = \phi - \phi_E \tag{7.6}$$

where ϕ_E is the surface work function in the presence of an accelerating field. At the maximum in the potential energy curve, where $r = r_o$

$$(dV/dr)_{r=r_o} = 0 \tag{7.7}$$

Hence

$$e/16\pi\mathscr{E}_o r_o^2 - F = 0 \tag{7.8}$$

which gives

$$r_o = (e/16\pi\mathscr{E}_o)^{\frac{1}{2}}F^{-\frac{1}{2}} \tag{7.9}$$

and the decrease in the work function is given by

$$\delta\phi = -V(r_o) = (eF/4\pi\mathscr{E}_o)^{\frac{1}{2}} \tag{7.10}$$

7.3   Polycrystalline surfaces

Having defined the work function of a single crystal surface, with and without applied electric fields, we must now consider the case of a polycrystalline surface. It is convenient to define in this instance a parameter μ, the chemical potential of electrons inside the metal, by the equation

$$\mu = \bar{\mu} - e\Phi_I \tag{7.11}$$

where Φ_I is the electrostatic potential inside the metal. Thus equation (7.2) becomes

$$\phi = -\Phi_o + \Phi_I - \mu/e \tag{7.12}$$

$$\phi = \Delta\Phi - \mu/e \tag{7.13}$$

The quantity $\Delta\Phi$ is the difference between the electrostatic potential of an electron inside the metal and the electrostatic potential of an electron outside the metal at a defined point. $\Delta\Phi$ is dependent on the surface conditions of the metal and is therefore a function of the structure of the metal surface and will vary also according to the presence of atoms or molecules adsorbed on that surface.

Many surfaces consist of a number of areas of different crystal orientations; these surfaces are said to be 'patchy' and the individual areas are therefore called patches.

The meaning and definition of the work function of such a surface will now be considered, firstly in the absence of applied electric fields. The electrostatic potential at a distance r_o (defined as in equation (7.9)) from the surface of patch i is Φ_{o_i}. It can be shown that, at a distance from the surface which is large compared with the patch dimensions, the electrostatic potential attains a constant value $\bar{\Phi}_o$, given by

$$\bar{\Phi}_o = \sum_i f_i \Phi_{o_i} \tag{7.14}$$

where f_i is the fractional area of the total surface occupied by patch i. Nearer the surface the electrostatic potential varies from place to place giving rise to patch fields.

The work function of patch i is defined by

$$\phi_i = -\Phi_{o_i} - \bar{\mu}/e \tag{7.15}$$

and the mean work function for the whole surface may be defined by

$$\bar{\phi} = -\bar{\Phi}_o - \bar{\mu}/e \tag{7.16}$$

$$= \sum_i f_i \Phi_{o_i} - \bar{\mu}/e \tag{7.17}$$

and therefore, as $\bar{\mu}$ is independent of i and $\sum_i f_i = 1$, equation (7.12) may be rewritten

$$\bar{\phi} = -\sum_i f_i (\Phi_{o_i} + \bar{\mu}/e) \tag{7.18}$$

Hence from equations (7.15) and (7.18) the mean work function of the surface $\bar{\phi}$ is given by

$$\bar{\phi} = \sum_i f_i \phi_i \tag{7.19}$$

A consequence of the patchiness of a metal surface is that some patches will be in an accelerating field (those with $\phi_i > \bar{\phi}$) whilst those with $\phi_i < \bar{\phi}$ will be in a retarding field. A zero field condition exists only for those patches where $\phi_i = \bar{\phi}$.

In the presence of an applied electric field two distinct cases must be considered: firstly where the applied field is small compared with the patch fields. In this case if the distance r_o, calculated from equation (7.9) is much larger than the patch dimensions then $\bar{\phi}$ will be reduced by an amount given by equation (7.6) just as in the case of a single crystal surface.

For larger values of the applied electric field the mean work function no longer has any meaning. If the accelerating field is sufficiently large that r_o is small compared with the patch dimensions then the passage of electrons across the surface depends solely on the individual patch work functions and not on any mean work function.

The work function of a metal surface has now been defined, but it is necessary to realise that even for a single crystal surface the work function must be considered to consist of two parts, one part which depends on the metal itself and the other which depends on the character of the electrical double layer at the surface. This latter component depends on the surface crystallographic orientation so that the work function must be expected to vary from crystal plane to crystal plane, which in fact it does (Smoluchowski, 1941). Fig. 7.2 shows this double layer at a metal surface for both (*a*) an atomically smooth surface and (*b*) an atomically stepped surface. If a metal surface is completely planar as in fig. 7.2(*a*), the electron cloud in the interior will not terminate there abruptly since this would correspond to an infinite kinetic energy. Instead, there will be a gradual decay with a Debye length of 5–10 nm where the wavefunction decays exponentially to zero with the result that the electron cloud extends outside the metal surface, leaving an electron deficiency within the metal and a consequent double layer or potential step with its negative end outermost. The potential across this step is automatically included in the measurement of ϕ but its presence has the effect of making the metal work function vary from crystal plane to crystal plane. Very close packed crystal faces will therefore have high work functions and atomically rough or loosely packed ones, low work functions.

Additional surface double layers arise from adsorption. Changes in the charge distribution occur in such a way that a dipole moment P_A can be associated with each adsorbate atom. The adsorbed layer will, therefore, contribute a term $\Delta\phi$ to the work function

$$\Delta\phi = 2\pi P_A N_s \theta \qquad (7.20)$$

where N_s is the maximum number of adsorption sites per unit area and θ the fraction of occupied sites. This equation implies a linearity of $\Delta\phi$ with θ which is seldom seen. The above equation is correct as written if the dipole is centred symmetrically about the (imaginary) surface of electroneutrality, fig. 7.3(*a*), since an emerging electron must do work against only half of the adlayer potential. However, if the dipole is

Fig. 7.2. Charge distribution at a metal surface (schematic): (*a*) an atomically smooth surface; (*b*) an atomically stepped surface.

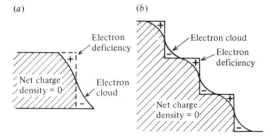

contained wholly within the adlayer, fig. 7.3(*b*) and (*c*), then the appropriate equation is

$$\Delta\phi = 4\pi P_A N_s \theta \tag{7.21}$$

Generally, it is safest to assume equation (7.20).

7.4 Work function measurements based upon the diode method

The diode method has been widely used for measuring the surface potentials of gases adsorbed on metal surfaces, particularly on evaporated metal films. The usual method of measurement relies on the fact that the anode current of a thermionic diode operated in either the retarding field or space charge limited mode may be expressed in the form

$$I_a = f(\phi_a - V_a) \tag{7.22}$$

where I_a is the anode current, V_a is the applied anode voltage, ϕ_a is the anode work function and f is a monotonic function. Now if I_a is plotted against V_a to give the anode current–voltage characteristic, both curves will be represented by:

$$I_a = f(\phi_a - V_a) \tag{7.23}$$
$$I'_a = f(\phi'_a - V'_a) \tag{7.24}$$

hence, provided the form of the function f does not change during the adsorption process at the diode anode, two parallel curves will be obtained. The separation between these curves in the V_a direction is then always given by $\Delta V = V_a - V'_a = \phi_a - \phi'_a$.

It is easier experimentally to hold I_a constant during the adsorption process and measure the change in V_a which ensues, rather than make repeated plots of the current–voltage curves, and this may be done by using a circuit arrangement of the type shown in fig. 7.4(*a*) (for manual control) or fig. 7.4(*b*) for automatic control. Generally, it is necessary to check the current–voltage characteristics for parallelism before and after the adsorption process.

Fig. 7.3. Dipoles arising in (*a*) covalent chemisorption, (*b*) ionic chemisorption and (*c*) physisorption on the surface of a metal. The distance between the charge centres is denoted *d*.

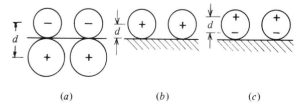

(*a*)   (*b*)   (*c*)

It was pointed out earlier that the diode may be operated in either the space charge limited or retarding field mode. In the space charge limited mode, the diode anode is slightly positive with respect to the cathode (see fig. 7.5(a)), and the anode current is controlled by a space charge between the diode anode and cathode. The relationship between the anode current and the anode voltage has been considered by a number of workers.

Simplified derivations give:

$$I_a = B(V_a + \phi_c - \phi_a)^n \tag{7.25}$$

where B is a constant which depends on the diode geometry and n has a value of 1.5 or slightly less. Thus the diode current is a monotonic function of $(V_a + \phi_c - \phi_a)$ and the method described in the preceding section holds true only if ϕ_c, the cathode work function, does not change during the adsorption process. This requirement will, in general, be satisfied as the filament is usually operated at temperatures in excess of

Fig. 7.4. Circuit arrangements suitable for surface potential measurements with the diode. D is the diode cell; E, electrometer; F, filament power supply; I, constant current source; P, potentiometer; R, 100 Ω resistor.

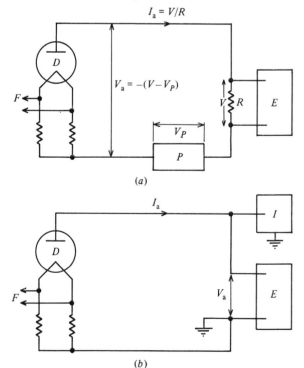

(a)

(b)

1500 K at which temperature adsorption on the filament is highly unlikely.

In the retarding field mode the anode of the diode is negative with respect to the cathode. The variation in the potential energy of an electron passing from the cathode to the anode of a retarding field diode is shown in fig. 7.5(*b*). The relationship between the anode current and anode voltage is given by:

$$I_a = AT_f^2 \exp\left[(V_a - \phi_a)/kT_f\right] \tag{7.26}$$

where A is a constant and T_f is the diode filament temperature. Again the anode current is a monotonic function of $(V_a - \phi_a)$ and the surface potential may be measured in the same way as described for the space charge limited diode.

The preceding analysis has considered essentially single crystal anodes, that is, anodes of uniform work function. Generally, however, the anode in a diode cell is formed from an evaporated metal film and is polycrystalline in structure. In this case the effect of the patchiness of the anode must be considered. The effect of anode patchiness in the

Fig. 7.5. Electron potential energies in (*a*) the space charge limited and (*b*) the retarding field diodes.

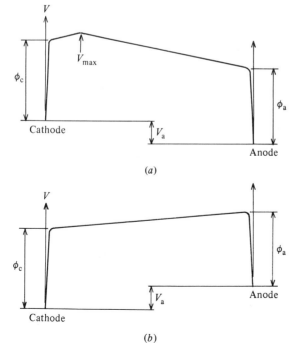

(*a*)

(*b*)

retarding field diode depends on the physical form of the diode. For the present purposes the diodes can be divided into two types, firstly those where any electron reflected from the diode anode will, in all probability, be collected by some electrode other than the anode (Anderson, 1941) and secondly those where an electron reflected from the anode will probably return to the anode at some other point, for example, the spherical diode of Pritchard (1965).

In the first type the anode current may be represented by

$$I_a = BT_f^2(1 - \bar{r}) \exp\left[(V_a - \bar{\phi}_a)/kT_f\right] \tag{7.27}$$

where $\bar{r}$ is the mean electron reflection coefficient for the surface. This form of diode is capable of measuring the true mean surface potential of an adsorbate on a patchy surface, a view upheld by the experiments of Anderson (1952), who found close agreement for the surface potential of Ba on Ag as measured by the vibrating capacitor and retarding field diode methods, the vibrating capacitor being widely taken to give the true mean surface potential.

In the second type of diode, namely that where electrons reflected from the anode have a large probability of approaching it elsewhere, the anode current will be

$$I_a = AT_f^2 \exp\left[(V_a - \bar{\phi})/kT_f\right] \tag{7.28}$$

This type of diode will therefore also be expected to give values of the mean surface potential of the adsorbate. This view is confirmed by the excellent agreement between the surface potentials of a number of gases at various coverages on Au and Ag films measured by Ford (1966) using both the retarding field diode and vibrating capacitor methods.

It should be noted that the diode methods will only give true mean surface potentials if the patch size of the anode patches is small compared with the anode–cathode separation. This results from the fact that the electrostatic potential outside the surface only attains a constant value Φ_o at distances from the surface which are large compared with the patch dimensions. It is only at such distances that the mean work function takes on any meaning. Hence, in the diode method, the electrons will 'see' only the mean work function of the surface rather than the individual patch work functions.

The presence of a hot filament and the emission of electrons from said filament leads to restrictions on the way the diode may be used. In particular it cannot be used (a) in the presence of high gas pressures; (b) where chemical reactions may occur at the hot filament; (c) where the emitted electrons can interact with the adsorbate. All of these objections can be overcome to some extent. One of the effects of high gas pressures is to cool the filament which, in turn, has a large effect on the diode characteristics. This may be overcome by maintaining the filament resistance and therefore its temperature constant. Chemical reactions at the hot filament represent a restriction on the range of gas pressures and types of gas for which surface potential measurements may be made by the diode method. Highly reactive gases such as O may corrode the diode filament thus changing its resistive and emissive properties. The use of materials such as Ir and Re for the filament allows this problem to be overcome. The adsorption of O on Au surfaces has been studied in this way (Ford & Pritchard, 1968). More sensitive molecules such as CH_4, C_2H_2, C_2H_4 and C_2H_6, for example, cannot be studied by the diode method as they pyrolyse at the hot filament. Equally, the study of the adsorption of H is difficult as the hot filament atomises the H. This problem can be reduced by the use of low work function cathodes such as LaB_6 coated Re, which allow filament operation at much lower temperatures.

The usefulness of the diode method lies in its essential simplicity and the absence of a reference electrode contamination problem. Its main disadvantages lie in the low maximum pressure at which it can be operated; the types of adsorbate which can be studied must be limited to those which do not interact chemically with the filament.

The practical conception of the diode has four main forms, which are (a) the electron beam method, (b) spherical or cylindrical diode, (c) crossed filament diode, (d) scanning diode.

In the electron beam method a beam of electrons from an electron gun impinge upon an anode at normal incidence (see fig. 7.6). The anode constitutes the sample upon which surface potential measurements are to be conducted. An example of this form of diode is described by Anderson (1941) and it finds its use mainly for small specimens in the form of foil, ribbon or single crystal.

In the spherical or cylindrical diode, so named by virtue of the geometry of their anodes, a central filament is surrounded by an anode generally, though not always, formed by the evaporation of a metal film. Diodes of this type have been described by Pritchard (1963) and

Fig. 7.6. The electron beam diode showing A, anode (surface under investigation) B_1 and B_2, beam defining apertures usually held at -10 V with respect to C, the cathode (Anderson, 1941).

Mignolet (1955), see fig. 7.7(a) and (b). It is convenient in diodes of this type to use the diode filament as the evaporation source.

The crossed filament diode (Hayes, Hill, Lecchini & Pethica, 1965), consists basically of two filaments mounted so that they are mutually perpendicular (see fig. 7.8). One of these filaments acts as the anode, the other as the cathode. This configuration is particularly useful for studies of the adsorption properties of refractory metals, such as W or Mo, since the surfaces can be cleaned by flashing. In addition the cell lends itself to flash filament measurements. Owing to the close proximity of the two filaments this type of cell cannot be used for low temperature studies owing to the effect of cathode heating on the anode.

A method of work function measurement which has not been widely used is the beam scanning diode described by Haas & Thomas (1966). The form of the diode is shown in fig. 7.9. It comprises an electron gun capable of producing a fine beam of electrons which then fall normally onto a fine mesh grid placed in front of the sample. This grid is maintained at the same potential as the final anode of the electron gun, around $+1000$ V with respect to the cathode. The cathode and anode form a retarding field diode, the anode current of which is controlled by the work function of the area of anode upon which the finely focussed electron beam is impinging.

The grid g_4 is placed close enough to the anode surface to produce a high accelerating field, of the order of 5000 V cm and the situation for

Fig. 7.7. The two principal configurations of diode cell, (a) spherical and (b) cylindrical. A, anode contact; F, functions both as evaporation source for films and as diode cathode; F_1, film evaporation source; F_2, diode cathode; V, pumping orifice; W, thermostatted water supply.

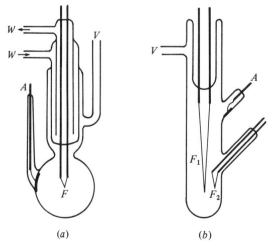

(a)   (b)

patches with work function differences of the order of one volt and of dimensions of the order of 10^{-5} m or greater corresponds to the situation of a patchy surface in an electric field which is large compared with the patch field. For electron beams around 10^{-5} m in diameter the method enables the relative individual work functions of patches of dimensions $> 10^{-5}$ m to be determined. For smaller patches the mean work function of the area under the electron beam is measured.

This method is capable of giving a work function map of the surface if the electron beam is scanned across the surface. The changes which ensue following the adsorption of gas on the surface can also be followed,

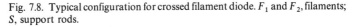

Fig. 7.8. Typical configuration for crossed filament diode. F_1 and F_2, filaments; S, support rods.

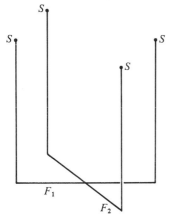

Fig. 7.9. Electron beam scanning diode. A, anode; C, cathode; D, beam defining aperture; E, electron beam; FF, field free region; SC, scan and focus coils, arranged to produce orthogonal scanning; Z, electrometer and signal processing circuits; g_1, g_2, g_3 and g_4, grids. $V_c = -V_a \sim -2\,\text{V}$; $V_{g1}0 \rightarrow -15\,\text{V}$; $V_{g2} = -500\,\text{V}$; $V_{g3} + V_{g4} = -800\,\text{V}$ (Haas & Thomas, 1966).

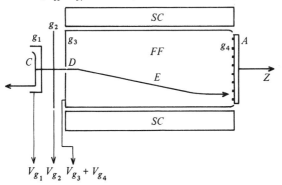

enabling the distribution of adsorbed gas between the various patches to be examined. Owing to the long electron path in the device it is not possible to use this method at high gas pressures.

7.5   Work function measurements based on CPD

A method of measuring work function changes, which does not impose any restriction on the experimental conditions which may be used, is the vibrating capacitor technique. This method depends on the measurement of the CPD which exists between two plates in electrical contact. If we consider, to begin with, two isolated metal plates then the energy diagram for the electrons in each plate is represented by fig. 7.10(a) where the vacuum levels correspond. When the two plates are connected electrically the condition for thermodynamic equilibrium is that the chemical potential per electron shall be uniform throughout the system. Since electrons obey Fermi–Dirac statistics, the chemical potential is also the Fermi energy E_F so that the equilibrium condition is equality of the Fermi energy for each metal (fig. 7.10(b)). Electrons just outside the metal surfaces now have different potentials since the equality of the Fermi levels is obtained by a flow of electrons from plate 2 to plate 1 leaving plate 2 positively charged and plate 1 negatively charged. The difference in the two potentials is just the difference between the respective work functions of the metals. If a potential V_{12} is introduced to restore the correspondence of the vacuum levels (fig. 10(c)) this potential will be equal, but opposite to, the work function difference.

Fig. 7.10. Potential energy relationships for electrons in two metal plates. (a) Plates isolated and charge free, E_{F_1} and E_{F_2} are the Fermi levels. (b) Plates connected electrically showing equivalence of Fermi levels. (c) Effect of applying a balancing external potential between the plates.

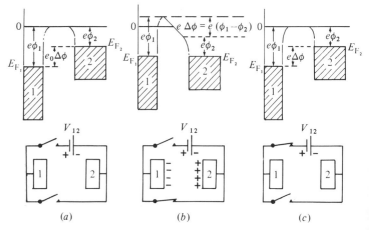

If the work function of one of the plates can be maintained constant (reference surface), then changes in the work function of the other plate will manifest themselves as a contact potential change, it is then merely a question of measuring this change. It is useful to calculate the number of electrons involved in this process, for example a capacitor formed from two parallel metal plates of area 10^{-4} m^2 and separated by 10^{-4} m will require to transfer only around 5×10^{11} electrons to set up a contact potential of 1 V.

There are two distinct approaches to the measurement problem and they both depend on detecting the charge flow between the plates of a capacitor. Historically, the first and most important approach is the vibrating capacitor technique based on the experimental approach of Kelvin (1898).

In this technique the capacitor plates are caused to vibrate with respect to each other whilst the potential between the plates is monitored. If no charge resides on the plates the potential is zero and remains zero as the capacity is altered. If, however, there is a charge on the plates, as a consequence of the contact potential, then varying the capacitance causes the voltage between the capacitor plates to vary in sympathy. If an external potential is introduced in series with the capacitor and adjusted so that it is equal in magnitude and opposite in sign to the contact potential, then the net charge on the plates becomes zero and no change in potential will occur when the plates are vibrated.

Originally Kelvin detected this state of affairs manually using his quadrant electrometer. Nowadays the process can be carried out electronically to provide an automatic recording of the contact potential and any changes which occur, fig. 7.11. There are many elegant and sensitive versions of this technique.

Fig. 7.11. Schematic diagram of an automatic contact potential measuring and recording circuit designed for the system of Delchar (1971) in which frequency doubling occurs in the driving system.

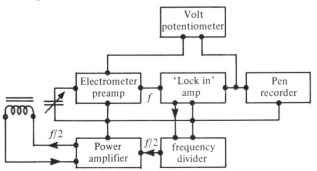

An alternative approach, the static capacitor technique, operates by detecting the charge flow which occurs during the change in contact potential between two plates set up to constitute a capacitor. Here it is necessary to apply the nulling potential in a time t_r (where t_r is the response time) very much less than the time constant formed by the capacitance C of the contact potential cell and the input resistance R of the measuring circuit. Ideally t_r/RC should be less than 0.01 to give accuracy greater than 1%.

The important distinction between the two capacitor techniques is that the vibrating capacitor will measure the contact potential very precisely whilst the static capacitor will measure only contact potential change. A further disadvantage of the static capacitor technique is the fact that it does not distinguish the charge source, only the flow of charge, so that any changes which cause charge flow are interpreted as a contact potential change.

The experimental manifestations of the vibrating capacitor technique may be distinguished principally by the different techniques used to introduce vibrating motion into the vacuum system. One of the earliest applications of the vibrating capacitor technique in surface studies is that of Mignolet (1950) where the vibration of the capacitor plate is achieved by using a natural resonance of the cell structure itself (see fig. 7.12). An extension of this approach is that due originally to Parker

Fig. 7.12. Vibrating condenser cell (after Mignolet, 1950).

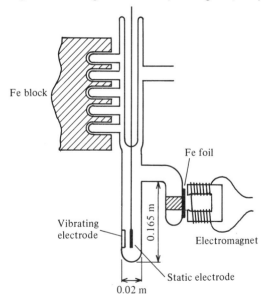

(1962) which uses a resonant bar, clamped at its midpoint, which is also the point of entry to the vacuum system. Magnetic excitation of the end of the bar lying outside the vacuum chamber causes a sympathetic vibration of the end within the vacuum chamber upon which is mounted the reference electrode (fig. 7.13). This technique has been used by Delchar (1971) to examine the adsorption of O and CO on Cu single crystals.

Whilst it is the usual convention to vibrate the capacitor plate in a plane perpendicular to the experimental surface this is not the only configuration which can be used. The moving plate can, for instance, be moved from side to side across the experimental surface so that the separation distance between the two planes is kept small and constant whilst the overlap is modulated at the vibration frequency. This configuration has the advantages that the electrical signal is produced as twice the drive frequency, the experimental surface is freely accessible to adsorbing gases and electron bombardment or Auger analysis can be carried out readily, merely by holding the vibrating surface to one side. One of the earliest examinations of the adsorptive properties of W single crystal surfaces was carried out by this method (Delchar & Ehrlich, 1965)

Fig. 7.13. Cut-away view of a contact potential cell based upon the resonant bar technique. *A*, Ar$^+$ ion gun for sample cleaning; *B*, reference surface (Au); *C*, sample surface; *D*, sample heater; *E*, resonant bar; *F*, thermocouple; *G*, alignment system; *H*, electrometer connection (Delchar, 1971).

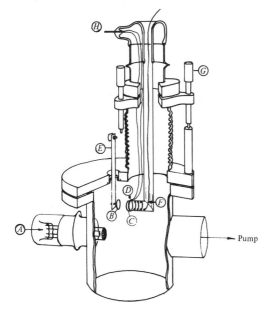

(see fig. 7.14). The static capacitor technique, fig. 7.15, has not been widely used since its first description in the literature (Delchar, Eberhagen & Tompkins, 1963). This is a consequence of the inherent defects outlined above. Notwithstanding these defects the method has recently reemerged with the old-fashioned servo-amplifier approach replaced by a modern integrated circuit (Pasco & Ficalora, 1980). Consequently, the response time and measurement accuracy is greatly improved.

Both of the above methods depend on the availability of a reference surface which will remain inert and unchanged during adsorption processes. A variety of surfaces have been used ranging from oxidised Ni (Mignolet, 1950) to SnO coated glass (Delchar, Eberhagen & Tompkins, 1963) and Au (Delchar, 1971).

7.6   Field emission measurements

The field emission microscope described in chapter 6 provides yet another way of measuring surface potentials since the tunnelling

Fig. 7.14. UHV contact potential system. *A*, W single crystal; *B*, Dewar and support assembly; *C*, electrometer connection; *D*, electron gun for outgassing; *E*, Pt reference surface; *F*, supermalloy driving slug; *G*, bellows; *H*, selective getters (Ni); *I*, vibrator support assembly; *J*, metal valves (after Delchar & Ehrlich, 1965).

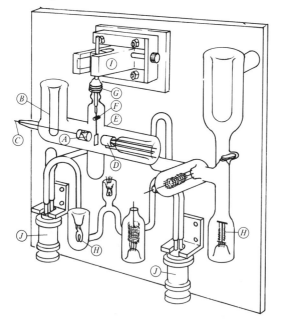

electron current is dependent on the work function of the emitter tip. It is not always possible to measure the work function of the emitter itself but one can measure the work function change or surface potential.

The Fowler–Nordheim (1928) equation describing the tunnelling of electrons through the emitter tip can be recast to the form

$$I/V^2 = a \exp\left(-b\phi^{\frac{3}{2}}/cV\right) \tag{7.29}$$

where a, b and c are constants and $F = cV$. A plot of $\ln I/V^2$ against $1/V$ yields a straight line of slope proportional to $\phi^{\frac{3}{2}}$. Thus successive measurements on an emitter, first clean and then gas covered, provide two plots whose slopes will be in the ratio of the work functions to the power $\frac{3}{2}$. In this method a value for the average work function of the

Fig. 7.15. Contact potential cell for measurements by the static capacitor technique on evaporated metal films. NFB, negative feedback connection; C, evaporation source; A, evaporated metal film (after Delchar, Eberhagen & Tompkins, 1963).

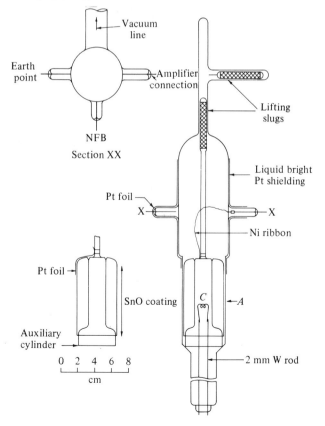

cleam emitter must be assumed before the work function of the gas covered surface can be determined. If the field F is known then the work function can be determined absolutely.

The current from the emitter is the combination of emission from all the crystal planes, but it is possible to ensure that only the emission from a particular crystal plane is selected by using the probe hole technique described by Engel & Gomer (1969), see fig. 7.16. In this approach the anode has a small aperture cut in its centre behind which lies a channeltron electron multiplier. Since the electrons leave the emitter following the field lines, it is possible to select a particular crystal plane for measurements merely by tilting or twisting the emitter until the

Fig. 7.16. Schematic diagram of a field emission tube with a probe hole used for isolating the work function changes on individual crystal planes (Engel & Gomer, 1969).

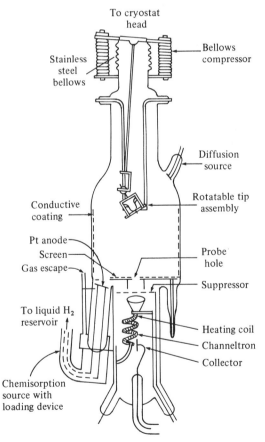

emission from the required plane is centred over the hole. Alignment is made easy by viewing the emission pattern on the fluorescent coating of the anode surface.

7.7 Photoelectric measurements

When radiation of frequency v is incident on a metal surface photoelectrons are produced provided that $hv > e\phi$ where ϕ is the metal work function. Theoretical analysis, due originally to Fowler (1931), shows that the quantum yield I (photoelectrons per light quantum absorbed) is related to v by the equation

$$I = bT^2F(\mu) \tag{7.30}$$

where b is a constant at least for the small range of frequencies near to the photoelectric threshold, F is expressible as a series in μ where $\mu = (hv - e\phi)/kT$.

In practice it is convenient to rearrange equation (7.30) in the form

$$\ln (I/T^2) - \ln b = \ln [F(hv - e\phi)kT] \tag{7.31}$$

then two plots are required, firstly $\ln (I/T^2)$ against hv/kT, based on the experimental observations, and secondly $\ln F(hv/kT)$ against hv/kT. The function F has been tabulated, and $e\phi/kT$ is a constant for a given surface. The horizontal displacement which is necessary to superimpose these plots is just $e\phi/kT$, the vertical displacement is $\ln b$.

In the region which is not too close to the threshold frequency v_0, defined by $hv_0 = e\phi$, we can approximate equation (7.30) by

$$I = bh^2(v - v_0)^2/2k^2 \tag{7.32}$$

It is now possible, and often more convenient, to plot $I^{\frac{1}{2}}$ against v to obtain b and v_0. This is effectively a zero temperature approximation and modifications of it have been suggested by Crowell, Kao, Anderson & Rideout (1972).

There have been many sets of measurements carried out by the photoelectric method, but typical of the required experimental arrangements is the apparatus of, for example, Baker, Johnson & Maire (1971), fig. 7.17.

7.8 Experimental results

It is not always possible to interpret changes in work function strictly according to the models of fig. 7.3. In some cases, though, there is no problem and, for example, the adsorption of the rare gases and the alkali metals is easy to explain.

The rare gases, with their closed electron shells, cannot bind to a metal surface by electron transfer, but, except for He, they are very polarisable

Fig. 7.17. A glass photocell for photoelectric measurements. *A*, monel block; *B*, mica substrate; *C*, quartz window; *D*, Ni evaporation source; *E*, cleaving arm; *F*, glass metal seal; *G*, pumping line; *H*, photocurrent collector lead (Baker *et al.*, 1971).

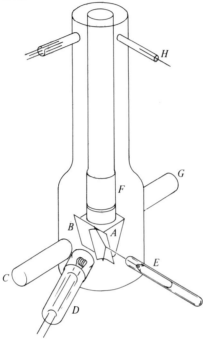

Fig. 7.18. Work function changes of W single-crystal planes after N adsorption at 300 K (after Delchar & Ehrlich, 1965).

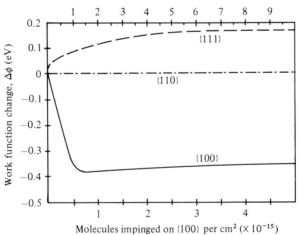

so that, at least at low temperatures, they can be held at a metal surface by van der Waals' type forces. The electron cloud around the rare gas atom is then distorted so that it lies mainly between the metal and the adatom leaving a resulting dipole set with its positive end out from the surface and a corresponding positive surface potential, see fig. 7.3(c).

In table 7.1 are set out some values, obtained by various methods, for the adsorption of the gases Ne, Ar, Kr and Xe on W. These results, although not particularly modern, display clearly the trend to increasing polarisation as one increases the size of the rare gas adatom. The size of the induced dipole and the resulting decrease in work function which accompanies it is revealed.

Ionic adsorption would be anticipated for the alkali metals, that is, a situation corresponding to fig. 7.3(b). Here there should be charge exchange leading to a large dipole, positive end out from the surface.

Fig. 7.19. Work function changes during O adsorption on evaporated Ni films at 178, 273 and 298 K. The arrows indicate the addition of doses of O to the N surface (after Delchar & Tompkins, 1967).

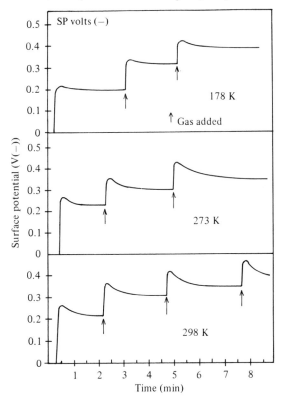

Table 7.1. *Work function changes following adsorption of inert gases and alkali metals on W*

System	Method	$\Delta\phi$ (eV)	Temp. (K)	Cover-age, θ	Reference
W–Ne	FEM	+0.15	4.2	1	(Gomer, 1958)
W–Ar	FEM	+0.87	20	1	(Ehrlich & Hudda, 1959)
W–Kr	FEM	+1.18	20	1	(Ehrlich & Hudda, 1959)
W–Xe	FEM	+1.40	4.2	1	(Gomer, 1958)
W–K	CPD	−2.77	300	0.6	(Fehrs & Stickney, 1971)
W–Cs	CPD	−2.88	300	1.0	(Fehrs & Stickney, 1971)

This in turn should lead to a large reduction in the work function indicated by a positive surface potential. The data from table 7.1 shows this very clearly, with work function decreases of up to 3 V.

Most other gases, including the electronegative gases such as O, fall into the category described by fig. 7.3(a). It is in this category that difficulties of interpretation emerge. One example of these difficulties is the behaviour of N on the three principal crystal planes of W, fig. 7.18. Here the sign of the surface potential changes for adsorption on the {100} and {111} planes, despite the fact that auxiliary flash desorption evidence suggests that the binding state and binding energy are much the same in both instances.

The adsorption of O on Ni when studied by the static capacitor method showed that not only adsorption but also incorporation can occur on the surface (Delchar & Tompkins, 1967) (fig. 7.19). The arrows indicate the addition of doses of O to the Ni surface. The essentially instantaneous change in work function is followed by a time dependent change which is more pronounced at the higher temperatures. This time dependent change marks the disappearance of O from the metal surface into the bulk, the first step in the build-up of an oxide layer. This sort of effect, where surface rearrangement occurs, is found for O on other transition metal surfaces using the vibrating capacitor techniques (Quinn & Roberts, 1964) and is probably an example of phonon assisted incorporation.

8   Atomic and molecular beam scattering

8.1   Introduction

Although many techniques have been developed to study surface properties, most of these techniques are not fully surface-specific and yield information about the surface properties, entangled with information on the first few atom layers. An interaction which can be surface-specific is that between a gas atom or molecule and a surface. This interaction spans a range of phenomena, from diffraction through inelastic scattering to irreversible chemisorption, depending on the nature of the gas–surface potential. Gas atoms or molecules of low kinetic energy (< 0.1 eV) act as very soft probes of the surface and, since they are physically unable to penetrate into the solid, exhibit an extreme sensitivity to the outermost atomic layer, a sensitivity which surpasses that of LEED or AES. Indeed, one of the attendant difficulties in the development of atomic and molecular beam scattering from surfaces has been the problem of obtaining surfaces which are sufficiently clean to show, for example, any diffraction features which may be present. Surfaces which on examination by AES show no impurities and give sharp, well-defined LEED patterns may, nevertheless, be insufficiently clean for atomic or molecular beam studies. A good example of this sensitivity is provided by the work of Lapujoulade, Lejay & Papanicolaou (1979).

Although diffractive scattering from surfaces has been known since the pioneering work of Esterman & Stern (1930), for a long time it was confined to ionic crystals, which possess very corrugated surfaces, most notably LiF. With improved techniques it has become possible, in recent years, to observe diffraction from clean metals and even semiconductor surfaces (Cardillo & Becker, 1978); initially diffractive effects were most readily observed from the rather corrugated surface planes with high Miller indices (Tendulkar & Stickney, 1971), but it is now possible to obtain diffraction from low index planes (Horne & Miller, 1977). In 1979, the first results on an adsorbate covered metal surface H_2 on Ni$\{110\}$ were published (Rieder & Engel, 1979), the latter demonstrated the usefulness of this technique in obtaining structural information, for example the H atom hard core radius and the Ni—H bond length.

Despite the surface specificity of atomic or molecular beam techniques it has usually been very much easier to work with charged particle beams when studying surfaces, since they are easy to generate, collimate and control; the scattered particles are easy to detect. In addition, essentially monoenergetic beams may be produced quite readily. As a consequence, the development of molecular beam techniques has lagged behind that of other surface techniques. However, it is now possible to produce neutral beams with characteristics suited to the performance of unambiguous surface experiments; in particular, beams can be made nearly mono-energetic and of high intensity. This development, together with advances in the techniques for beam detection, has reactivated a technique which was first used in 1911 (Dunnoyer, 1911). Despite these advances, work with atomic beams rather than molecular beams predominates and, in particular, He has been the most widely used beam gas.

8.2   The beam–surface interaction

When a neutral atom or molecule in a beam collides with a solid surface it can interact either elastically or inelastically (fig. 8.1). In an elastic collision there is no net energy transfer between the gas atom and the solid and one may see diffraction phenomena. Typically this occurs where the attractive well depth is less than ~ 40 meV. The repulsive portion of the atom–surface potential for closed shell atoms, such as He, arises from the overlap of the electron charge of the solid with that of the incoming atom so that scattering is from the electron charge distribution of the solid rather than from the ion cores. In fact, it has been shown (Esbjerg & Norskov, 1980) that the locus of classical turning points of a scattered He atom corresponds to a constant electron density contour in which the density is small compared with that found in say, a chemical bond. The periodic variations in amplitude, perpendicular to the surface, of a given electron density contour, will thus depend on the electron

Fig. 8.1. Schematic diagram showing the elastic and inelastic process which can occur in scattering from a surface under conditions where the surface periodicity is unimportant. $\mathbf{k}_i$ and $\mathbf{k}_f$ are the wavevectors of the incident and reflected atoms. $\mathbf{K}_i$ and $\mathbf{K}_f$ are the wavevector components parallel to the surface.

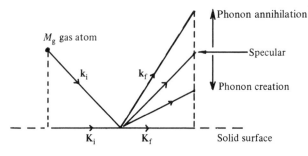

density, so that the corrugation surface and its amplitude are energy dependent.

In contrast to molecular beam scattering, LEED scattering processes occur from the ion cores of the first 3–5 atomic layers. Consequently, LEED information is complementary to that obtained in atomic/molecular beam experiments.

For greater well depths and for heavier atoms and molecules, inelastic scattering predominates. In an inelastic collision, energy exchange occurs and the interaction results in the creation or annihilation of surface phonons in the solid (fig. 8.1) or, additionally, the excitation or de-excitation of internal degrees of freedom in the case of molecules. To be useful in inelastic scattering, as a probe of surface dynamical properties, the momentum transfer between the incoming atom/molecule must range up to the dimensions of the Brillouin zone so that dispersion relations may be determined. Additionally, it is necessary to have beam energies in the range of the phonon energies in order to obtain high resolution.

Although the foregoing implies that the incoming atom merely 'bounces' off the surface in some way, there are, in fact, three possible outcomes of the inelastic collision between an atom and a surface. First, the atom can lose enough energy to become trapped or adsorbed on the surface. Adsorbed atoms will eventually desorb and contribute to the scattered signal. Since these atoms will then have equilibrated with the surface they are likely to desorb with a cosine spatial distribution and a Maxwellian velocity distribution characteristic of the surface temperature (fig. 8.2(c)). Second, the atom can lose some of its energy but still be scattered directly back into the gas phase (fig. 8.2(b)). It is this type of scattering which has been treated theoretically. The third case, fig. 8.2(a), represents purely elastic scattering. There is an intermediate situation, where the atom may lose insufficient energy to adsorb, but equally cannot scatter immediately and hops or diffuses over the surface before eventually desorbing.

Different scattering regimes exist which may be characterised as a function of the relative values of beam energy and mass, the surface atomic mass, temperature and available phonon energies, using a series of dimensionless parameters (Goodman, 1971). A measure of whether or not quantum effects are expected to be important is provided by $\varepsilon_{\theta_D} = E_i/k_B\theta_D$ where $k_B\theta_D$ is the maximum phonon energy of the solid, θ_D is the Debye temperature of the solid, E_i is the incident atom energy. Thus for $\varepsilon_{\theta_D} \gg 1$ many-phonon transitions will be expected in the scattering process and a classical treatment will be appropriate. If $\varepsilon_{\theta_D} \ll 1$ quantum mechanical, one phonon and diffraction (zero phonon) processes will be important. In fact classical theories seem to work quite well even when $\varepsilon_{\theta_D} < 1$. Another parameter which gives some indication of whether or not quantum effects are important is $\mathscr{F} = T_s/\theta_D$ where T_s is the surface temperature. Here, quantum effects are expected to be important or not according to whether $\mathscr{F} \ll 1$ or $\gg 1$. The ratio $E_i/W = \varepsilon_W$, where W is the attractive well depth, measures the importance of W so that for $\varepsilon_W \gg 1$ a purely repulsive interaction can be adopted, whilst for $\varepsilon_W \ll 1$ trapping and hopping will predominate. The mass ratio $\mu = M_g/M_s$, where M_g and M_s are the masses of gas and solid atoms respectively, gives an indication as to whether single collisions or multiple collisions with surface atoms are important. Thus for $\mu \ll 1$ we have light gas atoms striking relatively heavy surface atoms and a single collision is needed to reverse the component of momentum of a light gas atom normal to the surface; interaction times will be relatively short and the interaction near-elastic since $\Delta E/E_i \approx 4\mu$. Finally, a measure of the interaction distance is provided by the parameter $\mathscr{R} = R/R_c$ where R_c is a critical value of the gas

Fig. 8.2. Angular distributions of scattered particles illustrating: (*a*) elastic scattering, (*b*) weak inelastic scattering, (*c*) trapping.

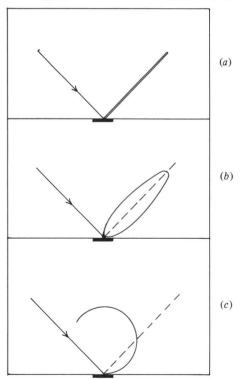

atom–surface interaction radius R, and R itself is the distance of closest approach of the centre of a gas atom to that of a surface atom during the collision. The critical value R_c is that value of R which just allows penetration of the incident gas atom through the surface layer.

8.3 Inelastic scattering, the classical view

The classical theories of inelastic scattering are valid over a wide range of values of T_s and T_i. They may be expected to apply, except for a relatively small quantum regime, to T_s up to 300 K and E_i up to ~ 100 eV. Above ~ 100 eV penetration and sputtering effects become important. The thermal scattering regime is characterised by relatively low incident beam energies, $E_i < 0.1$ eV, and relatively large gas–solid interaction distances R (no surface penetration), which results in scattering from a surface which 'appears' smooth or flat. This regime was first discussed by Oman (1968), and in this case the most important gas–solid interaction mechanism is through the thermal motion of the surface atoms and is, it transpires, most applicable to scattering from metals. Because of the apparently flat surface (at least as seen by the incoming atom), the thermal motion which is important during scattering is that lying in the direction normal to the surface.

The theoretical models incorporating a flat surface and only perpendicular surface atom motion are the so-called cube models. The first of these, the hard cube model, was developed by Logan & Stickney (1966) and is illustrated in fig. 8.3(*a*). Implicit in the model are the following assumptions:

(1) the interatomic gas–solid potential is such that the repulsive force is impulsive;
(2) the scattering potential is uniform in the plane of the surface (smooth surface) and since there is no motion of surface atoms parallel to the surface there is no change in the tangential component of the incident particle velocity;
(3) surface atoms are represented by hard cubes;
(4) a temperature-dependent velocity distribution is assigned to the surface atoms; there is no attractive part to the potential.

This hard cube model can be solved exactly and angular distributions of scattered atoms can be calculated if the velocity distribution of the incident beam is known. This model is, of course, very unrealistic, neglecting as it does the attractive part of the gas–solid potential in the low incident beam energy region, where it is most important. Also, the interaction between solid atoms is neglected, and tangential momentum exchange is not considered.

Some of these failings are corrected in the 'soft cube' model of Logan & Keck (1968), where an attractive stationary potential is introduced, which increases the normal component of the gas velocity before the repulsive collision and decreases it again afterwards, together with an exponential repulsive part. The surface atom involved in the collision is connected by a single spring to a fixed lattice. The ensemble of oscillators making up the surface has an equilibrium distribution of vibrational energies corresponding to the temperature of the solid. This model is

Fig. 8.3. (a) The hard cubes model of gas–surface scattering. (b) The soft cubes model of gas–surface scattering. The subscripts n and t refer to the normal and tangential components of velocity, the subscripts i and f refer to the incident and reflected beams respectively. The subscript c refers to the cube.

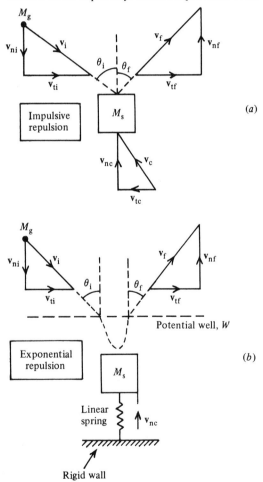

shown in fig. 8.3(*b*). The model introduces adjustable parameters for the potential well depth, range of interaction and lattice atom frequency. The solutions of the equations for angular distributions are now approximate. The soft cube model is most successful when applied to the scattering of heavy molecules where potential attractions would be expected to be largest.

The hard and soft cube models are, of course, single-particle models since they are restricted to energy transfer along the momentum component perpendicular to the surface. Three-dimensional models have been developed in which an ensemble of lattice points is constructed to correspond to a particular crystal plane. Classical trajectories for the scattered atoms are calculated for known incident velocities and angles, by solving the equations of motion of the gas atom and the lattice points. In some theories the gas–atom solid–atom potential is assumed to be a pairwise Morse interaction, whilst in others a pairwise Lennard–Jones 6–12 potential is used. A typical three-dimensional model is shown in fig. 8.4. All surface atoms are connected to nearest neighbours by harmonic springs. A large number of trajectories must be calculated to obtain reasonably reliable results but the approach is capable of reproducing experimental results (McClure, 1972).

The theoretical study of high incident beam energies $(0.1 < E_i < 100$ eV), shows that new features appear which may be attributed to the incident molecules 'seeing' the periodic surface lattice. This is the regime of large incident energies, short interaction distances and a large ratio of incident beam energy to the thermal energy of the solid. The flat surface,

Fig. 8.4. Oman's classical model of gas–surface scattering.

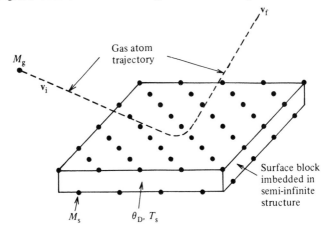

Solid atoms, either independent oscillators or oscillators coupled by 'springs'

cube models no longer apply in this regime and have been successfully replaced by a hard sphere model (Goodman, 1967). Associated with this structure scattering regime is rainbow scattering, which may be viewed as a classical mechanical result of the two-dimensional periodicity of the gas–solid interaction potential, in the same way as diffraction is a quantum mechanical result of this periodicity, with the difference that rainbow scattering does not require long range surface order. The formation of surface rainbows is illustrated in fig. 8.5 in which a two-dimensional scattering model (with one-dimensional periodicity) is used to illustrate the rainbow process. Fig. 8.5(a) shows eight different, but parallel, incident trajectories distributed over a unit cell of surface periodicity. As we pass from atom 1 to atom 8 we find that one or two of the atoms (2 and 3 here) have the maximum scattering angle θ_{max} and emerge essentially parallel whilst atoms 6 and 7 have the minimum scattering angle θ_{min}.

Fig. 8.5. Two-dimensional illustration of the origin of surface rainbows (a) scattering paths from the corrugated potential, (b) intensity maxima in the scattered beam as a function of scattering angle, (c) disposition of rainbow peaks with respect to the specular beam.

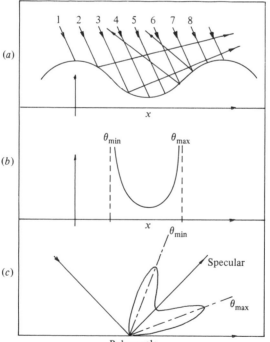

Polar angle

If the surface corrugation is represented by the function $Z = \mathscr{A}\cos(2\pi x/a)$ where $\mathscr{A}$ is the amplitude of the potential corrugation, x is the impact parameter and a is the atom spacing, then atoms 2 and 3 result from reflections where $x = a/4$ and atoms 6 and 7 from reflections where $x = 3a/4$. These reflections result in intensity maxima either side of the specular direction as shown in fig. 8.5(b). Indeed, in this simple model one would expect to find the two maxima set symmetrically about the specular direction, fig. 8.5(c), with an angular separation $\Delta\theta$ given by

$$\Delta\theta = 4\tan^{-1} 2\pi\mathscr{A}/a \tag{8.1}$$

Thus, in principle, it is possible to determine the surface corrugation depth $2\mathscr{A}$. Experiments which demonstrate rainbow peaks in scattering from metals have been carried out, for example, by Hulpke & Mann (1983). The theoretical basis of rainbow scattering was originally described by McClure (1970), who carried out calculations of the rainbow scattering of Ne from LiF, in excellent agreement with experiment.

8.3.1   *Inelastic scattering, the quantum mechanical view*

The quantum mechanical picture of gas–surface scattering is represented in fig. 8.6. Tangential components of the various wavevectors are indicated by upper-case letters; we see an N-phonon, inelastic scattering process about the diffraction peak denoted by $\mathbf{G}$ which is one of the reciprocal lattice vectors; the frequency of the nth phonon is ω_n and $\mathbf{Q}_n$ is the tangential component of its wavevector $\mathbf{q}_n$. All vectors are here assumed to lie in the plane of incidence.

Fig. 8.6. The quantum picture of gas–surface scattering showing an N-phonon inelastic scattering process about the diffraction peak denoted by $\mathbf{G}$, which is one of the surface reciprocal lattice vectors.

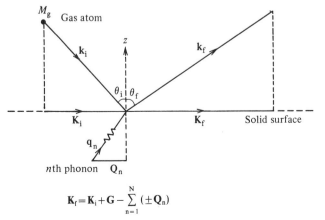

$$\mathbf{K}_f = \mathbf{K}_i + \mathbf{G} - \sum_{n=1}^{N}(\pm\mathbf{Q}_n)$$

The quantum laws of conservation of energy and tangential momentum are then, respectively,

$$k_f^2 = k_i^2 - (2M_g/\hbar) \sum_{n=1}^{N} (\pm\omega_n) \tag{8.2}$$

and

$$\mathbf{K}_f = \mathbf{K}_i + \mathbf{G} - \sum_{n=1}^{N} (\pm\mathbf{Q}_n) \tag{8.3}$$

$+$ signs are chosen if the nth phonon is created, i.e. gas atom loses energy to the solid and the $-$ sign if the nth phonon is annihilated, k_f is the final gas atom wavevector and k_i is the initial gas atom wavevector.

In the case of molecules it is not only phonon creation which can produce loss peaks in scattered molecule distribution but processes which are peculiar to molecules as distinct from atoms, and which result from the redistribution of energy between translational and internal (that is, vibrational and rotational) degrees of freedom during the collision with a crystal. Conservation of energy relates the magnitude k_f of the wavevector for scattered molecules to its initial magnitude k_i by the expression

$$k_f^2 - k_i^2 = 2m \, \Delta E/\hbar^2 \tag{8.4}$$

where m is the mass of the incident molecule and ΔE is the energy lost in the collision from internal degrees of freedom and converted into translational energy. The surface component $\mathbf{K}$ of the wavevector must also be conserved for coherent scattering from a planar periodic array so that

$$\mathbf{K}_f - \mathbf{K}_i = 2\pi\mathbf{G}(m, n) \tag{8.5}$$

where $\mathbf{G}(m, n)$ is a reciprocal lattice vector of the surface. For elastic encounters ΔE is zero and only specular and diffracted peaks appear. For collisions in which the energy of internal degrees of freedom is exchanged entirely with translational motion, ΔE assumes values characteristic of transitions in the free molecule. Generally speaking, the spacing of vibrational levels in molecules is such that they will not be involved in the redistribution process. Rotational levels can be more easily involved in this process and will manifest themselves as additional loss peaks in the angular distribution of the scattered molecules.

8.4   Elastic scattering

The primary importance of elastic scattering is that diffraction features can occur. Atom diffraction can then be used to determine the surface unit cell size as well as to carry out a structural analysis of the surface. In the same fashion though, as with LEED, we must consider the concept of a coherence length (or area) for an incident molecular beam. In other words, how large an area of surface is involved in the production of diffraction data? Intertwined with this question is the more practical one: how perfect must the periodic surface array be in order to produce good atom/molecular beam diffraction data? These questions have been considered by Comsa (1979), who showed that the measured diffraction pattern is ultimately the result of the coherent interference of each particle with itself so that the diffraction pattern results from the incoherent summation of the diffraction probability patterns of the individual atoms. The energy and angular spreads of the measuring system cause the individual probability patterns to be shifted against each other leading to a broadening of the measured diffraction peaks. For a perfect infinite crystal this angular broadening may be correlated to a length on the surface, the transfer width, w, which is a measure of the broadening effects of the measuring system. There are two broadening effects inherent in any beam measuring system, namely that due to instrumental aperture dimensions and that due to the incident beam energy spread. The latter is the more important of the two.

A real measuring system will be able to produce resolved diffraction beams in a direction θ_f only if the period of the perfect surface array, d, is smaller than w, otherwise the peaks will be smeared out. Alternatively, even the perfect measuring system cannot produce diffraction patterns unless there are at least two scatterers present within the length w.

Comsa (1979) has derived expressions for the contributions of the two broadening effects. For the angular broadening, w_θ, he obtains

$$w_\theta \approx \lambda/(|\Delta_\theta \theta_f| \cos \theta_f) \qquad (8.6)$$

When the more important effect of energy broadening is included he obtains

$$w \approx \lambda/[(\Delta_\theta \theta_f)^2 \cos^2 \theta_f + (\sin \theta_i - \sin \theta_f)^2 \overline{(\Delta E)^2}/E^2]^{\frac{1}{2}} \qquad (8.7)$$

where ΔE is the energy spread of the incident beam. It is worth noting that for $\theta_i = \theta_f$, the specular condition, the energy spread ΔE has no influence on the broadness of the specular beam. Using the Bragg condition we obtain

$$w_E \approx d/\{|n| [\overline{(\Delta E)^2}/E^2]^{\frac{1}{2}}\} \qquad (8.8)$$

where n is the diffraction order. Since $d \leqslant w_E$ if we are to resolve diffraction peaks, we can derive an upper limit for n given by

$$|n| \leqslant 1/[\overline{(\Delta E)^2}/E^2]^{\frac{1}{2}} \qquad (8.9)$$

where the limit is imposed by the beam energy spread alone, any angular imprecision leading to a further reduction in $|n|$.

The effect of these relations may be most readily seen by considering a Maxwellian beam from a Knudsen source where $\overline{(\Delta E)^2}/E^2 \approx 0.5$. Equation (8.9) limits the well-resolved diffracted beams to zero and first order, as is found experimentally. For supersonic nozzle sources where $\overline{(\Delta E)^2}/E^2$ can be $\sim 10^{-3}$ or less, very high order diffraction features are, in principle resolvable, again confirmed by experiment.

The size of the two-dimensional unit cell can be determined from the angular position of the diffraction peaks, whilst the detailed form of the surface determines the relative intensities of the diffraction peaks. In order to determine the structure the relative intensities of the individual beams must be calculated for a given structural model and the model parameters modified until the calculated and measured intensities agree. Generally, it is necessary to consider the case in which the surface is rather smooth and the vertical displacements in the electron charge distribution are small compared with the unit cell dimension. Additional simplification can be achieved if the energy of the incoming He atom is large compared with the depth of the attractive well. The surface can then be approximated to by a periodically modulated hard wall (corrugated hard wall), and the interaction potential can be written as

$$V(R, z) = \infty \quad \text{for } z < \zeta(R) \tag{8.10a}$$

$$= 0 \quad \text{otherwise} \tag{8.10b}$$

The corrugation function $\zeta(R)$ describes the periodically modulated surface and is classically the locus of the turning points of the scattered atoms for all impact points in the unit cell. As pointed out earlier, it is a replica of the surface–electron charge distribution. If the model of the interaction is refined to include a soft repulsive potential then the calculated diffraction intensities are changed slightly (Armand & Manson, 1979). The detailed quantum mechanical treatment of the interaction with a corrugated hard well has been described by Garcia, Goodman, Celli & Hill (1978).

Diffraction, of course, necessarily implies conservation of energy and momentum parallel to the surface, i.e. $k_i^2 = k_G^2$ and $\mathbf{K}_i + \mathbf{G} = \mathbf{K}_G$. Here $\mathbf{k}_i$ and $\mathbf{k}_G$ are the wavevectors of the incident and diffracted beams, $\mathbf{K}_i$ and $\mathbf{K}_G$ are the vector components parallel to the surface, $|\mathbf{G}| = 2\pi/a$ denotes the reciprocal lattice vector involved in the diffraction process, a is the lattice constant. For a beam incident normally, in addition to the specular beam, diffraction peaks will occur at final scattering angles θ_f given by the Bragg formula

$$k_i^2 \sin^2 \theta_f = G^2(m^2 + n^2) \tag{8.11}$$

m and n are the indices of the diffraction peak. Similarly, for beams incident at angle θ_i the diffraction condition becomes

$$\mathbf{k}_i(\sin \theta_f - \sin \theta_i) = \mathbf{G}(m, n) \tag{8.12}$$

Elastic scattering is typifed by fig. 8.2(a).

If the wavevectors $\mathbf{k}_i$ and $\mathbf{k}_G$ are separated into components parallel and perpendicular to the surface respectively so that $\mathbf{k}_i = (\mathbf{K}_i, \mathbf{k}_{iz})$ and $\mathbf{k}_G = (\mathbf{K}_G, \mathbf{k}_{Gz})$, i.e. upper-case symbols represent parallel components, we can write

$$\Psi(R, z) = \exp(i\mathbf{k}_i \cdot \mathbf{r}) + \sum_G A_G \exp(-i\mathbf{k}_G \cdot \mathbf{r}) \tag{8.13}$$

where the Rayleigh assumption has been applied, that is, the incoming and outgoing beams can be considered as plane waves up to the surface. If the corrugation is such that the ratio of the maximum height ζ_{max} to the lattice constant, a, is less than 0.2, for a two-dimensional corrugation, the Rayleigh approximation is valid. For diffracted beams F, $k_{Fz}{}^2 > 0$, and the intensity is given by $P_F = (\mathbf{k}_{Fz}/\mathbf{k}_{iz})|A_F|^2$ where $\sum P_F = 1$. The boundary condition

$$\Psi(R, \zeta(R)) = 0 \tag{8.14}$$

can be used with equation (8.13) to obtain an equation for the coefficients A_G, thus

$$\sum_G A_G \exp(i\mathbf{k}_{Gz}\zeta(\mathbf{R})) \exp(i\mathbf{G} \cdot \mathbf{R}) = -\exp(-i\mathbf{k}_{iz}\zeta(\mathbf{R})) \tag{8.15}$$

Equation (8.15) can be solved numerically to yield the coefficients A_G (Garcia, Ibanez, Solana & Cabrera, 1976).

This approach neglects the effects of thermal vibration of the surface atoms which leads to an angular dependent decrease in the elastic intensities. It is not easy to correct for this effect theoretically, although experimentally it may be done by measuring diffraction intensities over a wide range of temperatures and extrapolating to $T = 0$ K.

8.5 The production and use of molecular beams

The methods available for the production of molecular beams divide into two classes, effusive sources and nozzle sources. Historically, effusive sources were the first to be developed, starting with the apparatus of Dunnoyer (1911) for the production of a molecular beam of Na atoms and followed by the comprehensive molecular beam programme of Stern, which started in 1919 and included the now famous Stern–Gerlach experiment (Gerlach & Stern, 1921).

In the effusive source, gas merely effuses from an oven or source through an orifice, tube or array of tubes and is then allowed to pass through a collimating aperture or apertures to produce the molecular beam. In sources of this type the source pressure is adjusted to give a source Knudsen number greater than unity and thus provide free molecular flow through the source. (The source Knudsen number is the ratio of the gas mean-free-path to the orifice or source diameter.) There is no significant mass transport in the direction of the beam and the velocity distribution of molecules moving along the beam axis is necessarily Maxwellian, and characteristic of molecules having the orifice or oven temperature. The velocity distribution $I(v)$ in the beam is of the form

$$I(v) \propto (v^3/\alpha^2) \exp(-v^2/\alpha^2) \tag{8.16}$$

where $\alpha = 2kT/M$; M is the mass of the molecule. The beam flux at the reaction surface is then given by

$$I = 1.118 \times 10^{22} \, P_s A_s / l^2 (MT)^{\frac{1}{2}} \tag{8.17}$$

where P_s is the source pressure in Nm^{-2}, A_s (m^2) area of source aperture, l (m) source to surface distance, T the source temperature. The energy range obtainable from this type of source is 0.008–0.32 eV for source temperature 77–3000 K respectively. The upper limit is dictated by the lack of oven materials able to withstand higher temperatures. The effusing molecules have a cosine distribution in space.

The nozzle source, on the other hand, consists of a nozzle or jet (converging then diverging profile) aligned with a hollow truncated cone or skimmer (really a special collimator) which serves to 'skim off' all those molecules not directed along the beam axis, and followed by the usual collimating aperture. The two types of arrangement are shown schematically in fig. 8.7. In the nozzle source the Knudsen number is less than one, gas is expanded isentropically from a high pressure, through a nozzle into the vacuum chamber. As the gas expands into the lower pressure region, lower gas densities result with increasing distance from the nozzle exit and an eventual transition into free molecular flow from the continuum flow in the nozzle occurs as the collision rates decrease. It was first recognised that the transition from continuum to free molecular flow conditions could take place in supersonic nozzle flow, by Kantrowitz & Grey (1951) although this was merely a rediscovery of a suggestion made by Rodebush (1931).

In the nozzle source the average translational energy is $\frac{5}{2}kT$ (flux average), only slightly higher than for an oven source, but the continuum flow in the nozzle permits the use of seeding techniques in which a light gas accelerates heavier gas molecules to high velocities, and allows gas temperatures much higher than the nozzle wall temperatures. Further, in the nozzle source there is net mass transport through the nozzle and this involves the conversion of the total enthalpy into beam translational energy, yielding higher beam energies. As a result, the gas is cooled during the expansion through the nozzle and the velocity distribution of molecules along the beam axis, although still approximately Maxwellian, is that appropriate to a temperature T much lower than that of the nozzle. In this way rather narrow velocity distributions may be obtained. The most probable speed in a nozzle source is higher than that of an effusive source by a factor of $(2\gamma/(3\gamma-1))^{\frac{1}{2}}$ where γ is the specific heat ratio. The advantages of nozzle sources over effusive sources in providing high molecular beam intensities can be measured by comparing the centreline fluxes for each case. The centreline flux provided by a nozzle source is higher by a factor of $\gamma\mathcal{M}^2/2 + \frac{3}{2}$ for equal orifice flows F_s. Here $\mathcal{M}$ is the Mach number of the gas stream. The higher flux for the nozzle source is the result of the directed flow provided. If the comparison is based on identical gas flows through the nozzle providing the supersonic jet and the effusive source flow, then the relative advantage is less great, usually about a factor of 2. Additionally, much higher beam intensities are obtainable by this approach. A comparison of nozzle and effusive source velocity distributions is shown in fig. 8.8. In addition to the freezing of translational degrees of freedom, nozzle sources also exhibit freezing of rotational and vibrational degrees of freedom and in some instances clustering or condensation of the beam molecules.

Fig. 8.7. Comparison of effusive (*a*) and nozzle source (*b*) systems. Skimmer and source at similar locations.

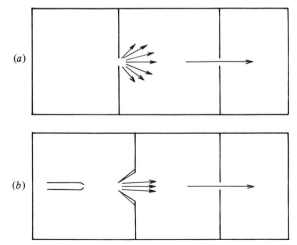

A theoretical description of nozzle sources can be made and the treatment of Kantrowitz & Grey (1951) will be followed here. It is based on several assumptions of ideal behaviour of the gas flow at the skimmer. Given a supersonic gas stream impinging on a core shaped skimmer as depicted in fig. 8.7(*b*), it is assumed that no shock waves are formed either in front of, or within, the skimmer. The absence of collisions due to molecules reflected from the skimmer surface is similarly assumed.

The molecular velocity distribution at the skimmer entrance is taken as that of a gas with Maxwellian velocity distribution at a temperature *T*, superimposed on an average flow velocity u_s. Across the area of the orifice the flow is assumed to be parallel to the axis. The velocity distribution at the skimmer entrance is then given by

$$\frac{dn}{n} = \left(\frac{m}{2\pi kT}\right)^{\frac{3}{2}} \exp\left\{-\frac{m}{2kT}\left[(u-u_s)^2 + v^2 + w^2\right]\right\} du\, dv\, dw \quad (8.18)$$

where *n* is the molecular number density, *m* is the molecular mass, *u* is the molecular velocity parallel to the axis, and *v* and *w* are those perpendicular to the axis.

The flux f_s at the skimmer may be obtained by integration of the differential flux *u* d*n* parallel to the axis over positive values of *u* and *all* values of *v* and *w*

$$f_s = n_s \int_{u>0_n} u\, dn = n_s \left(\frac{m}{2\pi kT}\right)^{\frac{1}{2}} \int_{-u_s}^{\infty} \exp\left[-\frac{m}{2kT}(u-u_s)^2\right] u\, dn$$

$$(8.19)$$

The flow through the skimmer F_s is given by the flux multiplied by the skimmer orifice area *A*.

For molecules with a fixed axial velocity *u*, the flux f_s on a plane perpendicular to the axis at an axial distance *l* from the skimmer orifice,

Fig. 8.8. Theoretical velocity distributions of effusive and nozzle source beams.

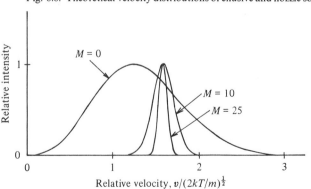

Relative velocity, $v/(2kT/m)^{\frac{1}{2}}$

and at a radial distance r from the axis, is given by

$$f_u = F_u \frac{1}{\pi l^2} \left(\frac{m}{2kT} \right) u^2 \exp \left(-\frac{mu^2 r^2}{2kTl^2} \right) \tag{8.20}$$

where F_u is the flow of molecules of velocity u through the skimmer. It is assumed that the dimension l is much larger than the diameter of the skimmer orifice.

The velocity distribution of molecules on the axis is given by a combination of equations (8.19) and (8.20) with r set to zero, as

$$df = An_s \frac{1}{\pi l^2} \frac{1}{\pi^{\frac{1}{2}}} \left(\frac{m}{2kT} \right)^{\frac{3}{2}} u^3 \exp \left[-\frac{m(u-u_s)^2}{2kT} \right] du \tag{8.21}$$

The equations outlined above apply for all positive values of u_s and for $u_s = 0$. For effusive sources, as already pointed out, $u_s = 0$ and the equations for skimmer flux, and centreline velocity distribution, reduce to the following:

$$f_s = \frac{1}{4} n_s \left(\frac{8kT_s}{\pi m} \right)^{\frac{1}{2}}$$

and

$$df = An_s \frac{1}{\pi l^2} \frac{1}{\pi^{\frac{1}{2}}} \left(\frac{m}{2kT} \right)^{\frac{3}{2}} u^3 \exp \left(-\frac{mu^2}{2kT} \right) du \tag{8.22}$$

The equations for fluxes and velocity distributions for nozzle source beams are often expressed in terms of the Mach number $\mathscr{M}$ of the gas stream at the skimmer. The Mach number may be expressed as u_s/a_s, the ratio of flow velocity to the local sonic velocity a_s, where $a_s = \gamma k T^{\frac{1}{2}}/m$ and γ is the specific heat ratio of the gas. Alternatively, we may talk of the 'speed ratio' $S = \mathscr{V}/\Delta\mathscr{V}$ which describes the velocity distribution in the beam. This quantity determines the resolution in an inelastic experiment and depends on the Knudsen number K

$$S = \text{const} \times K^{-(\gamma-1)/\gamma} \tag{8.23}$$

with the constant depending on γ (Anderson & Fenn, 1965).

It is clear from equation (8.23) that a high speed ratio can be achieved by increasing the gas pressure or orifice diameter. Unfortunately, both of these changes increase the total gas flow into the system so that the pumping capacity becomes the ultimate limitation. The gas flow varies with the square of the orifice radius, but only linearly with pressure so that it is usually advantageous to increase the speed ratio by working at higher pressures. Speed ratios as high as 200 have been achieved (Brusdeylins, Meyer, Toennies & Winkelmann, 1977; Campargue, Lebéhot & Lemonnier, 1977) which correspond to a velocity resolution of 0.5% or in energy terms of 1%. Thus with a beam energy of 20 meV (from a liquid nitrogen cooled source) energy resolution of 200 μeV is possible or, in terms familiar to optical spectroscopy, around 1.6 cm^{-1}. This resolution compares well with optical spectroscopy resolution and is achieved without reducing intensity since the nozzle source has the unique property of intensity and resolution increasing in parallel, limited ultimately by the pumping capacity of the vacuum system.

If the resolution requirements can be relaxed to a figure of, say, 10%, then the high pumping speed requirement can be removed by using a pulsed beam source in which pulses of gas are allowed to expand through the nozzle (Gentry & Giese, 1978). The gas pulses are produced by combining the nozzle with a magnetically operated valve which gives pulse lengths down to 10 μs duration and repetition rates of 10–20 Hz. The intensity in the beam, $\sim 1 \times 10^{21}$ atoms sr^{-1} s^{-1}, is about two orders of magnitude higher than with a continuous source and comes already chopped, removing the need for a beam chopper.

When the length of the nozzle source is reduced to zero, i.e. merely a plane circular orifice joining the high pressure source to the low pressure region, the flowing gas forms what is now called a free jet since the gas expands free of containing nozzle walls. Such a jet is often termed 'underexpanded' because the gas does not expand as rapidly as flow along the walls would require. The use of free jets has the primary advantage of simplicity, but the assumption of parallel gas flow at the skimmer is not valid. The divergence of a free jet is similar to that from a point source. For high values of $\mathcal{M}$ the divergence due to geometrical spreading of the jet may exceed that occurring as a result of the random thermal velocities of the molecules of the jet. Nevertheless, the Kantrowitz–Grey assumptions are entirely correct in many applications and the predictions based on them serve as an ideal with which the performance of experimental systems may be compared. It is probably true to say that the majority of nozzle sources in use today are really of the 'free jet' type.

In the same way that the geometric details of the nozzle may be simplified, so also can the structure of the skimmer. Again for many applications it is sufficient to use merely an aperture in a thin plate rather than the truncated cone structure of an orthodox skimmer (Moran, 1970).

8.6   Detectors

Once a molecular beam has been scattered from a target surface the angular distribution of scattered atoms or molecules must be determined, unfortunately neutral beams are not easy to detect. The beam detectors most commonly used either depend on ionisation by electron impact or measure temperature rise using a sensitive bolometer. Of these two, the first is the most widely used, generally in the form of a small quadrupole mass spectrometer, which in effect measures the partial pressure exerted by the scattered beam. Normally the ionisation region of the detector is designed either for fly-through or stagnation operation. In the former the resultant ion current is proportional to the number density n of the beam. The latter stagnation or accumulation mode yields an ion current proportional to the number density within the enclosed volume, which is in turn related to the flux f of the beam. Bolometer detectors, on the other hand, measure the beam intensity, which is proportional to the energy of the scattered beam. They can, for example, take the form of doped Si bolometers operated at temperatures down to 1.6 K (Boato, Cantini & Mattera, 1976). They suffer from the disadvantage that they do not respond sufficiently rapidly for time-of-flight measurements. For dissociated molecules a surface conductivity detector may be used (Nahr, Hoinkes & Wilsch, 1971) but its response time is not sufficiently fast for application in time-of-flight measurements either.

Systems utilising the pulsed jet source, with its higher intensity, can use an ion gauge as a detector. Its sensitivity is high and if built with scaled-down dimensions it can have a short response time. The real limitation on this type of detector lies in the current amplifier which it feeds. The amplifier must have a bandwidth matching the signal rise time.

In all detector arrangements differential pumping at the detector is necessary for good signal-to-noise ratio. This differential pumping not only reduces the background gas level, but also the residual level of the beam gas. The latter component cannot be filtered out in, say, a mass spectrometer and will limit the ultimate signal-to-noise ratio.

8.7 Experimental arrangements

A typical molecular beam system uses a high pressure gas source supplying a nozzle or free jet source. This source places a heavy load on the pumping arrangements so that very high pumping speeds are required in the nozzle chamber if the background gas pressure is to be reduced to a level low enough to avoid scattering of atoms/molecules out of the beam. The introduction of a secondary pumped chamber and collimator serves two purposes, namely to define further the beam geometry and also to reduce the effusive gas load on the final experimental chamber, where ultimate pressures of 10^{-9} torr will be required when the beam is on. The pumping speed at the final chamber must also necessarily be very high in order to achieve vacua of 10^{-10} torr or better. The introduction of an intermediate chamber is convenient since it allows the insertion of a chopper or velocity selector disk in the beam path. When the scattered beam intensity is very low the use of a chopped beam allows the application of phase sensitive detection to the detector signal and enables very low intensity signals to be extracted from the background noise.

An example of a 'free jet' system used to study elastic scattering (diffraction) from surfaces is shown schematically in fig. 8.9. Here, a dual nozzle system is used with three stage differential pumping. Under typical operating conditions the pressure behind each nozzle is 2 atm with the successive stages of pumping yielding a final, target chamber, pressure of 5×10^{-10} torr, with the beam on. The nozzles can be cooled to 80 K or heated resistively to 1200 K yielding He beam wavelengths of between 0.2 and 1.1 Å. The speed ratio is ~ 0.17–0.11 depending on nozzle temperature. These data are obtained with a 'free jet' nozzle 80 μm in diameter. Systems of this type are relatively compact compared with those used for inelastic scattering measurements and it is possible, indeed necessary, to mount the target on a standard manipulator so that angle of incidence and azimuthal angle can be varied. The quadrupole mass spectrometer can here be rotated around the sample in the scattering plane and also within $\pm 15°$ normal to the scattering plane.

An example of an experimental configuration which represents the 'state of the art' in systems designed for high resolution, inelastic scattering studies is shown in fig. 8.10. The beam source is a 5 μm-diameter free jet nozzle operated with He at 200 atm pressure. The first stage is pumped with a 5000 l s^{-1} diffusion pump. The beam chopper rotates at 9000 r.p.m. (150 Hz) and has slits 0.2 mm wide providing beam pulses of 2 μs length. The sample is mounted on a manipulator with three translational and three rotational degrees of freedom; the scattered atoms are detected by a quadrupole mass filter fitted with a head-on

Fig. 8.9. Schematic diagram of a molecular beam system designed for diffraction studies showing the essential features common to nozzle based systems (Engel & Rieder, 1981).

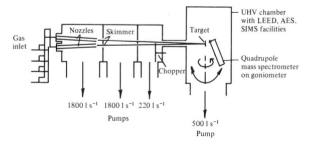

ioniser. Between the sample and detector are three stages of differential pumping. Overall, this system uses nine stages of differential pumping to produce the total pressure differential of 18 orders of magnitude between the nozzle, at 200 atm, and the detector at 10^{-12} torr. With the nozzle operated at 77 K giving a beam energy of 18 meV, this system has an energy resolution of 190 μeV. The high signal-to-noise ratio of this arrangement makes it possible to examine not only elastic scattering (a few per cent of incident beam), but in particular the inelastic peaks which have intensities 2–3 orders of magnitude lower than this. Owing to the size of this experimental arrangement it is necessary to operate with a fixed angle between source and detector using a rotatable sample.

In addition to improving the signal-to-noise ratio in beam experiments, the phase-sensitive detection system can also give information about the time-of-flight of molecules between the chopping point and the detector. This information is contained in the phase shift at maximum signal intensity. Furthermore, if both the phase shift and signal amplitude can be measured as a function of chopping frequency, then a complete velocity distribution can be obtained.

An alternative to the phase shift approach is the time-of-flight technique. In this method a narrow pulse of molecules is allowed to traverse the flight path and is detected by a multichannel signal averaging instrument where the signal intensity as a function of time is

Fig. 8.10. Cross-section through a molecular beam system designed for very high resolution inelastic scattering studies (Brusdeylins *et al.*, 1980).

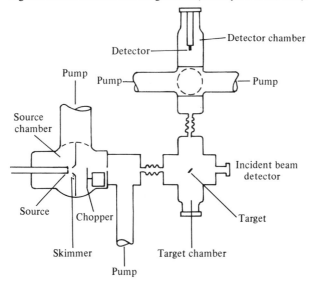

measured and stored. After a suitable number of cycles, a complete intensity versus time-of-flight curve is obtained.

An additional use of beam chopping should be mentioned here, namely, modulated beam relaxation spectrometry. It finds its use in determining the kinetics of desorption. The principle is quite straightforward; the intensity I_0 of the primary beam is modulated periodically by means of a chopper to produce a square waveform. The kinetics of the surface process cause a variation of the waveform of the scattered intensity I, see fig. 8.11, which is usually characterised by recording the amplitude I_1 and phase lag ϕ of the first Fourier component, by means of lock-in techniques.

If one considers a simple first order adsorption–desorption process, the variation of the surface concentration is described by

$$dn_s/dt = sI_0 - k_d n_s \qquad (8.24)$$

where n_s is the surface atom concentration, s the sticking coefficient and k_d the first order rate constant. The amplitude and phase lag of the signal are given by

$$I_1 = I_0/(1 + \omega^2/k_d^2)^{\frac{1}{2}} \qquad (8.25)$$

$$\tan \phi = \omega/k_d \qquad (8.26)$$

The rate constant for desorption, k_d, may be written as

$$k_d = v \exp(-E_d/kT) \qquad (8.27)$$

where v is the pre-exponential factor, equal to kT/h, and E_d is the activation energy for desorption. For a first order process $k_d = 1/\tau$, where τ is the mean residence time of the adsorbed atom on the surface and we can write $\tan \varphi = \omega \pi = \omega \tau_0 \exp(E_d/kT)$.

Figure 8.12 shows a plot of $\ln \tau$ versus $1/T$ for the system CO–Pt{111} which demonstrates very clearly the accessible time scale of this technique and also the vanishingly small coverages which may be examined by this method; coverages which are not accessible by other

Fig. 8.11. Basis of the modulated beam technique for studying the kinetics of desorption and of surface reactions.

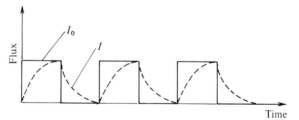

methods. In this instance, $E_d = 1.51$ eV molecule^{-1} and v_i is unusually large at 1.25×10^{15} s^{-1}.

8.8   Scattering studies

It was pointed out in earlier paragraphs that atomic or molecular beam diffraction is a very attractive technique for obtaining accurate surface structural data. Atomic de Broglie wavelengths are in the region of 0.5–1 Å and incident energies can be made low, typically less than 0.1 eV, so that there is no penetration into the surface as is found with LEED. Unfortunately, atomic beam scattering suffers from the fact that only a small fraction of the total scattering arises from coherent events (Mason & Williams, 1972) and this is particularly true

Fig. 8.12. The mean surface residence time τ for CO on Pt$\{111\}$ as a function of temperature, measured using the modulated beam technique (Campbell, Ertl, Kuipers & Segner, 1981).

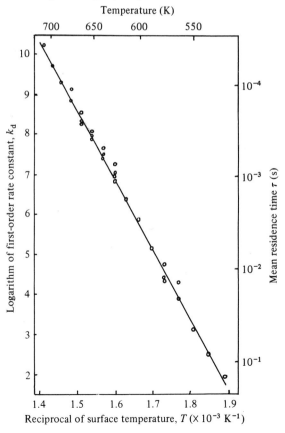

Temperature (K)

for metals where diffraction tends to be seen either only for relatively rough or corrugated surfaces such as the W{112} surface (Tendulkar & Stickney, 1971) or for surfaces at very low temperatures (Boato, Cantini & Tatarek, 1976). It is likely that the absence of diffraction features from metal surfaces is simply due to a combination of the very weak periodicity of the surface potential and the reduction in elastic intensity caused by the Debye–Waller factor (Beeby, 1971). Most experiments on metals have been carried out at temperatures higher than the characteristic Debye temperature. Although the diffraction of atoms from smooth metallic surfaces is quite difficult to observe, the diffraction of hydrogen has been seen from the {100} plane of Cu (Lapujoulade, Le Cruer, Lefort, Lejay & Maruel, 1980), see fig. 8.13.

Alkali halide crystals, on the other hand, yield intense, unambiguous diffraction patterns. For example, Williams (1971) used a nearly monoenergetic nozzle beam of helium to study the clean LiF{001} surface. He obtained sharp diffraction peaks essentially in agreement

Fig. 8.13. Normalised in-plane scattered intensity versus angle for H scattering from the Cu{100} surface. For the specular peak the intensities have been reduced by a factor of ten. The figures beneath the diffraction peak labels show changes in rotational quantum number j. Thus the peak labelled $\overset{(00)}{0\text{–}2}$ results from a (00) transition with an increase in rotational motion from $j=0$ to $j=2$ (Lapujoulade *et al.*, 1981).

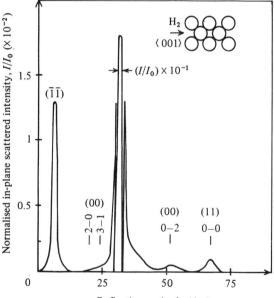

with theory; the first order peak about 10% of the specular {100} peak and the second order peak about 1%. Contrast the peak intensities of the data in fig. 8.13 for H_2 on Cu{100}. In an experiment with a nozzle beam of neon incident on the LiF{001} surface, diffraction was observed but the scattering intensity in this case was much weaker than for helium, indicating a greater fraction of inelastic collisions.

Very high resolution measurements of the inelastic scattering of He atoms from a LiF{001} surface have in fact yielded peaks which can be attributed to single surface phonons (Brusdeylins, Doak & Toennies, 1980), fig. 8.14.

Inelastic scattering, which aims to measure the intrinsic vibration excitations of the surface, requires that the experimental conditions be chosen to ensure quantum mechanical interaction. Thus, the incident atom or molecule must have a low mass and low velocity so that the particle wavelength is comparable with, or larger than, the surface periodicity. In addition, it is necessary, or at least desirable, to have a gas–solid interaction potential with a shallow minimum. This latter condition precludes gases which undergo chemisorption and the usual

Fig. 8.14. Time-of-flight spectrum for He atoms scattering from a LiF(001) surface along the [100] azimuth. The sharp peaks are due to single surface phonon interactions (Brusdeylins *et al.*, 1980).

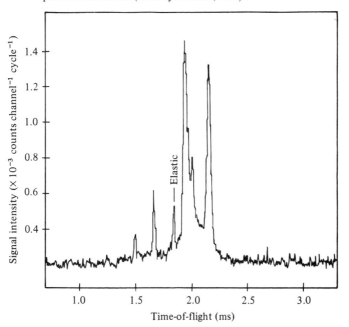

choice of gas atom is He. Inelastic effects have always been noticeable in gas–surface scattering although the earliest experiments were performed in a region well away from the quantum scattering regime, i.e. the thermal regime, where lobular rather than sharply directed scattering occurs, owing to multiphonon interactions.

The data of Brusdeylins *et al.* (1980), involving very high resolution time-of-flight measurements of the He atoms scattered along the $\langle 100 \rangle$ azimuth of LiF$\{100\}$, shows very sharp peaks. These peaks may be attributed to the creation and annihilation of phonons at the crystal surface. The sharp structure found is indicative of surface rather than bulk interactions. This experiment demonstrates that Inelastic Molecular Beam Scattering (IMBS) measurements are capable of measuring the excitation frequencies of surface localised vibration modes with resolution comparable with optical spectroscopy. It should be noted that these mode frequencies are observable through energy ranges inaccessible to electron spectroscopy, that is, transition energies where the spectra are obscured by the strong elastic peak. The sensitivity of IMBS to acoustic vibrations makes it complementary to both electron and photon spectroscopies, especially with its ability to examine vibration modes near to the Brillouin zone boundary. IMBS can measure the momenta of the surface vibration modes over the range covering the whole Brillouin zone. This makes it possible to determine the dispersion relations for surface modes.

Although the foregoing has dealt primarily with high resolution measurements of inelastic scattering in the quantum regime, it is interesting to examine wider aspects of inelastic scattering, thus for inelastic scattering there is no correlation between scattering intensity and microscopic surface roughness. The scattering is characterised by decreasing peak intensity with increasing angle of incidence and increasing surface temperature as in fig. 8.15 for Kr on W$\{110\}$ (Weinberg & Merrill, 1971).

There is also an inelastic regime in which substantial trapping occurs and atoms lose sufficient of their energy to become adsorbed. In consequence, large deviations from specularity occur evidenced by an increasingly cosine-like distribution. An example of this is shown in fig. 8.16 for H_2 incident on a Cu$\{111\}$ surface and shows how the build up of impurity, C in this case, causes a transition from essentially specular behaviour to a cosine-like distribution. This cosine component is from desorbed atoms which have equilibrated with the surface. A further feature of this regime is the increase in peak intensity with increasing surface temperature; this is due to the decreased trapping probability at high temperatures. The number of atoms trapped can be correlated to

Fig. 8.15. Kr scattering from W{110} surface at 45° incidence for a range of surface temperatures, ○ 375 K; △ 775 K; □ 1300 K (Weinberg & Merrill, 1971).

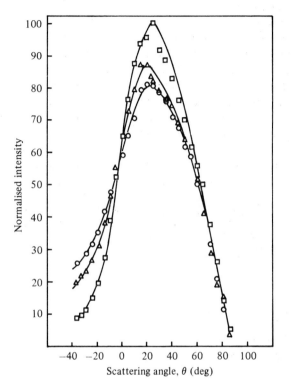

Fig. 8.16. Hydrogen scattering from a Cu{111} surface at 300 K (*a*) clean Cu surface, (*b*) same structure with traces of C impurity (Delchar, unpublished data).

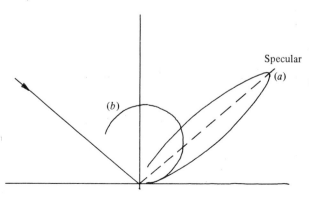

the estimated depth of the attractive potential well for various gas–metal systems. These observations of inelastic collisions are in qualitative agreement with the cube model.

An important measure of the translational energy transfer in an inelastic collision is the velocity distribution of the scattered beam. This has been measured, for instance, for an Ar nozzle beam scattered from the W{110} surface using the phase shift technique. It was found that the average velocity of the scattered gas increased as the angle of incidence increased. Also, the average energy decreased monotonically as a function of scattered angle away from the surface normal for all angles of incidence. The latter is predicted by the hard cube model.

Although the scattering of He atoms from metal surfaces is almost completely elastic, as can be demonstrated by the very narrow specular peaks obtained from clean, well-ordered, single crystal surfaces, for

Fig. 8.17. Attenuation of the intensity maximum in the scattering distribution for (a) He on Ag{111}, Pt{111} and W{110}; (b) Ne on Ag{111}, Pt{111} and W{110} (Sau & Merrill, 1973).

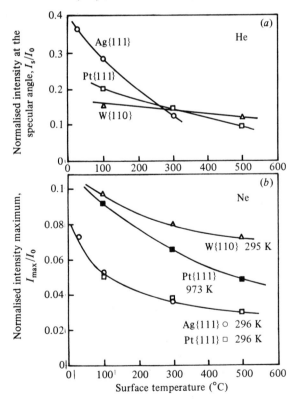

rough surfaces large non-specular components appear. A similar effect, a broadening of the beam width with increasing surface temperature, can be attributed to what might be called thermal roughening or increased mean square displacements of the vibrating surface atoms (Sau & Merrill, 1973), see fig. 8.17(*a*). A further indication of the sensitivity of a He beam to surface structure can be seen in a comparison of the specular intensity and peak width on going from the close-packed f.c.c. {111} face of silver to the slightly more open b.c.c. {110} face of W and to the still more open f.c.c. {100} face of Pt. The trend is a decrease in specular intensity, together with an increase in peak width, although the actual changes in surface structure are subtle and microscopic (fig. 8.17(*b*)).

Scattering of molecules from surfaces has additional possible complications as pointed out earlier. The scattering of NO, CO, N_2, O_2, H_2 and D_2 for example from Pt{100} produces a broad scattered peak set

Fig. 8.18. Comparison of H_2, D_2 and HD scattering from the same MgO(001) surface along the [010] azimuth; $T = 298$ K; angle of incidence, $\theta_i = 15°$ showing loss peaks due to exchange of translational energy with the rotational states (Grant-Rowe & Ehrlich, 1975).

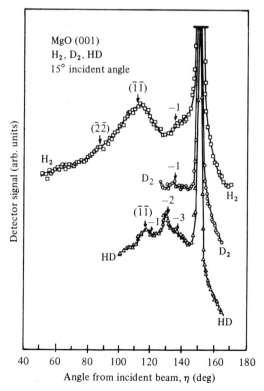

with its maximum at or near the specular angle. The width of the scattered beam here indicates substantial energy exchange or possibly surface roughness. For H_2, D_2 and HD beams incident on MgO at 300 K sharp diffraction peaks occur (fig. 8.18), together with additional loss peaks resulting from the exchange of translational energy with the rotational states of the H_2, D_2 and HD molecules (Grant-Rowe & Ehrlich, 1975). In going from H_2 to D_2 and then to HD, the intensity of specular scattering diminishes, with the height of the $(\bar{1}\bar{1})$ peak following this trend. The rotational loss peaks associated with $(\bar{1}\bar{1})$ show the opposite behaviour since they increase in the sequence H_2, D_2 and HD. This trend demonstrates that they are not tied to elastic diffraction. In this case, the efficiency of rotational transitions is found to be high; indeed, comparable to elastic scattering and to depend strongly upon the angle between the incident beam and the surface.

Elastic scattering and diffraction, as has been pointed out, can be used to determine surface structure. A nice example of this type of application can be found in the diffraction study of H_2 adsorption on Ni$\{110\}$ (Engel & Reider, 1981). Four homogeneous, ordered phases were observed as the coverage of H_2 was increased between $\theta = 0.7$ and $\theta = 1.6$ at a substrate temperature of 100 K. An irreversible reconstruction of the adlayer was observed upon heating to temperatures higher than 220 K. The diffraction structures are shown in fig. 8.19 for a He beam incident along the $\langle 001 \rangle$ azimuth. The adsorption of H_2 leads to a series of ordered adsorbed phases as the coverage is increased. For example, at $\theta = 0.7$ a (2×1) phase forms at 100 K, this was determined by both in-plane and out-of-plane measurements (fig. 8.20). The best fit corrugation function $\zeta(x, y)$ was determined as

$$\zeta(x, y) = -\tfrac{1}{2}\zeta(01) \cos (2\pi y/a_2) - \tfrac{1}{2}\zeta(02) \cos (2\pi 2y/a_2)$$
$$+ \zeta(\tfrac{1}{2}1) \sin (2\pi x/2a_1) \sin (2\pi y/a_2)$$
$$+ \zeta(\tfrac{1}{2}2) \sin (2\pi x/2a_1) \sin (2\pi 2y/a_1) \tag{8.28}$$

Here a_1 and a_2 denote the unit cell vectors of the clean Ni$\{110\}$ surface in $x = \langle 1\bar{1}0 \rangle$ and $y = \langle 001 \rangle$ directions, respectively. The Fourier coefficients corresponding to $\mathbf{G} = n\mathbf{a}_1 + m\mathbf{a}_2$ are denoted by $\zeta(nm)$. In this case the best fit parameters found were $\zeta(01) = 0.10 \pm 0.02$ Å, $\zeta(02) = 0.03 \pm 0.01$ Å, $\zeta(\tfrac{1}{2}1) = 0.070 \pm 0.01$ Å and $\zeta(\tfrac{1}{2}2) = 0.02 \pm 0.01$ Å. The calculated intensities obtained are plotted in fig. 8.21 for comparison with the experimental values. The best fit corrugation function given by equation (8.28) is plotted in fig. 8.21 as a corrugation map, or a map of surface charge density contours. The hard sphere model deduced from this map and the comparison with the clean surface contours is shown in fig. 8.22.

The scattering of atoms or molecules from surfaces can be divided into three regions determined largely by the energy of the incoming atom/molecules. We have considered examples of scattering with beam energies up to ~ 0.1 eV, where the scattering process can be described by quantum theory, or by the thermal scattering models; the energy span between 0.1 eV and 100 eV embraces structure scattering, for example, rainbow scattering, whilst, at incident energies above approximately 100 eV, penetration and sputtering occurs. This latter energy range is also an energy range in which very few experiments have been conducted, primarily owing to the difficulty of forming and detecting fast neutral atom beams. Computer models describing the interaction of fast atoms with deformable crystal surfaces have been developed by Garrison (1979) and Helbig, Linder, Morris & Steward (1982), actually with the primary aim of describing the spatial distribution of fast, 600 eV ion scattering. In ion scattering significant neutralisation occurs so that many ions are scattered from the surface as neutral atoms, otherwise the

Fig. 8.19. He diffraction traces for various coverages of H on a Ni(110) surface. $T_s = 100$ K, $\theta_i = 26°$, $\lambda = 1.08$ Å. The beam is incident along the [001] azimuth. The gain factor is a factor of two greater for the traces shown on the left-hand side of the figure (Engel & Rieder, 1981).

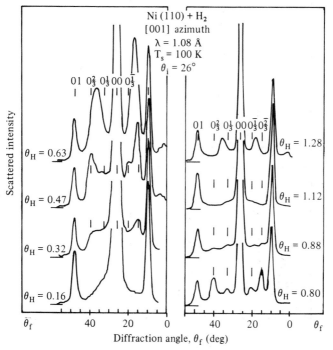

theory can be expected to describe the scattering of both species.

The principal prediction which arises from the computer models of fast atom/ion scattering from metal surfaces is the presence in the polar plot of backscattered atoms of a so-called 'rainbow' peak. In this instance the rainbow peak does not arise from the corrugation of the metal surface as seen by the incoming atom (true rainbow scattering), but from the presence of a second scattering centre (atom) in the surface, which places a lower limit on the deflection angle. The incoming atom can collide with a surface atom and be backscattered (single collision), or some of the scattered atoms may make a second collision with an adjacent surface atom (double collision). The intermediate condition where an atom almost makes a double collision, that is, where its exit from the surface is essentially defined by an adjacent atom, results in the

Fig. 8.20. Scattered He intensity as a function of the scattering angle θ_f for in-plane and out-of-plane ($\phi \neq 0$) detection for the 2×1 structure of the H_2 on Ni{110}. The (00) beam for the experimental (full line) and calculated intensity curves (dotted line) have been set equal. The total elastically scattered intensity is approximately 30% of the incoming intensity (Engel & Rieder, 1981).

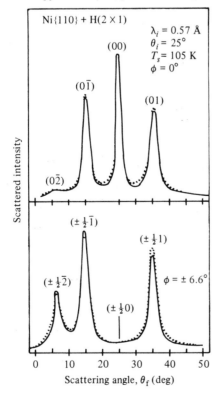

Fig. 8.21. Best fit corrugation map for the (2 × 1) phase, H_2 on Ni{110}. The surface unit cell is indicated (Engel & Rieder, 1981).

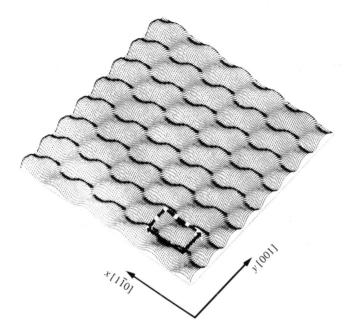

Fig. 8.22. Hard sphere model for the (2 × 1) adsorption phase. The small filled circles represent H atoms and the large circles the outermost layer of the Ni{110} substrate. The surface unit cell is indicated in the figure (Engel & Rieder, 1981).

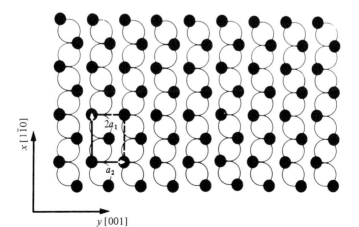

'rainbow' peak in the polar plot of backscattered intensity. The rainbow angle is thus primarily determined by the surface–atom spacing in the beam direction.

Experimental confirmation of the predictions of the computer model have been provided recently by Nielsen & Delchar (1984). They studied the spatial distribution of fast 150–1000 eV He atoms scattering from a W{100} surface along the ⟨110⟩ azimuth, see fig. 8.23. The fast neutral atoms were produced by a symmetrical charge exchange process between helium ions of the appropriate energy and thermal energy helium neutral atoms.

The data of fig. 8.23 show a sharp 'rainbow' peak for each of the four different incident energies, the peak occurring at the same polar angle for each energy. In addition, two further peaks are seen set symmetrically either side of the surface normal. These latter peaks result from He atoms which have penetrated the surface and re-emerged after scattering from

Fig. 8.23. Fast He atom scattering from a W{100} surface along the ⟨110⟩ azimuth showing the so-called 'rainbow' peak and two smaller peaks due to scattering from second and third layer W atoms for incident atom energies ranging from 150–1000 eV (Nielsen & Delchar, 1984).

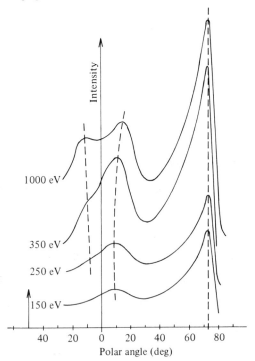

Fig. 8.24. Computer simulation of the polar distribution of backscattered, 350 eV, neutral He atoms incident on the W{100} surface along the ⟨110⟩ azimuth (Garrison & Chen, private communication).

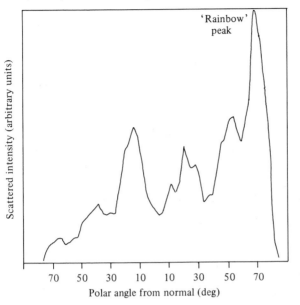

second and third layer atoms. Fig. 8.24 shows the prediction resulting from the computer simulation of the scattering process for 350 eV He atoms incident on the W{100} surface along the ⟨110⟩ azimuth (Garrison & Chen, 1984). The simulation of the backscattered atoms shows good agreement with the experimental data.

Further reading

The basis and wider application of molecular beam techniques are described in *Chemical Applications of Molecular Beam Scattering* (Fluendy & Lawley, 1973). Specific applications to surfaces are described in *Chemistry and Physics of Solid surfaces V* (Vanselow & Howe, 1984).

9 Vibrational spectroscopies

9.1 Introduction

The surface vibrations of adsorbates on crystal surfaces may be studied by Infrared Reflection–Absorption Spectroscopy (IRAS), Raman spectroscopy, High Resolution Electron Energy Loss Spectroscopy (HREELS) or molecular beam scattering. The last named is the subject of a separate chapter and will not be discussed further here. In all of these techniques, with the exception of molecular beam scattering, the inelastic process results from the same physical entity, the vibrating surface dipole, and to a certain extent similar selection rules apply for IRAS and HREELS; the selection rules for Raman activity complement those of IRAS.

One would expect that these techniques would yield much the same sort of structural information about adsorbed species on metal surfaces. This is generally true, though direct comparisons are not always possible since IRAS and Raman spectroscopy may be used at quite high pressures, akin to those encountered in catalytic systems, while HREELS may not. Balanced against this apparent disadvantage for HREELS is the fact that it can readily scan 1 eV, the whole infrared range, in one experiment, although the resolution attainable does not match that of IRAS. Historically, this type of experiment began with the transmission infrared experiments of Eischens & Pliskin (1958) which demonstrated the power of the technique, at least so far as supported metals were concerned. The above techniques between them allow access to a vast amount of information on surface structure and lateral interactions between adsorbates.

Raman spectroscopy has yet to prove a useful probe of surface vibrations, primarily due to the very high sensitivity required for the detection of the Raman signals. If one considers a monolayer of adsorbed molecules on a surface and assigns to these molecules the cross-section for Raman scattering typical of that found in the gas or liquid phase then, under the usual experimental conditions, the Raman signal will be between one and ten photons per second. This signal must be measured in the presence of a background of photons produced by the edge or wing of the exciting laser line, scattered elastically from the substrate surface defects or imperfections. Additionally, there will be fluorescence from the substrate.

Under certain circumstances the Raman cross-section is greatly enhanced (by many orders of magnitude), and this constitutes a so-called surface enhanced Raman spectroscopy. The conditions required for this enhancement are not fully understood, but at least one primary condition appears to be that the surface must be physically rough, with roughness on the scale of 50–500 Å.

On smooth surfaces, where the intrinsic Raman cross-section is small and the signal difficult to detect, schemes exist which enhance the signal by one or two orders of magnitude. One of these schemes involves coupling to surface electromagnetic waves (surface polaritons) on the metal to enhance the magnitude of the surface electric field associated with both the incident and scattered photon. Surface polaritons will be mentioned again in the context of IRAS.

Raman spectroscopy as a tool for studying surfaces and molecule–surface combinations is likely to become important if the enhancement effect can be extended to a wider range of examples. In view of the limited development which has so far taken place in this field, however, it does not seem appropriate to include a detailed examination of the technique.

In IRAS the absorption of infrared radiation due to excitation of surface vibrations of adsorbates is measured after reflection from a plane substrate surface, generally metal. Energy is extracted from the radiation field when the frequency of the light matches the eigenfrequency ω of the dipole-active oscillator and it is ultimately converted to heat via the anharmonic coupling of the infrared active oscillators to all eigenmodes of the system. The interaction between the radiation and the vibrating dipole is produced by the electric field of the light exerting a force on the charge e^* (the effective ionic charge of the oscillator). The wavelength of light is long compared with atomic distances so that the excitation will be almost completely in phase for neighbouring dipoles. For a surface adsorbate lattice this is equivalent to the statement that the wavevector $\mathbf{k}_\parallel$ of the surface wave is large, where $\mathbf{k}_\parallel$ is the component of the wavevector parallel to the surface. We can write $\mathbf{k}_\parallel$ in terms of the wavevector $\mathbf{k}_L$ of the incident light and the angle of incidence θ_i (with respect to the normal)

$$\mathbf{k}_\parallel = \mathbf{k}_L \sin \theta_i \tag{9.1}$$

A reflection infrared spectrum of an adsorbed layer can be obtained from direct measurement of the reflection losses as absorption bands are scanned; this method is known as reflection–absorption infrared spectroscopy. Alternatively, we can measure the changes in ellipticity of reflected plane polarised radiation. This ellipticity is a consequence of the different phase and amplitude changes following reflection, for the components of the radiation whose electric vectors lie parallel to the plane of incidence (*p*-polarisation) and perpendicular to it (*s*-polarisation). Infrared spectroscopy based on measurements of this type is called infrared ellipsometry (Stobie, Rao & Dignam, 1976).

The absorption of infrared radiation by thin films at a metal surface is markedly enhanced at high angles of incidence and is effectively limited to *p*-polarised radiation. This can be seen from a simple consideration of the electric fields produced by the radiation at a bare metal surface.

Fig. 9.1 illustrates the incident and reflected vectors of *s*- and *p*-polarised radiation. At all angles of incidence the *s*-polarised component is practically reversed in phase upon reflection, and as the reflection coefficient is near unity the resultant of the incident and reflected vectors is close to zero at the surface. Consequently, the *s*-component cannot interact significantly with surface dipoles. The *p*-component, however, suffers a phase change which varies strongly with angle of incidence. It can be seen from fig. 9.1 that at grazing incidence the *p*-component can be enhanced to yield a nearly double, resultant electric vector $\mathbf{E}_{p_{\perp}}$ perpendicular to the surface, whilst the tangential vector $\mathbf{E}_{p_{\parallel}}$ is very much weaker. These two vectors form the major and minor axes of an elliptical surface standing wave. The *p*-component can, therefore, interact strongly with vibrational modes of adsorbed species which give a dipole derivative perpendicular to the surface. Fig. 9.2 shows the angular dependence of the resultant amplitude of the electric field components at a bare metal surface relative to E_0, the amplitude of the incident ray, and demonstrates the pronounced maxima near grazing incidence. For a given incident beam width, the area of metal surface over which the enhanced field is effective increases as $1/\cos\theta$ and the intensity of absorption by a surface layer experiencing this field can be expected to depend on θ as $E^2 \sec\theta$. This intensity function is shown in fig. 9.3 and

Fig. 9.1. The incident and reflected electric vectors of the *p*- and *s*-polarized radiation at a metal surface. The plane of incidence is the *xz* plane.

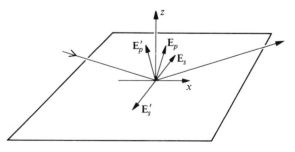

reproduces the angular dependence of the accurate absorption function deduced theoretically by Greenler (1966). Thus, to summarise, only perpendicular dipole components can be detected by IRAS. The considerations above for the photon—dipole interaction are quite straightforward; the position for the electron–dipole interaction is much more complex, but can be described as follows.

When an electron approaches the surface, the electric field of the electron exerts a force on dipole-active oscillators. Just as in IRAS and for the same reason, the electric field is practically normal to the surface. The same selection rule with respect to the orientation of the surface dipole oscillator then applies. Owing to the long range nature of the Coulomb field, the most significant contributions to the total interaction arise during the time when the electron is still many lattice spacings away from the surface. Under these conditions the field is nearly homogeneous

Fig. 9.2. Angular dependence of the resultant amplitude of the electric field components at a bare metal surface, relative to E_0 the amplitude of the incident beam for a surface with refractive index $n = 3$ and absorption index $k = 30$; k is a measure of the attenuation per vacuum wavelength λ in a path d, see for example equation (9.7).

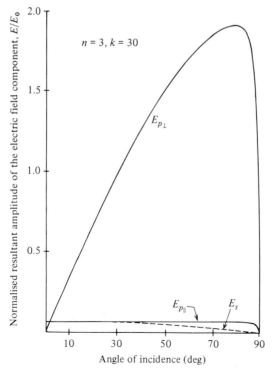

on the atomic scale and thus mostly long wavelength surface waves are excited. As the electron approaches the surface, the lateral extension of the electric field, being a function of the distance, decreases and for this reason a continuous distribution of surface wavevectors is excited, unlike IRAS. In IRAS the field is periodic in time, therefore only the fundamental frequency of the harmonic oscillator is excited. In electron energy loss spectroscopy, however, the total interaction time is of the order of an oscillator period and from the standpoint of the oscillator all frequencies can be excited, with a Poisson distribution in the excitation probabilities. In practice, for a single adsorbate layer, multiple excitations of this sort are not generally prominent, as the intensities of the corresponding electron losses are too low. An alternative viewpoint of the electron–surface wave interaction is provided by considering that the electric field of the dipole-active surface wave exerts a force on the incident electron. Then the maximum interaction occurs when the electron velocity parallel to the propagation direction of the wave

Fig. 9.3. Surface intensity function $(E/E_0)^2 \sec \theta$ for the electric field components at a bare metal surface of refractive index $n = 3$ and absorption index $k = 30$.

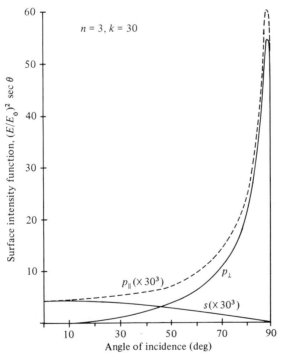

matches the phase velocity, $\omega_s/\mathbf{k}_\parallel$, where ω_s is the eigenfrequency of the surface wave, or the electron behaves like a 'surfer'.

$$v_{e_\parallel} = \frac{\omega_s}{k_\parallel} \tag{9.2}$$

For both IRAS and HREELS, $\mathbf{k}_\parallel$ is small compared with a reciprocal lattice vector and thus phonon dispersion has a negligible effect on the observed frequency. The interaction described above produces small-angle scattering, typically, substantially more intense than that observed at large deflection angles. The result is a 'lobe' of inelastic scattering events sharply peaked about the specular direction. The mechanism described above is known as dipole scattering and the information is obtained from the 'dipole lobe' around the specular direction. Unfortunately, if long range surface order is not maintained, then this 'dipole lobe' is broadened and greatly reduced in magnitude.

Although for small scattering angles about the specular direction the dipole scattering mechanism is usually dominant, outside this 'dipolar lobe' or at large scattering angles we have what is known as impact scattering. In practice these two regions are not distinct, but merge smoothly as one scans the angular distribution of scattered electrons. The impact scattering component results from the short range portion of the electron–molecule interaction and the angular distribution of the inelastically scattered electrons is relatively broad. The physical picture of the scattering event in impact scattering is qualitatively different from that discussed in the small-angle (dipole scattering) regime. There the electron does not 'see' the electric field until it is ~ 1 Å from the metal surface which it then, of course, penetrates, typically to a depth of around 10 Å. This suggests that the inelastic event is initiated primarily while the electron is inside the metal rather than when it is in the vacuum above it.

For dipole allowed modes, examination of the cross-section in the regime of large-angle deflection shows that the cross-section does not fall off, but levels out to a value substantially lower than that found near the specular direction, but which is nonetheless detectable. Whilst this part of the scattering regime has received little attention until recently, the area is potentially very rich in information since the energy and angle variation of the loss cross-section contains detailed information on the surface geometry; one is not confined to dipole moments perpendicular to the surface.

A fine example of the information which can be retrieved from this impact scattering regime is to be seen in the study of the formate ion on Pt{110} by Hofmann, Bare, Richardson & King (1983).

To summarise, scattering into a narrow angle about the specular beam direction shows electron energy losses arising from the vibrational excitation of modes that possess a net change in dipole moment perpendicular to the metal surface. In the 'impact' scattering regime we see wide angle scattering from adsorbate modes vibrating parallel as well as perpendicular to the surface, due to the short range part of the electron scattering potential and this provides information on the structure of the surface–adsorbate system.

There remains a further scattering mechanism to consider, namely inelastic scattering via an intermediate negative ion resonance. Negative ion resonances are often observed in elastic collisions between electrons and molecules in the gas phase where, for certain electron energies, the electron and molecule form a compound state with a lifetime between 10^{-10} and 10^{-15} s. In the case of a chemisorbed molecule, the resonance is quenched by the coupling of the molecule to the surface and the sharp resonance characteristic found for the gas phase is broadened. Thus, on a surface, the lifetime of such a resonance is likely to be very short when a chemisorbed molecule is coupled to the substrate electron states and, in fact, very few instances of such resonances have been seen (Ibach, 1981). Physisorbed molecules have been found to exhibit negative ion resonances and an example is N_2 physisorbed on Ag (Demuth, Schmeisser & Avouris, 1981).

The vibrational cross-sections for resonance scattering are rather high, often two orders of magnitude higher than impact scattering cross-sections at the same energy. To set this in perspective, the attainable sensitivity for excitation via impact scattering allows the detection of $\sim \frac{1}{100}$ monolayer so that physisorbed molecules, excited via resonance scattering, may be detected at between 10^{-3} and 10^{-4} of a monolayer by this means.

The selection rule which operates in negative ion resonance may be summarised briefly as follows. The vibrational, normal modes strongly excited on resonance are those which appear in the decomposition of the direct product of the irreducible representation of the wavefunction of the negative ion with itself. Consequently, if the symmetry group of the molecule on the surface allows only non-degenerate representations, it follows that, just as in dipole scattering, only totally symmetric vibrational modes are strongly excited on resonance.

9.2   IRAS

In the preceding section a qualitative view has been presented of the photon–dipole interaction. In this section a quantitative approach is adopted, but not one based on the treatment of Greenler (1966). Rather, a simpler expression is used, but one which is still generally valid for surface spectroscopy on metals.

The starting point is the linear approximation theory of McIntyre & Aspnes (1971), this approximation is justified because the change in reflectivity due to adsorbates $\Delta R/R$, is always small. In addition, only terms which are linear in the concentration of adsorbates are considered since, experimentally, a linear relationship is found to exist between ΔR and coverage (Bradshaw & Pritchard, 1970). When the film thickness d is much less than wavelength λ, linear approximations lead to equations (9.3) and (9.4) where θ_i is the angle of incidence from the surrounding medium, ε_1, ε_2 and ε_3 are the complex dielectric constants of the ambient atmosphere (vacuum), the thin isotropic film and of the substrate respectively: n_1 is the refractive index of the ambient (see fig. 9.4).

$$(\Delta R/R)_\perp = (8\pi dn_1 \cos \theta_i/\lambda) \operatorname{Im} (\varepsilon_2 - \varepsilon_3)/(\varepsilon_1 - \varepsilon_3) \tag{9.3}$$

$$(\Delta R/R)_\parallel = (8\pi dn_1 \cos \theta_i/\lambda)$$
$$\times \operatorname{Im} \left\{ \left(\frac{\varepsilon_2 - \varepsilon_3}{\varepsilon_1 - \varepsilon_3} \right) \left[\frac{1 - (\varepsilon_1/\varepsilon_2\varepsilon_3)(\varepsilon_2 + \varepsilon_3) \sin^2 \theta_1}{1 - (1/\varepsilon_3)(\varepsilon_1 + \varepsilon_3) \sin^2 \theta_i} \right] \right\} \tag{9.4}$$

In the infrared region ε_3 is usually much larger than ε_2 for metals and $(\Delta R/R)_\perp$ is then negligible. Equation (9.4) can be simplified for the case of highly reflecting metals and angles of incidence less than the optimum values where $\cos^2 \theta_i > |1/\varepsilon_3|$ and $n_1 = 1$

$$(\Delta R/R)_\parallel = (8\pi d \sin \theta_i \tan \theta_i/n_2^3 \lambda) \operatorname{Im} (-1/\varepsilon_2) \tag{9.5}$$

The process of simplification can be taken a step further for the case where the absorption index k_2 is small, when equation (9.5) reduces to

$$(\Delta R/R)_\parallel = (4 \sin \theta_i \tan \theta_i/n_2^3) 4\pi k_2 d/\lambda \tag{9.6}$$

The first set of terms in equation (9.6) represents the enhancement of absorption in angular reflection at a metal surface whilst the second term is the Beer's law attenuation coefficient that would be expected in transmission at normal incidence where I_0 is the incident intensity and

Fig. 9.4. Labelling system for the combination ambient atmosphere, adsorbate and substrate showing the respective refractive indices and dielectric constants.

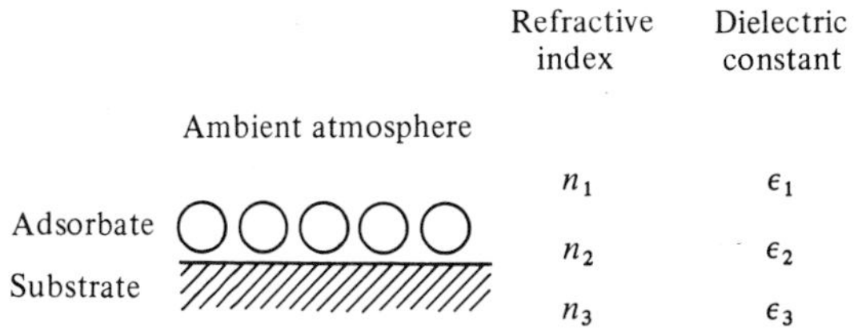

	Refractive index	Dielectric constant
Ambient atmosphere		
	n_1	ϵ_1
Adsorbate ⭘⭘⭘⭘⭘	n_2	ϵ_2
Substrate ▨▨▨▨▨▨	n_3	ϵ_3

the transmitted intensity is given by

$$I = I_0 \exp(-4k_2 d/\lambda) \tag{9.7}$$

In practice, the curve in fig. 9.3 deviates from the function $4 \sin \theta_i \tan \theta_i$ only when $\theta_i > 80°$.

Ibach (1977b) used equation (9.5) together with an harmonic oscillator model for ε_2 to derive a value for the effective ionic charge e^* associated with the surface dipole derivative. The expression he obtained is

$$\left(\frac{\Delta R}{R}\right)_\parallel d\omega = \frac{32\pi^3}{\lambda} \sin \theta_i \tan \theta_i \, Nd \frac{e_\perp{}^*}{\mu\omega_0} \tag{9.8}$$

where Nd is the surface concentration of oscillators which may be replaced by $N_s \theta$ where N_s is the number of surface sites and θ the fractional coverage. $e_\perp{}^*$ is the component of e^* in the direction of the polarisation of the light, i.e. perpendicular to the surface, μ is the reduced mass of the harmonic oscillator and ω_0 the oscillator frequency. Since infrared experiments are usually carried out with unpolarised light, at large angles of incidence $R_\parallel$ is roughly equal to $R_\perp$ and therefore

$$\Delta R/R = \tfrac{1}{2}(\Delta R/R)_\parallel \tag{9.9}$$

Equation (9.8) is accurate enough for most practical purposes and offers greater convenience in use than those previously proposed. Application of this expression, of course, yields values for e^*, the effective ionic charge. This parameter is of considerable importance, for a layer of atoms e^* provides information about the derivative of surface potential ϕ with respect to the thickness r of the adsorbate layer

$$d\phi/dr = \varepsilon_2{}^{-1} e^* N_s \tag{9.10}$$

In self-consistent, first principle theories of chemisorption, $d\phi/dr$ may be calculated, see for example Applebaum & Haman (1975). Comparison between theory and experiment can thus be achieved.

An entirely different basis for IRAS at metal surfaces is provided by surface electromagnetic waves or surface polaritons (Bell, Alexander, Ward & Tyler, 1975). A surface electromagnetic wave can propagate over macroscopic distances. Its associated electric and magnetic fields decay exponentially in both directions normal to the interface, but can interact strongly with surface layers. Whilst it cannot couple directly with free photons it can be excited and detected through the evanescent fields at total internal reflection elements close to the metal surface (Schoenwald, Burstein & Elson, 1973). Calculations by Bell *et al.* (1975) predict strong signals from a monolayer of CO on Pt where propagation distances are several millimetres in the 10 μm wavelength region.

The surface electromagnetic wave technique requires intense, highly collimated sources, i.e. tunable lasers, and the spectra mentioned above have been obtained in the tuning range of CO_2 lasers. Because the surface evanescent wave has a limited range normal to the metal surface this confers selective surface sensitivity akin to that of IRAS.

9.3   Electron energy loss spectroscopy

It is convenient to discuss the inelastic scattering of electrons from an ordered surface lattice in the dipole scattering regime using the treatment of Evans & Mills (1972). According to their treatment the intensity of an energy loss normalised by the elastic intensity is, for the case of specular reflection:

$$S = 8\pi \hat{E}_0^{-1} \hat{p}_\perp^2 \hat{N}_s \theta \cos^{-1} \theta_i F(\alpha, \theta_i) \tag{9.11}$$

Quantities labelled by $\hat{}$ are in atomic units, E_0 is the primary energy, $p_\perp$ the perpendicular component of the dipole moment and $F(\alpha, \theta_i)$ an angular term given by

$$F(\alpha, \theta_i) = [(\sin^2 \theta_i - 2 \cos^2 \theta_i)/(1 + \alpha^2)]$$
$$+ (1 + \cos^2 \theta_i) \ln (1 + 1/\alpha^2) \tag{9.12}$$
$$\alpha = \vartheta_E/\vartheta_c, \quad \vartheta_E = \hbar\omega/2E_0 \tag{9.13}$$

which contains the angle ϑ_c which is the maximum angle over which the spectrometer accepts reflected electrons scattered slightly off the specular direction, the acceptance angle. In practice this is the angle where the intensity of the elastic peak has fallen to half its maximum value. It is assumed also that the spectrometer aperture is circular. It is assumed that ϑ_c is the same in and out of the scattering plane. It is possible to show that the relative intensity S is

$$S = 4\pi (1836 \hbar\hat{\omega} \hat{E}_0)^{-1} \hat{N}_s \theta (\hat{e}^{*2}/\hat{\mu}) \cos^{-1} \theta_i F(\alpha, \theta_i) \tag{9.14}$$

this equation holds well when the specular peak is sharp, i.e. for well-ordered surface lattices. For disordered lattices and even for coverages where a full regular surface lattice is not completed the situation is more complicated. In general, the intensity in HREELS is not simply linear in coverage, even if e^* is independent of coverage.

The electrons, as has been pointed out, excite surface phonons with a continuous distribution of wavevectors and this distribution has to be different parallel and perpendicular to the plane of incidence. Consequently, the angular distribution of electrons around the specular angle depends on both the polar angle ϑ and the azimuthal angle, ψ. The angular dependence of the differential cross-section averaged over ψ is given by

$$\frac{\overline{dS}}{d\Omega} \propto \frac{(1 + \cos^2 \theta_i)}{\cos \theta_i} \vartheta \frac{(\vartheta^2 + \gamma^2 \vartheta_E^2)}{(\vartheta^2 + \vartheta_E^2)^2} \tag{9.15}$$

where $\gamma^2 = 2 \sin^2 \theta_i/(1 + \cos^2 \theta_i)$ and Ω is the solid angle $\vartheta \, d\vartheta/d\psi$.

In fig. 9.5, the averaged cross-section normalised at $\vartheta = 0$ is plotted against ϑ/ϑ_E. It can be seen that $dS/d\Omega$ falls rather rapidly to begin with and tails off smoothly for higher ϑ. The dashed lines indicate the limit to which a typical spectrometer ($\vartheta_c = 1.5°$, $E_0 = 5$ eV) accepts the inelastic intensity for phonon energies of 100 and 300 meV respectively. While $\vartheta_c > \vartheta_E$, i.e. most of the elastic intensity is accepted by the analyser, disorder affects the elastic and inelastic intensity by approximately the same factor and the ratio S of equation (9.14) remains essentially unchanged. If the condition $\vartheta_c > \vartheta_E$ is not fulfilled then S for disordered surfaces will be larger than calculated. More detailed consideration of the theoretical aspects of electron energy loss spectroscopy is given by Ibach & Mills (1982).

9.4   Experimental methods in IRAS

The early work using reflection–absorption methods generally employed multiple reflections at high angles of incidence obtained with closely spaced parallel mirrors in the form of evaporated films deposited on glass or on metal foils (see Pritchard (1975)). This approach, while useful for room temperature measurements on highly reflecting metals, is

Fig. 9.5. Differential cross-section for inelastic scattering of electrons from a metal surface, averaged over the azimuth versus ϑ/ϑ_E. The curves are normalised to one for $\vartheta/\vartheta_E = 0$. The dashed lines indicate the limit up to which a typical spectrometer will accept the inelastic intensity of phonon losses.

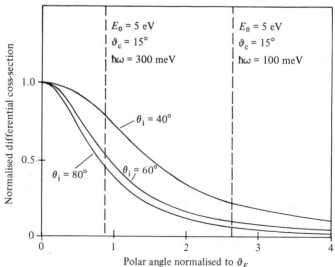

not normally used, primarily because multiple reflections are not advantageous in general. Although the absorption in a single reflection is very small, reaching, for example, 3% for a monolayer of CO on Cu, and the prospect of increasing the relative intensity of the absorption band in a multiple reflection is attractive, each additional reflection is accompanied by an energy loss due to the less than perfect reflectivity of real metals. Consequently, too many reflections ultimately reduce the absolute magnitude of the absorption signal and, indeed, are thoroughly inconvenient for single crystal studies. For a given wavelength the optimum number of reflections is a minimum when the angle of incidence corresponds to the maximum reflection–absorption in a single reflection. In most cases the change in reflectance ΔR in a single reflection is at least half the optimum value. The situation for many metals has been covered in detail by Greenler (1974, 1975). This conclusion has an important bearing on the design of experimental systems primarily because a single reflection is much more convenient for single crystal experiments, in addition it allows other surface techniques to be deployed.

An experimental arrangement used in studies on Cu single crystals (Horn & Pritchard, 1975) is shown in fig. 9.6. In the design of experimental arrangements for IRAS the mechanical stability of the

Fig. 9.6. Horizontal section through a cell used for IRAS and surface potential measurements on Cu single crystals. The surface potential measurements are made with a movable Kelvin probe (Pritchard, 1979a).

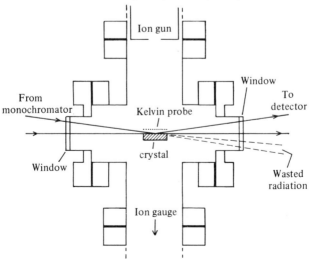

sample mounting is very important because any fluctuation in the interception of radiation by the crystal appears as noise in the recorded spectrum. In an ideal world the noise would be only the shot noise of the radiation, but in practice detector noise is the limiting factor, added to which is the stability of the source and spectrometer atmosphere. The usual mode of operation is to record the reflection spectrum intensity both before and after adsorption and to measure the difference; clearly here the stability of source and atmosphere are critical. In principle, therefore, better stability should be given by double beam spectrometers. Probably the best approach is through the technique of wavelength modulations and methods based on polarisation modulation.

Wavelength modulation is obtained by a variety of means each superimposing a small sinusoidal wavelength modulation as the spectrum is scanned. The modulation can be produced by oscillating the position of the monochromator slit, or of a mirror before the slit, or of a dispersive element. The modulation generates at the detector a signal of the same frequency that is proportional to the derivative of the conventional spectrum. The derivative signal changes rapidly as the adsorbate peaks are scanned and enhances the weak but relatively sharp absorption peaks superimposed on a large but smooth background due to the metal surface itself. By this means the sensitivity of single beam reflection spectra can be increased to less than 0.01% (Horn & Pritchard, 1975). In practice the derivative spectrum of the clean surface is subtracted from that of the adsorbate covered surface and the resulting difference integrated. Integration reduces the noise level and enables high sensitivity to be achieved without the use of spectrum averaging. It should be noted however, that if the background spectrum possesses much detailed structure as a consequence of absorbing species in the spectrometer atmosphere, the technique of wavelength modulation is less useful and needs to be replaced by polarisation modulation.

Polarisation modulation relies on the very different interaction of the *s*- and *p*-polarised components of radiation incident on a surface film which can serve to generate a double beam spectrometer with a single optical path (Bradshaw & Hoffman, 1978). The *s*- and *p*-polarised components from the source are separated and passed alternately through the monochromator before reflection from the crystal. Since the *s*- and *p*-components interact identically with the surrounding atmosphere, the technique is less susceptible to the spectrometer atmosphere than wavelength modulation.

An alternative way of using the *s*- and *p*-polarised beams resides in the ellipsometric spectroscopy developed primarily by Dignam, Rao, Moskovits & Stobie (1971). Here the relative absorbence of the *p*-

components is determined from the ellipsometric parameter Δ, the relative phase retardation, and tan ψ, the relative amplitude attenuation of the two components. A sensitivity of 0.005% has been achieved by this method.

9.4.1 *Applications of IRAS*

One of the adsorbates most thoroughly studied by IRAS is CO; it possesses two important experimental advantages, namely, that it gives intense absorption bands and is easily handled in a UHV system. The above advantages have been coupled with the easy reversibility of the absorption on Cu surfaces to yield a wide range of experimental data ranging initially from evaporated metal films to single crystal surfaces. An overview of the type of result obtainable from these measurements is shown in fig. 9.7 which combines results obtained from experiments on single crystal planes with those produced by evaporated metal films of Cu.

One of the earliest observations of a reflection spectrum of CO on a metal was made on Pt foil by Low & McManus (1967) using Fourier transform spectroscopy. The combination CO–Pt is of considerable interest because it has been studied by HREELS techniques also, thus affording a comparison between the two techniques.

Fig. 9.7. Comparison of spectrum of CO on a polycrystalline Cu film with band shapes and positions found on individual single crystal planes (Pritchard, 1979b).

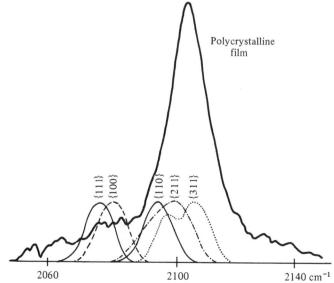

Shigeishi & King (1976), using a single reflection from a recrystallised Pt ribbon, believed to expose essentially {111} surfaces, found an exceptionally intense (6% absorption) spectrum with a single band. This band shifted smoothly from 2065 cm^{-1} at low coverage to 2101 cm^{-1} at saturation at 300 K; no other peaks were observed in this experiment or indeed in those carried out subsequently (Horn & Pritchard, 1977; Crossley & King, 1977). However, electron energy loss spectra for CO on Pt{111} obtained by Froitzheim, Ibach & Lehwald (1977) indicated a second peak at 1870 cm^{-1}. The above shifts have been attributed to one of two possible factors, either dipole coupling between CO molecules aligned parallel to each other on the surface (Hammaker, Francis & Eischens, 1965) or a reduction of dπ* backbonding (into the antibonding 2π*-orbital of CO) (Blyholder, 1964).

Using the theory developed by Hammaker *et al.* (1965), it has been shown that a ^{12}CO molecule couples very weakly into a ^{13}CO environment owing to the large difference in the isolated molecule stretching frequency for the two isotopes (Crossley & King, 1977). Thus, using variable ^{13}CO/^{12}CO mixtures at constant total coverage they demonstrated that the entire frequency shift can be reproduced as the ^{12}CO composition increases from 0 to 100%, and must, therefore, be attributable to coupling effects and not chemical bonding effects.

Additional insights into the surface arrangements of CO adsorbed on Pt{111} surfaces have been obtained (Crossley & King, 1980) resulting in a model based on 'gaseous' and 'island' sites. At 300 K and low coverage a band due to the C—O stretching frequency from a linearly chemisorbed species is observed, first at 2065 cm^{-1}, shifting to 2070 cm^{-1} at 0.3×10^{14} molecules cm^{-2}. In the range $(0.4–1.6) \times 10^{14}$ molecules cm^{-2}, however, the band has two overlapping components, the second one appearing at 2083 cm^{-1} when the first has shifted to 2074 cm^{-1}. The first band becomes less intense (disappearing at 1.8×10^{14} molecules cm^{-2}) and the second one grows with coverage reaching 2101 cm^{-1} at saturation: ^{12}CO/^{13}CO data at saturation coverage yield a singleton frequency of 2065 cm^{-1}. At coverage up to 0.4×10^{14} molecules cm^{-2} CO adsorption on Pt{111} can be considered to occur into a random or 'gaseous' state with isolated species. At higher coverages, these free species are in equilibrium with 'islands' or clumps producing the second high frequency band. At high coverages the free species are replaced or consumed at the expense of the islands.

Although the above accounts for the bands seen in the range 2065–2101 cm^{-1} it does not explain the band seen at 1870 cm^{-1}. Recent work by Hayden & Bradshaw (1983) has shown two bands in this region (attributed to bridge structures) at 1840–1857 cm^{-1} and at

~ 1810 cm^{-1}. These measurements were made at 150 K. At 95 K only one band was seen, at 1840–1857 cm^{-1}, whilst at 300 K the two bands perceived at 150 K were almost unresolved with an accompanying upward shift of the 1810–1825 cm^{-1} band contributing to the convergence. These data bring out an important difference between the resolution of IRAS and HREELS (the latter seeing only a single band in the bridging region) namely the resolution of HREELS is usually not better than 50 cm^{-1} while for IRAS it can be around 10 cm^{-1}, or ~ 1 meV. Thus, in this instance, the two bands separated at most by 47 cm^{-1} cannot be resolved by HREELS.

The interpretation applied to the above data is based on the general rule that increasing the coordination number of the CO site decreases the C—O stretch frequency. Thus, the band at 1840–1857 cm^{-1} is attributed to the C—O stretch of the molecule adsorbed in the two-fold bridge sites. The band at 1810 cm^{-1} is then attributed to the C—O stretch of molecules in three-fold hollow sites.

An early application of IRAS was the identification of the weakly bound α-CO state on W at 295 K as a molecularly adsorbed form (Yates & King, 1972). This was demonstrated using a single reflection from a W ribbon whereby the infrared data could be correlated with thermal desorption peaks from a second similar ribbon, fig. 9.8. Although a single reflection involves a large loss in sensitivity (60–70%) with Cu, the loss is much less severe with the transition metals. Their lower reflectivities have the consequence that a single reflection usually gives two-thirds or more of the optimum absorption signal. Thus, in the case of CO on W,

Fig. 9.8. UHV reflection cell for use with W ribbon samples (Yates & King, 1972).

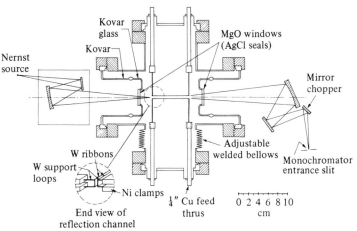

the infrared spectra served to show that the α-binding state is a non-dissociated molecular state. The appearance and growth of infrared bands at 2128 cm^{-1} and 2020 cm^{-1} coincided with CO adsorption in the α_1- and α_2-states, these latter were distinguished by electron impact desorption behaviour. No infrared bands could be detected for the more abundant and strongly bound β-states.

The adsorption of carbon monoxide on several faces of Cu has been studied by single reflection. Some results are gathered in fig. 9.7. For Cu{100} CO adsorption showed a narrow band at 2085 cm^{-1}, shifting on saturation to 2094 cm^{-1} without broadening. LEED studies have indicated corresponding c(2 × 2) and out-of-registry compression structures. The sharpness of the band at 2094 cm^{-1} suggests that the bond between CO and the Cu surface varies little with position on the surface, and similar behaviour has been found with other faces of Cu. The band observed on Cu{100} is quite distinct from that consistently found with polycrystalline Cu films, or supported Cu, which lie in the range 2100–2103 cm^{-1}; similarly with Cu{111} where there is an even bigger discrepancy with the band appearing at 207 cm^{-1}. Since the results for the higher index planes {110}, {211} and {311} are in much closer agreement with typical polycrystalline behaviour, this has led to the conclusion that stepped or higher index planes predominate in films and supported copper surfaces. It is interesting to note that just as Cu gives a single intense band with CO, so does Au, a measurable band being produced for 5% of monolayer coverage (Kottke, Greenler & Tomkins, 1972).

9.5   Experimental methods in HREELS

In order to perform experiments on vibrational spectroscopy involving the scattering of low energy electrons from surfaces, it is necessary to use spectrometers with an overall resolution higher than 30 meV. A good example of such a spectrometer is shown in fig. 9.9 and is due to Froitzheim, Ibach & Lehwald (1975). The main parts of this spectrometer are the two cylindrical deflectors with one of them fed by an electron gun. This cylindrical deflector and electron gun combination feeds an accelerating electron lens system to form a source producing highly monoenergetic electrons; the second deflector works as an analyser. The overall resolution of this combination can be varied between 7 and 150 meV independently of the primary energy which can be varied between 1 and 70 eV. The angular resolution is ∼1.5°. This latter resolution is important as the current dipole scattering theory describes only small-angle scattering. In the same fashion as described for IRAS experiments, it is generally useful to combine the HREELS measurements with other surface techniques, primarily those of LEED, AES and flash filament techniques; this combination of techniques is necessary to enable the HREELS data to be interpreted.

9.5.1 *Experimental results from HREELS*

In HREELS (dipole scattering regime) there exists the same kind of selection rules as in IRAS, in this instance because the incident electrons interact primarily with the long range fields set up in the vacuum by the oscillating dipoles, we find that fields created by the oscillation of a dipole are screened by a factor of $1/\varepsilon$ if the dipole is orientated parallel to the surface and not at all if the dipole is orientated perpendicular to the surface. To summarise, then, only those vibrations can be excited which create dipoles with a component perpendicular to the surface. This condition is illustrated in fig. 9.10 for a set of imaginary adsorption states.

State (*a*), an atom joined by a single bond to the surface atom immediately below it, would yield a single loss peak.

State (*b*), an atom joined in a symmetric bridge position, has only one

Fig. 9.9. Schematic diagram of a high resolution electron energy loss spectrometer (Froitzheim *et al.*, 1975).

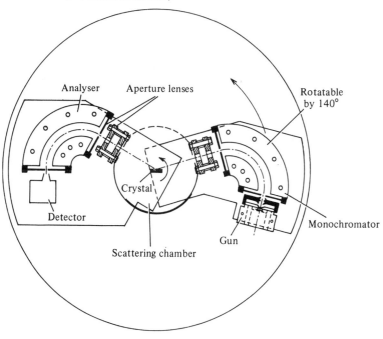

normal vibration with a perpendicular component, and again yields a single loss peak.

State (c), an atom joined asymmetrically in a bridge bond. Here a second vibration mode with a perpendicular component appears.

State (d), an adsorbed molecule in a stretched position, gives two loss peaks. The low frequency mode originates from the vibration of the molecule as a whole against the surface, the moving mass is large. The higher frequency loss peak originates in the stretching vibration of the molecule.

State (e), an adsorbed molecule in a bridge position, gives a loss spectrum with two peaks. The low frequency mode derives from the vibration of the isolated atoms perpendicular to the surface; the high frequency mode originates from the stretching vibration of the quasi-

Fig. 9.10. Set of imaginary adsorption states showing the expected loss peaks associated with each structure.

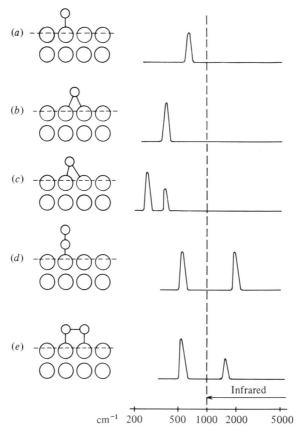

molecule. The frequency split is proportional to the square root of the mass ratio. For simplicity the above spectra assume the mass of the adsorbate atoms to be much less than the mass of the substrate atoms.

Examples of the above structures found in experimental data from HREELS experiments are numerous and it suffices to illustrate the observed structures by reference to a limited number of results.

The adsorption of H on W{100}, carried out by Froitzheim *et al.* (1976), illustrates many of the features set out above and the loss peaks can be compared with data from LEED and ESD. Fig. 9.11 shows the energy loss spectra of H on W{100}. The two losses at 155 and 130 meV correspond to atomic H adsorbed in on-top and bridge sites respectively. The loss at 155 meV occurs for low coverage, $\theta = 0.4$; at higher coverages the 130 meV loss appears in addition and reaches a maximum at $\theta = 2$ while the 155 meV loss disappears. Comparison with other techniques suggests that the 155 meV loss corresponds to H adsorbed in an on-top position and the 130 meV loss to a bridge position. In fact, H adsorption on the W{100} plane is greatly complicated by the surface reconstruction which accompanies the adsorption process and the fact that

Fig. 9.11. Electron energy loss spectra for different coverages of H on the W{100} surface. Instrumental resolution shown as inset (Froitzheim *et al.*, 1976).

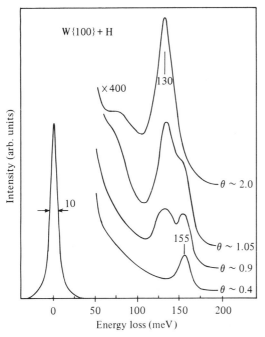

the configurations are temperature dependent. The mechanism for the reconstruction of both clean and H covered W{100} surfaces remains in doubt, with somewhat conflicting evidence from auxiliary techniques such as LEED. Models have been proposed that predict reconstructions involving a vertical shift, a lateral shift, or a combination of both. A more recent example of a study of H adsorption on the W{100} surface may be found in the work of Didham, Allison & Willis (1983), who have concluded that H adsorption induces commensurate and incommensurate displacements. At low coverages the H atoms occupy a bridge site between two adjacent W atoms which are pulled closer together than their equilibrium separation in the bulk crystal. As the coverage increases the W atom spacing changes to its bulk value. This change is apparently not gradual, but represents transfer between two states, a lower energy state (pinched surface atoms) and a higher energy state (normal spacing).

It is possible to correlate HREELS and LEED measurements on this system to form a coherent picture of the changes involved in going from the commensurate to the incommensurate phases. Thus, to begin with, the bridge H atom is joined to W atoms lying flat, but pinched together. At higher coverage the 'pinched' W atoms involved in the bridge tilt out of the surface plane and display extra vibrational modes in the specular direction as modes parallel to the surface gain a perpendicular component (Willis, 1979). Finally, at higher coverage, the surface becomes disordered with the tilted W atoms relaxing into the surface plane at normal bulk lattice spacing; the surface is now flat.

The adsorption of CO on W{100} (fig. 9.12) reveals, as might be expected, additional complexities resulting from the possibility of dissociative and molecular adsorption. For small coverages of CO, corresponding to exposures of less than 1 L, two loss peaks are observed at ≈ 70 meV and 78 meV. Both frequencies are identical to the vibration frequencies of isolated C and O atoms respectively. This spectrum is clear evidence of dissociative adsorption of the β-CO state and agrees with evidence derived from XPS and isotope mixing studies. On increasing the CO exposure two additional loss peaks are observed whose intensity is dependent on the CO partial pressure. The high frequency is close to the frequency of the free molecule vibration and it is reasonable to assume that these losses are related to the characteristic vibrations of the undissociated α-CO state and correspond to a molecule in the upright position, probably directly over the top of a W atom. The data shown in due to Froitzheim *et al.* (1977).

The results described thus far apply to relatively simple molecules; it is of some interest to examine the information which can be obtained about more complex molecules on a more 'catalytic' type of surface. A system which has received a good deal of attention is that of Pt{111} with ethylene or acetylene as the adsorbate. Initially, LEED experiments conducted with acetylene were interpreted on the assumption that acetylene comprised the adsorbed species (Kesmodel, Stair, Baetzhold & Somorjai, 1976). Electron energy loss measurements (Ibach, Hopster & Sexton, 1977) have shown that this assumption is incorrect. Indeed, later work (Ibach & Lehwald, 1978) has shown that the surface species seen by Kesmodel *et al.* can be formed only from ethylene and is not seen with acetylene unless H is added, either as pre-adsorbed atoms or as atomic H from the gas phase (fig. 9.13). This conclusion has been confirmed by further LEED measurements.

A variety of surface structures have been proposed to account for the experimental findings including

(*a*) ethylidene CH_3—CH

Fig. 9.12. Electron energy loss spectra of CO on W{100} at 300 K. The spectra are recorded with a primary electron beam energy of 5 eV incident at an angle of 75°; $\Delta\phi$ is the measured increase in work function (Froitzheim *et al.*, 1977).

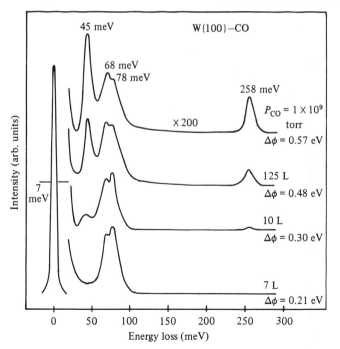

W{100}–CO

45 meV

68 meV
78 meV

258 meV

× 200

$P_{CO} = 1 \times 10^9$ torr
$\Delta\phi = 0.57$ eV

125 L
$\Delta\phi = 0.48$ eV

10 L
$\Delta\phi = 0.30$ eV

7 meV

7 L
$\Delta\phi = 0.21$ eV

Intensity (arb. units)

Energy loss (meV)

(b)   ethylidyne CH_3-C-

and

(c) $-CH_2-CH$

These structures must be considered together with an additional

Fig. 9.13. Electron energy loss spectra of acetylene on Pt{111} at 140 K and 2 L exposure following temperature cycling in vacuum and in a H atmosphere. Acetylene plus H and ethylene produce essentially the same room temperature spectrum (Ibach & Lehwald, 1978).

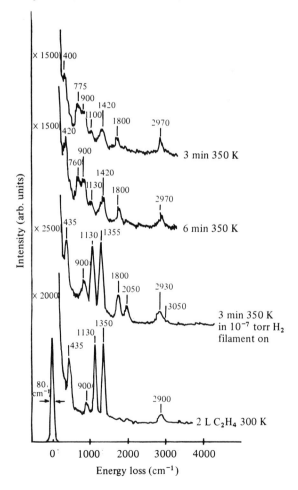

observation, namely that H is released from an ethylene covered surface when it is annealed at 300 K. The amount of H released is approximately one-quarter of the total amount of H on the surface suggesting a surface species of composition C_2H_3, i.e. species (*b*) or (*c*). The identity of the structure on the Pt{111} surface has been fixed by comparison with an organometallic compound $CH_3CCo_3(CO)_9$, for which the infrared absorption spectrum has been measured. Frequency and symmetry assignments for the fundamentals of the above compound compare very closely with these for the Pt{111} surface species and establish the adsorbed species as

ethylidyne $CH_3 \!\!-\!\! C \!\!\!<$

This is a particularly interesting example of a surface problem since it has required the interplay of several different surface techniques to obtain the solution.

Further reading
A review of IRAS with extensive views of the results obtained by this technique can be found in Pritchard (1979a). A comprehensive overview of HREELS is to be found in *Electron Spectroscopy for Surface Analysis* (Ibach, 1977c). Additional material and a comprehensive theoretical treatment of HREELS is presented in *Electron Energy Loss Spectroscopy* (Mills & Ibach, 1982).

References

Abrahamson, A. A. (1969) *Phys. Rev.*, **178**, 76.

Adams, D. L., Germer, L. H. & May, J. W. (1970) *Surface Sci.*, **22**, 45.

Alferieff, M. E. & Duke, C. B. (1967) *J. Chem. Phys.*, **46**, 923.

Allyn, C. L., Gustafsson, T. & Plummer, E. W. (1977) *Chem. Phys. Letters*, **47**, 127.

Almén, O. & Bruce, G. (1961) *Nucl. Instrum. Methods*, **11**, 257.

Andersen, C. A. & Hinthorne, J. R. (1973) *Anal. Chem.*, **45**, 1421.

Andersen, H. H. (1979) *Appl. Physics*, **18**, 131.

Anderson, J. B. & Fenn, J. B. (1965) *Physics of Fluids*, **8**, 780.

Anderson, P. A. (1941) *Phys. Rev.*, **59**, 1034.

Anderson, P. A. (1952) *Phys..Rev.*, **88**, 655.

Andersson, S. & Pendry, J. B. (1980) *J. Phys. C.-Solid State Phys.*, **13**, 3547.

Antinodes, E., Janse, E. C. & Sawatsky, G. A. (1977) *Phys. Rev. B*, **15**, 1669.

Applebaum, J. & Haman, D. A. (1975) *Phys. Rev. Letters*, **34**, 806.

Argile, C. & Rhead, G. E. (1975) *Surface Sci.*, **53**, 659.

Armand, G. & Manson, J. R. (1979) *Phys. Rev. Letters*, **43**, 1839.

Armitage, A. F., Woodruff, D. P. & Johnson, P. D. (1980) *Surface Sci.*, **100**, L483.

Baker, B. G., Johnson, B. B. & Maire, G. L. C. (1971) *Surface Sci.*, **24**, 572.

Barber, M., Vickerman, J. C. & Wolstenholme, J. (1977) *Surface Sci.*, **68**, 130.

Beeby, J. L. (1971) *J. Phys. C.*, **4**, L395.

Bell, R. J., Alexander, R. W., Ward, C. A. & Tyler, I. L. (1975) *Surface Sci.*, **48**, 253.

Benninghoven, A. (1970) *Z. Phys.*, **230**, 403.

Benninghoven, A. (1975) *Surface Sci.*, **53**, 596.

Bergland, C. N. & Spicer, W. E. (1964a) *Phys. Rev.*, **136**, A1030.

Bergland, C. N. & Spicer, W. E. (1964b) *Phys. Rev.*, **136**, A1044.

Bethe, H. (1928) *Ann. Phys.*, **87**, 55.

Blaise, G. &Slodzian, G. (1970) *J. Physique*, **31**, 93.

Blyholder, G. (1964) *J. Phys. Chem.*, **68**, 2772.

Boato, G., Cantini, P. & Mattera, L. (1976) *Surface Sci.*, **55**, 141.

Boato, G., Cantini, P. & Tatarek, R. (1976) *J. Phys. F*, **9**, L237.

Boiziau, C., Garot, C., Nuvolone, R. & Roussel, J. (1980) *Surface Sci.*, **91**, 313.

Bozso, F., Yates, J. T. Jr., Arias, J., Metiu, H. & Martin, R. M. (1983) *J. Chem. Phys.*, **78**, 4256.

Bradshaw, A. M. & Hoffmann, F. M. (1978) *Surface Sci.*, **72**, 513.

Bradshaw, A. M. & Pritchard, J. (1970) *Proc. Roy. Soc. (London)*, **A316**, 169.

Brochard, D. & Slodzian, G. (1971) *J. Physique*, **32**, 185.

Brusdeylins, G., Doak, R. B. & Toennies, J. P. (1980) *Proc. IVth Int. Conf. on Solid Surfaces*, **2**, 842. Suppl. Le Vide, les Couches Minces, No. 201.

Brusdeylins, G., Meyer, H. D., Toennies, J. P. & Winkelmann, K. (1977) *Progr. Astronaut. Aeronaut.*, **51**, 1047.

Brutschy, B. & Haberland, H. (1977) *J. Phys. E.*, **10**, 90.

Buck, T. M. (1977) in *Inelastic Ion–Surface Collisions*, ed. N. H. Tolk, J. C. Tully, W. Heiland & C. W. White (Academic Press, London), p. 47.

Buck, T. M., Wheatley, G. H., Miller, G. L., Robinson, D. A. H. & Chen, Y.-S. (1978) *Nucl. Instrum. Methods*, **149**, 591.

Burdick, G. A. (1963) *Phys. Rev.*, **129**, 138.

Campargue, R., Lebéhot, A. & Lemonnier, J. C. (1977) *Progr. Astronaut. Aeronaut.*, **51**, 1033.

Campbell, C. T., Ertl, G., Kuipers, H. & Segner, J. (1981) *Surface Sci.*, **107**, 207.

Cardillo, M. J. & Becker, G. E. (1978) *Phys. Rev. Letters*, **42**, 508.

Cardona, M. & Ley, L. (Eds) (1978) *Topics in Applied Physics* (Springer-Verlag, Berlin), **26**.

Citrin, P. H., Eisenberger, P. & Hewitt, R. C. (1980) *Phys. Rev. Letters*, **45**, 1948.

Comsa, G. (1979) *Surface Sci.*, **81**, 57.

Conrad, H., Ertl, G., Küppers, J., Wang, S. W., Gérard, K. & Haberland, H. (1979) *Phys. Rev. Letters*, **42**, 1082.

Crossley, A. & King, D. A. (1977) *Surface Sci.*, **68**, 528.

Crossley, A. & King, D. A. (1980) *Surface Sci.*, **95**, 131.

Crowell, C. R., Kao, T. W., Anderson, C. L. & Rideout, V. L. (1972) *Surface Sci.*, **32**, 591.

Davenport, J. W. (1976) *Phys. Rev. Letters*, **36**, 945.

Davenport, J. W. (1978) *J. Vac. Sci. Technol.*, **15**, 433.

Davis, L. E., MacDonald, N. C., Palmberg, P. W., Riach, G. E. & Weber, R. E. (1976) *Handbook of Auger Electron Spectroscopy*, 2nd edition (Physical Electronics Inc., Eden Prarie, Minn).

Dawson, P. H. (1977) *Phys. Rev. B*, **15**, 5522.

Dearnaley, G. (1969) *Reports on Prog. in Phys.*, **32**, 405.

Delchar, T. A. (1971) *Surface Sci.*, **27**, 11.

Delchar, T. A., Eberhagen, A. & Tompkins, F. C. (1963) *J. Sci. Instr.*, **40**, 105.

Delchar, T. A. & Ehrlich, G. (1965) *J. Chem. Phys.*, **42**, 2686.

Delchar, T. A., MacLennan, D. A. & Landers, A. M. (1969) *J. Chem. Phys.*, **50**, 1779.

Delchar, T. A. & Tompkins, F. C. (1967) *Proc. Roy. Soc. (London)*, **A300**, 141.

Demuth, J. E. & Eastman, D. E. (1974) *Phys. Rev. Letters*, **32**, 1123.

Demuth, J. E., Schmeisser, D. S. & Avouris, Ph. (1981) *Phys. Rev. Letters*, **47**, 1166.

den Boer, M. L., Einstein, T. L., Elam, W. T., Park, R. L., Roelofs, L. D. & Laramore, G. E. (1980) *J. Vac. Sci. Technol.*, **17**, 59.

De Wit, A. G. J., Bronckers, R. P. N. & Fluit, J. M. (1979) *Surface Sci.*, **82**, 177.

Didham, E. F. J., Allison, W. & Willis, R. F. (1983) *Surface Sci.*, **126**, 219.

Dignam, M. J., Rao, B., Moskovits, M. & Stobie, R. W. (1971) *Can. J. Chem.*, **48**, 1115.

Doniach, S. & Winick, H. (1980) *Synchrotron Radiation Research* (Plenum, New York).

Dunnoyer, L. (1911) *Le Radium*, **8**, 142.

Dupp, G. & Scharmann, A. (1966) *Z. Phys.*, **192**, 284.

Dyke, W. P. & Trolan, J. K. (1953) *Phys. Rev.*, **89**, 799.

Ehrlich, G. (1961a) *J. Chem. Phys.*, **34**, 29.

Ehrlich, G. (1961b) *J. Chem. Phys.*, **34**, 39.

Ehrlich, G. (1961c) *J. Appl. Phys.*, **32**, 4.

Ehrlich, G. & Hudda, F. G. (1959) *J. Chem. Phys.*, **30**, 493.

Ehrlich, G. & Hudda, F. G. (1961) *J. Chem. Phys.*, **35**, 1421.

Ehrlich, G. & Kirk, C. F. (1968) *J. Chem. Phys.*, **48**, 1465.

Eischens, R. P. & Pliskin, W. A. (1958) *Advan. Catal.*, **10**, 1.

Engel, T. & Gomer, R. (1969) *J. Chem. Phys.*, **50**, 2428.

Engel, T. & Rieder, K. H. (1981) *Surface Sci.*, **109**, 140.

Esbjerg, N. & Norskov, J. K. (1980) *Phys. Rev. Letters*, **45**, 807.

Estel, J., Hoinkes, H., Kaarmann, H., Nahr, N. & Wilsch, H. (1976) *Surface Sci.*, **54**, 393.

Esterman, I. & Stern, O. (1930) *Z. Phys.*, **61**, 95.

Evans, E. & Mills, D. L. (1972) *Phys. Rev. B*, **5**, 4126.

Fadley, C. S. (1978) in *Electron Spectroscopy: Theory, Techniques and Applications*, ed. C. R. Brundle & A. D. Baker (Academic Press, London).

Fano, U. & Cooper, J. W. (1968) *Rev. Mod. Phys.*, **40**, 441.

Fehrs, D. L. & Stickney, R. E. (1971) *Surface Sci.*, **24**, 309.

Feibelman, P. J. & Knotek, M. L. (1978) *Phys. Rev. B*, **18**, 6531.

Feibelman, P. J. & McGuire, E. J. (1977) *Phys. Rev. B*, **15**, 3575.

Feibelman, P. J. & McGuire, E. J. (1978) *Phys. Rev. B*, **17**, 690.

Feibelman, P. J., McGuire, E. J. & Pandey, K. C. (1977) *Phys. Rev. B*, **15**, 2202.

Feldman, L. C., Kauffman, R. L., Silverman, P. J., Zuhr, R. A. & Barrett, J. H. (1977) *Phys. Rev. Letters*, **39**, 1411.

Feuerbacher, B., Fitton, B. & Willis, R. F. (eds.) (1978) *Photoemission and the Electronic Properties of Surfaces* (Wiley, Chichester).

Feurstein, A., Grahmann, H., Kalbitzer, S. & Oetzmann, H. (1975) in *Ion Beam Surface Layer Analysis*, ed. O. Meyer, G. Linker & F. Kappeler (Plenum Press, New York), p. 471.

Fluendy, M. A. D. & Lawley, K. P. (1973) *Chemical Applications of Molecular Beam Scattering* (Chapman & Hall, London).

Ford, R. R. (1966) Ph.D. thesis, University of London.

Ford, R. R. & Pritchard, J. (1968) *Chem. Commun.*, 362.

Fowler, R. H. (1931) *Phys. Rev.*, **38**, 45.

Fowler, R. H. & Nordheim, L. W. (1928) *Proc. Roy. Soc.*, **A119**, 173.

Froitzheim, H., Ibach, H. & Lehwald, S. (1975) *Rev. Sci. Instrum.*, **46**, 1325.

Froitzheim, H., Ibach, H. & Lehwald, S. (1976), *Phys. Rev. Letters*, **36**, 1549.

Froitzheim, H., Ibach, H. & Lehwald, S. (1977) *Surface Sci.*, **63**, 56.

Gadzuk, J. W. (1974) *Phys. Rev. B*, **9**, 1978.

Garcia, N., Goodman, F. O., Celli, V. & Hill, N. R. (1978) *Phys. Rev. B*, **19**, 1808.

Garcia, N., Ibanez, J., Solana, J. & Cabrera, N. (1976) *Surface Sci.*, **60**, 385.

Garrison, B. J. (1979) *Surface Sci.*, **87**, 683.

Gentry, W. R. & Giese, C. F. (1978) *Rev. Sci. Instrum.*, **49**, 595.

Gerlach, W. & Stern, O. (1921) *Z. Phys.*, **8**, 110 & **9**, 349, 353.

Glupe, G. & Mehlhorn, W. (1967) *Phys. Letters*, **25A**, 274.

Godfrey, D. J. & Woodruff, D. P. (1979) *Surface Sci.*, **89**, 76.

Gomer, R. (1958) *J. Chem. Phys.*, **29**, 441, 443.

Gomer, R. (1961) *Field Emission and Field Ion Microscopy* (Harvard University Press).

Goodman, F. O. (1967) *Surface Sci.*, **7**, 391.

Goodman, F. O. (1971) *Surface Sci.*, **26**, 327.

Goymour, C. G. & King, D. A. (1973) *Trans. Faraday Soc.*, **69**, 736.

Grant-Rowe, G. & Ehrlich, G. (1975) *J. Chem. Phys.*, **63**, 4648.

Greenler, R. G. (1966) *J. Chem. Phys.*, **44**, 310.

Greenler, R. G. (1974) *Japan J. Appl. Phys. Suppl. 2*, part 2, 265.

Greenler, R. G. (1975) *J. Vac. Sci. Technol.*, **12**, 1410.

Guseva, M. I. (1960) *Soviet Phys; Solid State*, **1**, 1410.

Gustafsson, T. & Plummer, E. W. (1978) in *Photoemission and the Electronic Properties of Surfaces*, ed. B. Feuerbacher, B. Fitton & R. F. Willis (Wiley, Chichester), p. 353.

Haas, G. A. & Thomas, R. E. (1966) *Surface Sci.*, **4**, 64.

Haas, T. W., Grant, J. T. & Dooley, G. J. (1972) *J. Appl. Phys.*, **43**, 1853.

Hagstrum, H. D. (1954) *Phys. Rev.*, **96**, 336.

Hagstrum, H. D. (1961) *Phys. Rev.*, **123**, 758.

Hagstrum, H. D. (1966) *Phys. Rev.*, **150**, 495.

Hagstrum, H. D. (1977) in *Inelastic Ion–Surface Collisions*, ed. N. H. Tolk, W. Heiland & C. W. White (Academic Press, London), p. 1.

Hagstrum, H. D. & Becker, G. E. (1971) *J. Chem. Phys.*, **54**, 1015.

Hagstrum, H. D. & Becker, G. E. (1972) *Proc. Roy. Soc.*, **331**, 395.

Hammaker, R. A., Francis, S. A. & Eischens, R. P. (1965) *Spectrochimica Acta*, **21**, 1295.

Hayden, B. E. & Bradshaw, A. M. (1983) *Surface Sci.*, **125**, 787.

Hayes, F. H., Hill, M. P., Lecchini, M. A. & Pethica, B. A. (1965) *J. Chem. Phys.*, **42**, 2919.

Hayward, D. O. & Trapnell, B. M. (1964) *Chemisorption* (Butterworth, London).

Heiland, W. & Taglauer, E. (1972) *J. Vac. Sci. Technol.*, **9**, 620.

Heiland, W. & Taglauer, E. (1977) in *Inelastic Ion–Surface Collisions*, ed. N. H. Tolk, J. C. Tully, W. Heiland & C. W. White (Academic Press, London), p. 27.

Heiland, W., Iberl, F., Taglauer, E., Menzel, D. (1975) *Surface Sci.*, **53**, 383.

Heine, V. (1966) *Phys. Rev.*, **151**, 561.

Helbig, H. F., Linder, M. W., Morris, G. A. & Steward, S. A. (1982) *Surface Sci.*, **114**, 251.

Henzler, M. (1977) *Topics in Current Physics*, ed. H. Ibach (Springer-Verlag, Berlin), **4**, 117.

Hofmann, P., Bare, S. R., Richardson, N. V. & King, D. A. (1981) *Solid State Commun.*, **42**, 645.

Hofmann, P., Bare, S. R., Richardson, N. V. & King, D. A. (1983) *Surface Sci.*, **133**, L459.

Hofmann, S. (1976) *Appl. Phys.*, **9**, 56.

Holland, B. W. & Woodruff, D. P. (1973) *Surface Sci.*, **36**, 488.

Holland, S. P., Garrison, B. J. & Winograd, N. (1979) *Phys. Rev. Letters*, **43**, 220.

Holloway, S. & Beeby, J. L. (1978) *J. Phys. C.—Solid State Phys.*, **11**, L247.

Horn, K. & Pritchard, J. (1975) *Surface Sci.*, **52**, 437.

Horn, K. & Pritchard, J. (1977) *J. Physique*, **38**, C4, 164.

Horne, J. M. & Miller, D. R. (1977) *Surface Sci.*, **66**, 365.

Hulpke, E. & Mann, K. (1983) *Surface Sci.*, **133**, 171.

Ibach, H. (ed.) (1977a) *Topics in Current Physics* (Springer-Verlag, Berlin), **4**.

Ibach, H. (1977b) *Surface Sci.*, **66**, 56.

Ibach, H. (ed.) (1977c) *Topics in Current Physics* (Springer-Verlag, Berlin), **4**, p. 205.

Ibach, H. (1981) *Proc. EUCMOS XV* (Elsevier, Amsterdam).

Ibach, H., Hopster, H. & Sexton, B. (1977) *Appl. Surface Sci.*, **1**, 1.

Ibach, H. & Lehwald, S. (1978) *J. Vac. Sci. Technol.*, **15**, 407.

Ibach, H. & Mills, D. A. (1982) *Electron Energy Loss Spectroscopy* (Academic Press, London).

Inghram, M. G. & Gomer, R. (1954) *J. Chem. Phys.*, **22**, 1279.

International Tables for X-Ray Crystallography (Kynoch Press, Birmingham, England), 1952.

Ishitani, T. & Schimizu, R. (1974) *Phys. Letters*, **46A**, 487.

Jaeger, R., Feldhaus, J., Haase, J., Stöhr, J., Hussain, Z., Menzel, D. & Norman, D. (1980) *Phys. Rev. Letters*, **45**, 1870.

Johnson, P. D. & Delchar, T. A. (1977) *J. Phys. E*, **10**, 428.

Jones, R. G. & Woodruff, D. P. (1981) *Vacuum*, **31**, 411.

Jorgensen, W. L. & Salem, L. (1973) *The Organic Chemist's Book of Orbitals* (Academic Press, London).

Joyes, P. (1973) *Rad. Effects*, **19**, 235.

Kane, E. O. (1964) *Phys. Rev. Letters*, **12**, 97.

Kantrowitz, A. & Grey, J. (1951) *Rev. Sci. Instrum.*, **22**, 328.

Kelvin (1898) *Phil. Mag.*, **46**, 82.

Kennedy, D. J. & Manson, S. T. (1972) *Phys. Rev. A*, **5**, 227.

Kesmodel, L. L., Stair, P. C., Baetzhold, R. C. & Somorjai, G. A. (1976) *Phys. Rev. Letters*, **36**, 1316.

Kincaid, B. M., Meixner, A. E. & Platzman, P. M. (1978) *Phys. Rev. Letters*, **40**, 1296.

King, D. A. (1975) *Surface Sci.*, **47**, 384.

King, D. A. & Woodruff, D. P. (eds.) (1982) *The Chemical Physics of Solid Surfaces and Heterogeneous Catalysis, Vol. 4, Fundamental Studies of Heterogeneous Catalysis* (Elsevier, Amsterdam).

Kirschner, J. (1977) in *Topics in Current Physics*, ed. H. Ibach (Springer-Verlag, Berlin), **4**, 59.

Kirschner, J. & Etzkorn, H. W. (1985) in *Topics in Current Physics*, ed. H. E. Oechsner (Springer-Verlag, Heidelberg) (in press).

Kliewer, K. L. (1978) in *Photoemission and the Electronic Properties of Surfaces*, ed. B. Feuerbacher, B. Fitton & R. F. Willis (Wiley, Chichester), p. 45.

Knapp, J. A., Himpsel, F. J. & Eastman, D. E. (1979) *Phys. Rev. B*, **19**, 4952.

Knotek, M. L. & Feibelman, P. J. (1978) *Phys. Rev. Letters*, **40**, 964.

Knotek, M. L., Jones, V. O. & Rehn, V. (1979) *Phys. Rev. Letters*, **43**, 300.

Koch, E. E. (1982) *Handbook on Synchrotron Radiation* (North-Holland, Amsterdam).

Kono, S., Goldberg, S. M., Hall, N. F. T. & Fadley, C. S. (1978) *Phys. Rev. Letters*, **41**, 1831.

Kottke, M. L., Greenler, R. G. & Tompkins, H. G. (1972) *Surface Sci.*, **32**, 231.

Koyama, R. Y. & Smith, N. V. (1970) *Phys. Rev. B*, **2**, 3049.

Krause, M. O. & Ferreira, J. G. (1975) *J. Phys. B*, **8**, 2007.

Lapujoulade, J., Le Cruer, Y., Lefort, M., Lejay, Y. & Maruel, E. (1980) *Surface Sci.*, **103**, L85.

Lapujoulade, J., Lejay, Y. & Papanicolaou, N. (1979) *Surface Sci.*, **90**, 133.

Laramore, G. E. (1981) *Phys. Rev. A*, **24**, 1904.

Larsen, P. K., Chiang, S. & Smith, N. V. (1977) *Phys. Rev. B*, **15**, 3200.

Lichtman, D. (1965) *J. Vac. Sci. Technol.*, **2**, 70.

Logan, R. M. & Keck, J. C. (1968) *J. Chem. Phys.*, **49**, 860.

Logan, R. M. & Stickney, R. E. (1966) *J. Chem. Phys.*, **44**, 195.

Low, M. J. D. & McManus, J. C. (1967) *Chem. Comm.*, 1166.

Lu, T. M. & Lagally, M. G. (1980) *Surface Sci.*, **99**, 695.

MacLennan, D. A. & Delchar, T. A. (1969) *J. Chem. Phys.*, **50**, 1772.

Madey, T. E. (1972) *Surface Sci.*, **33**, 355.

Madey, T. E. (1980) *Surface Sci.*, **94**, 483.

Madey, T. E., Czyzewski, J. J. & Yates, J. T. Jr (1975) *Surface Sci.*, **49**, 465.

Madey, T. E. & Yates, J. T. Jr (1969) *J. Chem. Phys.*, **51**, 1264.

Madey, T. E. & Yates, J. T. Jr (1970) *Actes Colloq. Int. Structure Proprietes Surfaces Solides, Paris*, 155.

Madey, T. E. & Yates, J. T. Jr (1971) *J. Vac. Sci. Technol.*, **8**, 39.

Madey, T. E., Yates, J. T. Jr, King, D. A. & Uhlaner, C. J. (1970) *J. Chem. Phys.*, **52**, 5215.

Madix, R. J. (1979) *Surface Sci.*, **89**, 540.

References 443

Marcus, P. M., Demuth, J. E. & Jepsen, D. W. (1975) *Surface Sci.*, **53**, 501.
Mason, B. F. & Williams, B. R. (1972) *J. Chem. Phys.*, **56**, 1895.
Mathieu, H. J. & Landolt, D. (1975) *Surface Sci.*, **53**, 228.
Maul, J. (1974) thesis, Technische Universität, München.
Melmed, A. J. (1965) *J. Appl. Phys.*, **36**, 3585.
Melmed, A. J. & Gomer, R. (1959) *J. Chem. Phys.*, **30**, 586.
Menzel, D. & Fuggle, J. C. (1978) *Surface Sci.*, **74**, 321.
Menzel, D. & Gomer, R. (1964) *J. Chem. Phys.*, **41**, 3311.
Mignolet, J. C. P. (1950) *Disc. Faraday Soc.*, **8**, 326.
Mignolet, J. C. P. (1955) *Rec. Trav. Chim. Pay Bas*, **74**, 685.
Mills, D. L. & Ibach, H. (1982) *Electron Energy Loss Spectroscopy* (Academic Press, London).
Moore, G. E. (1961) *J. Appl. Phys.*, **32**, 1241.
Moran, J. P. (1970) *AIAA Journal*, **8**, 539.
Morgan, A. E. & Werner, H. W. (1977) *Anal. Chem.*, **49**, 927.
Müller, E. W. (1936) *Z. Phys.*, **37**, 838.
Müller, E. W. (1951) *Z. Phys.*, **131**, 136.
Müller, E. W. (1955) *Ann. Meeting Electron Microscopy Soc. of America*, The Pennsylvania State University.
Müller, E. W., Krishnaswamy, S. V. & McLane, S. B. (1970) *Surface Sci.*, **23**, 112.
Müller, E. W. & Tsong, T. T. (1969) *Field Ion Microscopy* (Elsevier, New York).
McClure, J. D. (1970) *J. Chem. Phys.*, **52**, 2712.
McClure, J. D. (1972) *J. Chem. Phys.*, **52**, 2823.
McIntyre, J. D. E. & Aspnes, D. E. (1971) *Surface Sci.*, **24**, 417.
Nahr, H., Hoinkes, H. & Wilsch, H. (1971) *J. Chem. Phys.*, **54**, 3022.
Netzer, F. P. & Madey, T. E. (1982) *Surface Sci.*, **119**, 422.
Niehus, H. & Krahl-Urban, B. (1981) *Rev. Sci. Instrum.*, **52**, 56.
Nielsen, H. B. & Delchar, T. A. (1984) *Surface Sci.*, **141**, 487.
Nishijima, M. & Propst, F. M. (1970) *Phys. Rev. B*, **2**, 2368.
Nordheim, L. W. (1928) *Proc. Roy. Soc.*, **A121**, 628.
Oechsner, H. (1973) *Z. Phys.*, **261**, 37.
Oliphant, M. L. E. (1929) *Proc. Roy. Soc.*, **A124**, 228.
Oman, R. A. (1968) *J. Chem. Phys.*, **48**, 3919.
Onderdelinden, D. (1968) *Can. J. Phys.*, **46**, 739.
Oppenheimer, J. R. (1928) *Phys. Rev.*, **31**, 67.
Park, R. L., Houston, J. E. & Schreiner, D. G. (1971) *Rev. Sci. Instrum.*, **42**, 60.
Park, R. L. & Madden, H. H. Jr (1968) *Surface Sci.*, **11**, 188.
Parker, J. H. (1962) *Rev. Sci. Instrum.*, **33**, 948.
Pasco, R. W. & Ficalora, P. J. (1980) *Rev. Sci. Instrum.*, **51**, 246.
Pendry, J. B. (1974) *Low Energy Electron Diffraction* (Academic Press, London).
Pendry, J. B. (1980) *J. Phys. C—Solid State Phys.*, **13**, 937.
Peng, Y. K. & Dawson, P. T. (1971) *J. Chem. Phys.*, **54**, 950.
Perdereau, J. & Rhead, G. E. (1971) *Surface Sci.*, **24**, 555.
Pessa, M., Lindroos, M., Asonem, H. & Smith, N. V. (1982) *Phys. Rev. B*, **25**, 738.
Plummer, E. W. (1977) *Proc. 7th Intern. Vac. Cong. & 3rd Intern. Conf. Solid Surfaces* (Vienna), ed. R. Dobrozemsky, F. Rüdenauer, F. P. Viehbäck & A. Breth, p. 647.
Plummer, E. W. & Young, R. D. (1970) *Phys. Rev. B*, **1**, 2088.
Powell, B. D. & Woodruff, D. P. (1976) *Phil. Mag.*, **34**, 169.
Powell, C. J. (1974) *Surface Sci.*, **44**, 29.
Preuss, E. (1980) *Surface Sci.*, **94**, 249.

Pritchard, J. (1963) *Trans. Faraday Soc.*, **59**, 437.

Pritchard, J. (1975) in *Moderne Verfahren der Oberflächenalyse*, Decherma Monographien, 1975, **78**, 231.

Pritchard, J. (1979a) in *Chemical Physics of Solids and Their Surfaces*, Vol. 7, ed. M. W. Roberts & J. M. Thomas (Chemical Society), p. 166.

Pritchard, J. (1979b) in *Chemical Physics of Solids and Their Surfaces*, Vol. 7, ed. M. W. Roberts and J. M. Thomas (Chemical Society), p. 171.

Prutton, M. & Peacock, D. C. (1982) *J. Microscopy*, **127**, 105.

Quinn, C. M. & Roberts, M. W. (1964) *Trans. Faraday Soc.*, **61**, 1775.

Quinn, J. J. (1962) *Phys. Rev.*, **126**, 1453.

Redhead, P. A. (1962) *Vacuum*, **12**, 203.

Redhead, P. A. (1964) *Can. J. Phys.*, **42**, 886.

Redhead, P. A. (1967) *Nuovo Cimento Suppl.*, **5**, 586.

Reid, R. J. (1972) *Surface Sci.*, **29**, 623.

Rieder, K. H. & Engel, T. (1979) *Phys. Rev. Letters*, **43**, 373.

Roberts, R. W. & Vanderslice, T. A. (1963) *Ultrahigh Vacuum and Its Applications* (Prentice-Hall, Englewood Cliffs, New Jersey).

Robinson, N. W. (1968) *The Physical Principles of Ultra-High Vacuum Systems and Equipment* (Chapman & Hall, London).

Rodebush, W. H. (1931) *Rev. Mod. Phys.*, **3**, 392.

Rork, G. D. & Consoliver, R. E. (1968) *Surface Sci.*, **10**, 291.

Roy, D. & Carette, J. D. (1977) in *Topics in Current Physics*, ed. H. Ibach (Springer-Verlag, Berlin), **4**.

Rundel, R. D., Dunning, F. B. & Stebbings, R. F. (1974) *Rev. Sci. Instrum.*, **45**, 116.

Rusch, J. W. & Erickson, R. L. (1977) in *Inelastic Ion–Surface Collisions*, ed. N. H. Tolk, J. C. Tully, W. Heiland & C. W. White (Academic Press, London), p. 73.

Rye R. R. & Barford, B. D. (1971) *Surface Sci.*, **27**, 667.

Rye, R. R., Houston, J. E., Jennison, D. R., Madey, T. E. & Holloway, P. H. (1979) *Ind. Eng. Chem. Prod. Rev. Dev.*, **18**, 2.

Sandstrom, D. R., Leck, J. H. & Donaldson, E. E. (1968) *J. Chem. Phys.*, **48**, 5683.

Saris, F. W. (1982) *Nucl. Instrum. Methods*, **194**, 625.

Saris, F. W. & van der Veen, J. F. (1977) *Proc. 7th Intern. Vac. Cong. & 3rd Intern. Conf. Solid Surfaces* (Vienna), ed. R. Dobrozemsky, F. Rüdenauer, F. P. Viehbäck & A. Breth, p. 2503.

Sau, R. & Merrill, R. P. (1973) *Surface Sci.*, **34**, 268.

Sawatsky, G. A. & Lenselink, A. (1980) *Phys. Rev. B*, **21**, 1790.

Schmidt, L. D. (1974) *Catal. Rev.*, **9**, 115.

Schoenwald, J., Burstein, E. & Elson, J. M. (1973) *Solid State Comm.*, **12**, 185.

Scofield, J. H. (1976) *J. Elect. Spect.*, **8**, 129.

Seah, M. P. & Dench, W. A. (1979) *Surface Interface Analysis*, **1**, 2.

Sevier, K. D. (1972) *Low Energy Electron Spectrometry* (Wiley, New York).

Shigeishi, R. A. & King, D. A. (1976) *Surface Sci.*, **58**, 379.

Shih, H. D., Jona, F., Jepsen, D. W. & Marcus, P. M. (1976) *Surface Sci.*, **60**, 445.

Shimizu, H., Ono, M. & Nakayama, K. (1973) *Surface Sci.*, **36**, 817.

Shirley, D. A. (1973) *Advan. Chem. Phys.*, **23**, 85.

Siegbahn, K., *et al.* (1967) *ESCA; Atomic, Molecular and Solid State Structure Studied by Means of Electron Spectroscopy* (Almqvist & Wiksells, Uppsala).

Sigmund, P. (1969) *Phys. Rev.*, **184**, 383.

Sigmund, P. (1977) in *Inelastic Ion–Surface Collisions*, ed. N. H. Tolk, J. C. Tully, W. Heiland & C. W. White (Academic Press, London), p. 121.

Smith, N. V., Benbow, R. L. & Hurych, Z. (1980) *Phys. Rev. B*, **21**, 4331.

Smoluchowski, R. (1941) *Phys. Rev.*, **60**, 661.

Southern, A. L., Willis, W. R. & Robinson, M. T. (1963) *J. Appl. Phys.*, **34**, 153.

Stobie, R. W., Rao, B. & Dignam, M. J. (1976) *Surface Sci.*, **56**, 334.

Streater, R. W., Moore, W. T., Watson, P. R., Frost, D. C. & Mitchell, K. A. R. (1978) *Surface Sci.*, **72**, 744.

Taglauer, E., Englert, W., Heiland, W. & Jackson, D. P. (1980) *Phys. Rev. Letters*, **45**, 740.

Taglauer, E. & Heiland, W. (1976) *Appl. Phys.*, **9**, 261.

Taglauer, E., Heiland, W. & Beitat, U. (1979) *Surface Sci.*, **89**, 710.

Tarng, M. L. & Wehner, G. K. (1971) *J. Appl. Phys.*, **42**, 2449.

Taylor, N. J. (1969) *Rev. Sci. Instrum.*, **40**, 792.

Tendulkar, D. V. & Stickney, R. E. (1971) *Surface Sci.*, **27**, 516.

Thapliyal H. V. (1978) Ph.D. thesis, Cornell University.

Thiry, P., Chandesris, D., Lecante, J., Guillot, C., Pinchaux, R. & Petroff, Y. (1979) *Phys. Rev. Letters*, **43**, 82.

Tolk, N. H., Traum, M. M., Tully, J. C. & Madey, T. E. (eds.) (1983) *Desorption Induced by Electronic Transitions, DIET I* (Springer-Verlag, New York).

Tolk, N. H., Tully, J. C., Heiland, W. & White, C. W. (eds.) (1977) *Inelastic Ion–Surface Collisions* (Academic Press, London).

Traum, M. M. & Woodruff, D. P. (1980) *J. Vac. Sci. Technol.*, **17**, 1202.

Tromp, R. M., Smeenk, R. G. & Saris, F. W. (1981) *Phys. Rev. Letters*, **46**, 939.

Tully, J. C. (1977) *Phys. Rev. B*, **16**, 4324.

Tully, J. C. & Tolk, N. H. (1977) in *Inelastic Ion–Surface Collisions*, eds. N. H. Tolk, J. C. Tully, W. Heiland & C. W. White (Academic Press, London), p. 105.

Turner, D. W., Baker, C., Baker, A. D. & Brundle, C. R. (1970) *Molecular Photoelectron Spectroscopy* (Interscience, New York).

Urbach, F. (1930) *Sitzber. Akad. Wiss. Wien, Math-Naturw. Kl. Abt. IIa*, **139**, 363.

van der Veen, J. F., Smeenk, R. G., Tromp, R. M. & Saris, F. W. (1979) *Surface Sci.*, **79**, 219.

van Hove, M. A. & Tong, S. Y. (1979) *Surface Crystallography by LEED*, Springer Series on Chemical Physics, Vol. 2 (Springer-Verlag, Berlin).

van Oostrum, A. G. J. (1966) *Philips Res. Rept.*, Suppl. **1**, 1.

Vanselow, R. & Howe, R. (eds.) (1984) *Chemistry and Physics of Solid Surfaces, V* (Springer-Verlag, New York), p. 257.

Venkatachalam, G. & Sinha, M. K. (1974) *Surface Sci.*, **44**, 157.

Vrakking, J. J. & Meyer, F. (1975) *Surface Sci.*, **47**, 50.

Wang, G. C. & Lagally, M. G. (1979) *Surface Sci.*, **81**, 69.

Wehner, G. K., Stuart, R. V. & Rosenberg, D. (1961) *General Mills Annual Report of Sputtering Yields*, Report No. 2243.

Weijsenfeld, C. H. (1966) thesis, University of Utrecht.

Weinberg, W. H. & Merrill, R. P. (1971) *J. Chem. Phys.*, **56**, 2881.

Wenass, E. & Howsmon, A. (1968) *The Structure and Chemistry of Solid Surfaces* (Proc. IV International Materials Symposium, Berkeley), ed. G. A. Somorjai (Wiley, New York), p. 1969.

Werner, H. W. (1978) *Electron and Ion Spectroscopy of Solids*, ed. L. Fiermans, J. Vennik & W. Dekeyser, NATO Advanced Study Institute Series, Physics B (Plenum Press, New York), **32**, 324.

Wertheim, G. K. (1978) in *Electron and Ion Spectroscopy of Solids*, ed. L. Fiermans, J. Vennik & W. Dekeyser (Plenum Press, New York), p. 192.

White, C. W., Thomas, E. W., van der Weg, W. F. & Tolk, N. H. (1977) in *Inelastic Ion Surface Collisions*, ed. N. H. Tolk, J. C. Tully, W. Heiland & C. W. White (Academic Press, London), p. 201.

Williams, A. R. & Lang, N. D. (1977) *Surface Sci.*, **68**, 138.

Williams, B. R. (1971) *J. Chem. Phys.*, **55**, 3220.

Willis, R. F. (1979) *Surface Sci.*, **89**, 457.

Wittmaack, K. (1975) *Surface Sci.*, **53**, 626.

Wittmaack, K. (1977) in *Inelastic Ion–Surface Collisions*, ed. N. H. Tolk, J. C. Tully, W. Heiland & C. W. White (Academic Press, London), p. 153.

Wittmaack, K. (1978) *Proc. 8th Intern. Conf. on X-Ray Optics and Microanalysis*, ed. P. Beaman, R. Ogilvie & D. Wittney (Science Press, Princeton).

Wittmaack, K. (1979) *Surface Sci.*, **89**, 668.

Wood, E. A. (1964) *J. Appl. Phys.*, **35**, 1306.

Wood, R. W. (1899) *Phys. Rev.*, **5**, 1.

Woodruff, D. P. (1976) *Disc. Faraday Soc.*, **60**, 218.

Woodruff, D. P. (1982) *Nucl. Instrum. Methods*, **194**, 639.

Woodruff, D. P. (1983) *Surface Sci.*, **124**, 320.

Woodruff, D. P. & Godfrey, D. J. (1980) *Solid State Commun.*, **34**, 679.

Woodruff, D. P., Johnson, P. D., Traum, M. M., Farrell, H. H., Smith, N. V., Benbow, R. L. & Huryzch, Z. (1981) *Surface Sci.*, **104**, 282.

Woodruff, D. P., Norman, D., Holland, B. W., Smith, N. V., Farrell, H. H. & Traum, M. M. (1978) *Phys. Rev. Letters*, **41**, 1130.

Yates, J. T. Jr & King, D. A. (1972) *Surface Sci.*, **30**, 601.

Yates, J. T. & Madey, T. E. (1971) *J. Chem. Phys.*, **54**, 4969.

Yin, L., Tsang, T. & Adler, I. (1976) *J. Elect. Spect.*, **9**, 67.

Yonts, O. C., Normand, C. E. & Harrison, D. E. (1960) *J. Appl. Phys.*, **31**, 447.

Yu, M. L. (1978) *J. Vac. Sci. Technol.*, **15**, 668.

Zanazzi, E. & Jona, F. (1977) *Surface Sci.*, **62**, 61.

Index